Lecture Notes in Electrical Engineering

Volume 1027

The book series *Lecture Notes in Electrical Engineering* (LNEE) publishes the latest developments in Electrical Engineering—quickly, informally and in high quality. While original research reported in proceedings and monographs has traditionally formed the core of LNEE, we also encourage authors to submit books devoted to supporting student education and professional training in the various fields and applications areas of electrical engineering. The series cover classical and emerging topics concerning:

- Communication Engineering, Information Theory and Networks
- Electronics Engineering and Microelectronics
- Signal, Image and Speech Processing
- Wireless and Mobile Communication
- Circuits and Systems
- Energy Systems, Power Electronics and Electrical Machines
- Electro-optical Engineering
- Instrumentation Engineering
- Avionics Engineering
- Control Systems
- Internet-of-Things and Cybersecurity
- Biomedical Devices, MEMS and NEMS

For general information about this book series, comments or suggestions, please contact leontina.dicecco@springer.com.

To submit a proposal or request further information, please contact the Publishing Editor in your country:

China

Jasmine Dou, Editor (jasmine.dou@springer.com)

India, Japan, Rest of Asia

Swati Meherishi, Editorial Director (Swati.Meherishi@springer.com)

Southeast Asia, Australia, New Zealand

Ramesh Nath Premnath, Editor (ramesh.premnath@springernature.com)

USA, Canada

Michael Luby, Senior Editor (michael.luby@springer.com)

All other Countries

Leontina Di Cecco, Senior Editor (leontina.dicecco@springer.com)

**** This series is indexed by EI Compendex and Scopus databases. ****

Ashwani Kumar Dubey · Vijayan Sugumaran ·
Peter Han Joo Chong
Editors

Advanced IoT Sensors, Networks and Systems

Select Proceedings of SPIN 2022

Springer

Editors
Ashwani Kumar Dubey
Department of Electronics
and Communication Engineering
Amity School of Engineering
and Technology
Amity University
Noida, Uttar Pradesh, India

Vijayan Sugumaran
Department of Decision and Information
Sciences
Center for Data Science and Big Data
Analytics
Oakland University
Rochester, MI, USA

Peter Han Joo Chong
School of Engineering, Computer
and Mathematical Sciences
Auckland University of Technology
Auckland, New Zealand

ISSN 1876-1100 ISSN 1876-1119 (electronic)
Lecture Notes in Electrical Engineering
ISBN 978-981-99-1311-4 ISBN 978-981-99-1312-1 (eBook)
https://doi.org/10.1007/978-981-99-1312-1

Technically Co-Sponsored by: Springer

This Springer imprint is published by the registered company Springer Nature Singapore Pte Ltd.
The registered company address is: 152 Beach Road, #21-01/04 Gateway East, Singapore 189721, Singapore

Foreword by Dr. Jon M. Jenkins

Advanced IoT Sensors, Networks and Systems contains a collection of high-quality, peer-reviewed research papers that were curated and presented at the *9th International Conference on Signal Processing and Integrated Networks (SPIN 2022)* held virtually on August 25 and 26, 2022. SPIN 2022 was organized by the Department of Electronics and Communication Engineering, Amity School of Engineering and Technology, Amity University Uttar Pradesh, Noida, UP, India. This collection explores the latest advances in the fields of Machine Learning, Wireless Communication, Cloud Computing, Robotics and Embedded Systems and Sensors and Sensor Technology. The collected papers were selected based on their originality, technical strength, novelty, scientific contribution, readability, and application to the field of Internet of Things. Selected topics include applications of machine learning to the classification of Alzheimer's disease, breast cancer, and to the detection of DDoS attacks, advances in wireless, wearable sensor signal processing and data analytics in healthcare applications, advances in lightweight data security systems for medical communications using two-factor authentication in cloud computing, trajectory optimization for satellites and UAVs, and modeling of Martian launch vehicles, to name a few. The diversity and depth of the topics addressed by these collected papers demonstrate the energy and fast-paced evolution of these vital technologies which are transforming our world and human experience for the better in ways both obvious and inconspicuous.

The readers are sure to find engaging papers of interest to them whether they are established researchers in the field seeking to keep up with the state of the art or they are newcomers seeking an introduction to these fields and the diversity of application areas in IoT Sensors, Networks and Systems.

September 2022

Dr. Jon M. Jenkins
NASA Ames Research Center
Moffett Field, CA, USA

Foreword by Dr. Vivek Lall

Economic growth in India has gained momentum for a large part of the last several years. The growth charter continues today also and will do so in the near and distant future. However, what India needs is advancements in technology much more than pure play capital investments. There is a dire need to invest exponentially in technological innovations. What India has to its advantage is its demographic dividend and this should be utilized in the most optimal manner. There is development across sectors viz. energy sector is evolving with innovations focusing on hydrogen and other renewables, and automobiles are witnessing the "e-revolution". The lines between virtual and physical are blurring by the day. Such technological advancements will impact every stratum of society—from economy to culture to environment.

To imagine an India in the next 25 years by 2047, it would be easier to analyze each segment of technology from communication technology to electronics to materials to energy to chemicals to aerospace to bioscience, among others. What India requires are institutional measures to restructure and reform the entire RD framework and have 5/10/15/20/25—year targets with evaluation and monitoring at each milestone period and scope for course correction where required. There need to be multi-institutional network projects/programs with extensive collaborations between industry and academia.

While India currently has to focus on the hard technology part of the subject in terms of infrastructure and services, there is a need to parallelly focus on the softer aspects of technological innovation. There is a lot of dependence between the two, and given India's plans for the coming 25 years focusing on both will be important.

There is a need to think beyond immediate technologies and innovations and imagine what could happen in the next 25 years and how India needs to be prepared for the same. Metaverse is the most recent revolution which is converging virtual and augmented reality—something that was possibly not thought of by the larger public even 5 years ago. The virtual world is transforming into a state of new normal. Online interactions involve immersive participation by varied avatars—connecting buyers and sellers, enabling socializing in different scenarios while also allowing distributed collaborations. This is much different from the existing norms of socializing, interactions and collaborations and hence has its own challenges. Personal privacy issues and risk of hacking of non-accessed devices may result in distrust issues while also putting reputation and identity issues at stake. That apart, Metaverse also requires a touch and feel engagement and due to time and space perceptions, the effectiveness in a Metaverse may get diluted. These are however early days and I am sure the pace of innovations will continue to evolve.

Health care, transportation and AI are potentially the three areas which are going to see the maximum innovations in technology. From mind mapping robots to 3D printed bones to hydrogen aircraft and flying taxis—it is a whole world out there being explored and researched. Technologies talking to each other and having a certain amount of autonomous behavior is what AI is bringing to the table. A decision-making ability up to a certain level based on pre-defined algorithms is what will define the communication between technologies going forward. The developments in unmanned systems are an example of how communication is going to move forward between technologies. Brain-computer interfaces are another area of developments in communication that will allow people to transmit thoughts directly to a computer. Digital is being blended with the real world, and there are new ways to visualize work and share ideas.

Gradually, we are witnessing a decreasing human touch in communications and technology and as things get automated further, the human touch will possibly remain only that—an interface to monitor the developments. Such rapid technological changes have their consequences with the primary one being policy making. Exponentially evolving technology framework does not beget an equally fast-paced policy framework. There are systems and processes for policy formulation and execution, and hence the risk is technology development may outpace policy making and the latter would only be playing catch up. The second major challenge is the absorption of such evolution. The majority section of the world would possibly be a decade behind in understanding, using and implementing the technological advancements at the pace of origination. Systems that need to be put in place to ensure maximum population are able to reap the benefits of such technological advancements and not result in few sections benefiting disproportionately.

Advanced IoT Sensors, Networks and Systems are at the heart of the next paradigm shift in technology, and we look forward to our collective future.

September 2022

Dr. Vivek Lall
Chief Executive
General Atomics Global Corporation
San Diego, USA

Foreword by Prof. Harry E. Ruda

Advanced IoT Sensors, Networks and Systems is comprised of a collection of peer-reviewed, high-quality research papers that were selected and presented at the *9th International Conference on Signal Processing and Integrated Networks (SPIN 2022)*. The conference was held virtually on August 25 and 26, 2022, organized by the Department of Electronics and Communication Engineering, Amity School of Engineering and Technology, Amity University Uttar Pradesh, Noida, UP, India. The book provides insights into the fields of Artificial Intelligence, Wireless Communication, Cloud Computing, Embedded Systems and Sensor Technology. The content should appeal to both beginners and experts in these fields as the coverage of the topics is comprehensive and current. Moreover, the authors have placed their contributions within the context of the state of the art. The book covers applications including electronic health monitoring, AI-based trajectory planning for UAVs, use of smart antenna in military, malicious node detection in heterogeneous IoT, trajectory optimization for Satellites, and QUIC protocol.

I am sure readers of this book will find this fascinating and educating text on perspectives in IoT Sensors, Networks and Systems.

September 2022

Prof. Harry E. Ruda
University of Toronto
Toronto, Canada

Preface

The book "Advanced IoT Sensors, Networks and Systems—Select Proceedings of SPIN 2022" presents high-quality peer-reviewed and presented research papers from the *"9th International Conference on Signal Processing and Integrated Networks (SPIN 2022)"* held virtually during **August 25–26, 2022**, organized by the *"Department of Electronics and Communication Engineering, Amity School of Engineering and Technology, Amity University Uttar Pradesh, Noida, UP, India"*.

The contents of the book are broadly divided into five technical tracks, namely (i) *Artificial Intelligence, Machine Learning and Data Sciences*, (ii) *Wireless Communication, WSN and Adhoc Networks*, (iii) *Cloud Computing, 5G, loT and IIoT*, (iv) *Robotics and Embedded Systems* and (v) *Sensors and Sensor Technology*. A few advanced key topics included in the chapters under different tracks of the book are electronic health records, QUIC protocol, smart antenna applications for military uses, secure importing of files through crypto wallet, malicious node detection in heterogeneous IoT, AI-based trajectory planning for UAVs, trajectory optimization for Satellites, design of 9T1R nvSRAM Cell, etc. The chapters of the book are the selected research papers based on their originality, technical strength, novelty, scientific contribution, relevance to the book theme, readability and application to the field of Internet of Things. This book features the latest advances in the field of IoT Sensors, Networks and Systems for diverse applications. This book serves as a definitive reference resource for researchers, professors and practitioners working in the field of IoT and related design and development of the sensors and systems. This book includes the chapters based on the following topics: Alzheimer's Disease Classification, Clustering for Ordinal Survey Data; Machine Learning Algorithms for Breast Cancer, Deep Learning Analysis of Electronic Health Records; Channel Assignment Method in Cognitive Radio System; Frequency Beam Scanning Antenna Using SIW for mmwave; QUIC Protocol; Prevention of Data Leakage using TDM and DES Encryption and Decryption; Free Space Optics with Intelligent Reconfigurable Surface; Dual Band Conformal Antenna for Military Applications; IEEE 802.11n Devices for Vehicular Networks; Application of Genetic Algorithm and Flower Pollination Algorithm in cloud computing; Malicious Node Detection in Heterogeneous Internet of Things; HSRP and GLBP over OSPF and RIP Routing

Protocols; Load Balancing Algorithm in Cloud Computing; Cryptowallet; Trajectory Optimization of a Satellite; Artificial Intelligence Aided Trajectory Planning for UAV; 9T1R nvSRAM Cell with Improved Read Delay and Margin; Active Inductor; and Martian Ascent Launch Vehicle.

Noida, India
Rochester, USA
Auckland, New Zealand

Prof. (Dr.) Ashwani Kumar Dubey
Prof. (Dr.) Vijayan Sugumaran
Prof. (Dr.) Peter Han Joo Chong

SPIN 2022

Conference Organization

Patrons-in-Chief

Dr. Ashok K. Chauhan, Founder President, Ritnand Balved Education Foundation (RBEF), India
Dr. Atul Chauhan, Chancellor, Amity University Uttar Pradesh, Noida, and President, RBEF, India

Chief Patron

Prof. (Dr.) Balvinder Shukla, Vice Chancellor, Amity University Uttar Pradesh, Noida, and President, RBEF, India

General Chair

Prof. (Dr.) Manoj Kumar Pandey, Joint Head, Amity School of Engineering and Technology (ASET), Amity University Uttar Pradesh, Noida, India

Conference Chair

Prof. (Dr.) J. K. Rai, Professor and HoD (ECE), Amity School of Engineering and Technology (ASET), Amity University Uttar Pradesh, Noida, India

Organizing Chair(s)

Prof. (Dr.) Pradeep Kumar, Professor and Dy. HoD (ECE), Amity School of Engineering and Technology (ASET), Amity University Uttar Pradesh, Noida, India
Prof. (Dr.) Ashwani Kumar Dubey, Professor (ECE), Amity School of Engineering and Technology (ASET), Amity University Uttar Pradesh, Noida, India

International Coordination Committee Chair(s)

Dr. M. R. Tripathy, ASET, Amity University Uttar Pradesh, Noida, India
Dr. Sindhu Hak Gupta, ASET, Amity University Uttar Pradesh, Noida, India

Editorial Committee

Faculty

Dr. Anil Kumar Shukla, ASET, Amity University Uttar Pradesh, Noida, India
Dr. Rinki Gupta, ASET, Amity University Uttar Pradesh, Noida, India

Student

Sagar C. V., ECE, ASET, Amity University Uttar Pradesh, Noida, India
Shubham Singh, ECE, ASET, Amity University Uttar Pradesh, Noida, India
Gowri Srinivasan, ECE, ASET, Amity University Uttar Pradesh, Noida, India
Vedanta Bhattacharya, ECE, ASET, Amity University Uttar Pradesh, Noida, India
Pranavi Madan, ECE, ASET, Amity University Uttar Pradesh, Noida, India
Yashasvi Roy, ECE, ASET, Amity University Uttar Pradesh, Noida, India
Nikita Agarwal, ECE, ASET, Amity University Uttar Pradesh, Noida, India
Ishaan Srivastava, ECE, ASET, Amity University Uttar Pradesh, Noida, India
Tina Kalita, ECE, ASET, Amity University Uttar Pradesh, Noida, India
T. Venkatasai, ECE, ASET, Amity University Uttar Pradesh, Noida, India

Finance Chair

Dr. Anil Kumar Shukla, ASET, Amity University Uttar Pradesh, Noida, India

Technical Session Committee

Dr. Rinki Gupta, ASET, Amity University Uttar Pradesh, Noida, India
Dr. Sanmukh Kaur, ASET, Amity University Uttar Pradesh, Noida, India
Dr. Shubhra Dixit, ASET, Amity University Uttar Pradesh, Noida, India
Dr. Nidhi Gaur, ASET, Amity University Uttar Pradesh, Noida, India
Dr. Richa Sharma, ASET, Amity University Uttar Pradesh, Noida, India
Ms. Anupama Bhan, ASET, Amity University Uttar Pradesh, Noida, India
Mr. Lala Bhaskar, ASET, Amity University Uttar Pradesh, Noida, India

Sponsorship Committee

Dr. Rinki Gupta, ASET, Amity University Uttar Pradesh, Noida, India
Dr. Sunny Anand, ASET, Amity University Uttar Pradesh, Noida, India

Website and Media Management Committee

Dr. Richa Sharma, ASET, Amity University Uttar Pradesh, Noida, India

Registration Committee

Ms. Anupama Bhan, ASET, Amity University Uttar Pradesh, Noida, India
Dr. Shubhra Dixit, ASET, Amity University Uttar Pradesh, Noida, India

Technical Program Committee

Chair

Dr. Shubhra Dixit, ASET, Amity University Uttar Pradesh, Noida, India

Members

Dr. Afaq Ahmad, Sultan Qaboos University, Muscat, Oman
Dr. Carlos M. T. González, University of Las Palmas de Gran Canaria, Spain
Dr. Rinki Gupta, Amity University Uttar Pradesh, India

Dr. Sindhu Hak, Amity University Uttar Pradesh, India
Dr. Rajeev Ratan, MVN University, Haryana, India
Dr. Yogesh Kumar, Chitkara University, Punjab, India
Dr. Yaduvir Singh, Harcourt Butler Technical University, UP, India
Dr. Partha Mangipudi, Amity University Uttar Pradesh, India
Dr. Alok Srivastava, MVN University, Haryana, India
Dr. Smriti Agarwal, Motilal Nehru National Institute of Technology, UP, India
Dr. Supriya Goel, University of Hyderabad Campus, Hyderabad, India
Dr. Abhishek Kashyap, Jaypee Institute of Information Technology, Noida, India
Dr. Deepti Mehrotra, Amity University Uttar Pradesh, India
Dr. Ajay Roy, Lovely Professional University, Punjab, India
Dr. Divya Upadhyay, Amity University Uttar Pradesh, India
Dr. Irshad Ahmad Ansari, Indian Institute of Information Technology, MP, India
Dr. P. Banerjee, Former Chief Scientist, NPL, New Delhi, India
Dr. Gourav Bathla, University of Petroleum and Energy Studies, UK, India
Dr. Digvijay S. Chauhan, Feroze Gandhi Institute of Engineering and Technology, UP, India
Dr. Ritu Gupta, Bhagwan Parshuram Institute of Technology, New Delhi, India
Dr. Sanmukh Kaur, Amity University Uttar Pradesh, India
Dr. Ravi Kumar Arya, National Institute of Technology, Delhi, India
Dr. Sakshi Bangia, YMCA University of Science and Technology, Haryana, India
Dr. Rashmi Gupta, G. G. Singh Indraprastha University, New Delhi, India
Dr. Basant Kumar, Motilal Nehru National Institute of Technology, UP India
Dr. Sachin Kumar, Amity University Uttar Pradesh, India
Dr. Kamlesh Kr. Singh, Amity University Uttar Pradesh, Lucknow, India
Dr. M. D. Upadhayay, Shiv Nadar University, UP, India
Dr. Rashmi Vashisth, Amity University Uttar Pradesh, India
Dr. Seema Verma, Amity University Uttar Pradesh, India
Dr. Tanu Wadhera, Indian Institute of Information Technology, Una, India
Dr. Bhalke Bhalke, AISSMS College of Engineering, Maharashtra, India
Dr. Dinesh Bhatia, North-Eastern Hill University, Meghalaya, India
Dr. Monica Kaushik, Amity University Uttar Pradesh, India
Dr. Sumit Kushwaha, Chandigarh University, Punjab, India
Dr. Shalini Sah, Amity University Uttar Pradesh, India
Dr. Subodh Wairya, Institute of Engineering and Tech, Lucknow, UP, India
Dr. Pawan Whig, V. I. P. S., New Delhi, India
Dr. Shamim Akhter, Jaypee Institute of Information Technology, Noida, India
Dr. Shakti Raj Chopra, Lovely Professional University, Punjab, India
Dr. Urvashi Garg, Haryana College of Technology and Management (HCTM), India
Dr. Pooja Jha, Amity University Jharkhand, Jharkhand, India
Dr. Alok Joshi, Jaypee Institute of Information Technology, Noida, India
Dr. Raj Kamal Kapur, Amity University Uttar Pradesh, India
Dr. J. K. Rai, Amity University Uttar Pradesh, India
Dr. Anil Kumar Shukla, Amity University Uttar Pradesh, India
Dr. Kuldeep Singh, G. University of Science and Technology, Haryana, India

Dr. Kuldeep Singh, Malaviya National Institute of Technology, Jaipur, India
Dr. Mayank Srivastava, Amity University Uttar Pradesh, India
Dr. Prashant Upadhyaya, Chandigarh University, Punjab, India
Dr. R. L. Yadava, Galgotias College of Engineering and Technology, UP, India
Dr. Mohammad Z. Khan, Integral University, Lucknow, India
Dr. Sunny Anand, Amity University Uttar Pradesh, India
Dr. Shailendra Aswale, S. R. Institute of Engineering and Information Technology (SRIEIT), Goa, India
Dr. Surender Kr. Grewal, D. Chhotu Ram University of Science and Technology, Haryana, India
Dr. Pragya Hejib, University of Delhi, New Delhi, India
Dr. Pradeep Kumar, Amity University Uttar Pradesh, India
Dr. Sunil Kumar, SCIT, Manipal University, Jaipur, India
Dr. Abhimanyu Nain, G. J. University of Science and Technology, Haryana, India
Dr. Seema Narwal, Dronacharya College of Engineering, Gurugram, India
Dr. Anurag Singh, NIT Delhi, New Delhi, India
Dr. Gurpreet Singh, Chitkara University, Rajpura, Punjab, India
Dr. Rajeev Tiwari, University of Petroleum and Energy Studies, UK, India

International Advisory Committee

Prof. Kazuya Kobayashi, Chuo University, Tokyo, Japan
Prof. Cher Ming Tan, Chang Gung University, Taipei, Taiwan
Dr. Teruaki Hayashi, The University of Tokyo, Japan
Dr. Ahmed Abdelgawad, Central Michigan University, USA
Dr. Annalisa Bruno, ERI, Nanyang Technical University, Singapore
Dr. Hien Quoc Ngo, ECIT, Queen's University Belfast, UK.
Dr. Paramita Guha, CSIR-National Physical Laboratory, New Delhi, India
Dr. Vijayan Sugumaran, Oakland University, USA
Dr. Lotte N. S. A. Struijk, Aalborg University, Denmark
Prof. Komal Bhatia, YMCAUST, Faridabad, Haryana, India
Dr. Peter Han Joo Chong, Auckland University of Technology, New Zealand
Dr. Álvaro Rocha, ISEG, University of Lisbon, Portugal
Dr. Sunil Vadera, University of Salford, UK
Dr. Carlos M. T.-González, University of Las Palmas de Gran Canaria (ULPGC), Spain
Prof. Z. A. Jaffery, Department of Electronics Engineering, Jamia Millia Islamia, ND, India
Dr. Boris Novikov, HSE University, St. Petersburg, Russia
Dr. Miguel López-Benítez, University of Liverpool, UK
Dr. Gaurav Sharma, University of Rochester, USA
Prof. Rajiv Kapoor, Delhi Technological University, New Delhi, India
Dr. Yury Shestopalov, University of Gävle, Gävle, Sweden

Dr. C. J. Reddy, Americas Altair Engineering, Inc, USA
Prof. Mridula Gupta, University of Delhi, New Delhi, India
Dr. Afaq Ahmad, Sultan Qaboos University, Sultanate of Oman, UAE
Dr. Thuy T. Le, San José State University, USA
Prof. Nandana Rajatheva, University of Oulu, Finland
Prof. Neeta Pandey, Delhi Technological University, New Delhi, India

Organizing Committee

Dr. M. R. Tripathy, Amity University Uttar Pradesh, India
Dr. Anu Mehra, Amity University Uttar Pradesh, India
Dr. Sindhu Hak Gupta, Amity University Uttar Pradesh, India
Dr. Rinki Gupta, Amity University Uttar Pradesh, India
Dr. Neeraj Khera, Amity University Uttar Pradesh, India
Dr. Sanmukh Kaur, Amity University Uttar Pradesh, India
Dr. Anil Kumar Shukla, Amity University Uttar Pradesh, India
Dr. Richa Sharma, Amity University Uttar Pradesh, India
Dr. Nidhi Gaur, Amity University Uttar Pradesh, India
Dr. Shalini Sah, Amity University Uttar Pradesh, India
Dr. Shubhra Dixit, Amity University Uttar Pradesh, India
Dr. Sunny Anand, Amity University Uttar Pradesh, India
Mr. Lala Bhaskar, Amity University Uttar Pradesh, India
Mr. Rahul Kr. Verma, Amity University Uttar Pradesh, India
Ms. Shikha Bathla, Amity University Uttar Pradesh, India
Ms. Anupama Bhan, Amity University Uttar Pradesh, India
Mr. Haneet Rana, Amity University Uttar Pradesh, India

Amity University at a Glance

Ritnand Balved Education Foundation (RBEF), a non-profit organization registered under the Societies Act-1860, is the umbrella body for all Amity institutions. Amity is home to over 1,25,000 brilliant students and is a pioneer in providing quality education in multiple domains offered by over 300 programs. Driven by the dream to create a world of knowledge par excellence, Amity University is spearheading the field of education. Amity has set a new standard of academic excellence in India by becoming Asia's only not-for-profit University to have been awarded US Regional Accreditation by WASC Senior College and University Commission, USA—considered to be the Gold Standard of Accreditations globally.

Amity University was founded with a focus on research and innovation. At Amity, there is a strong impetus to research which has led to Amity scientists and faculty filing more patents than any other Indian University, besides developing more than 2,120 management case studies bought by over 490 institutions worldwide including Harvard, Stanford, Oxford, McKinsey and KPMG.

In this Knowledge Era, Amity University has emerged as a Research and Innovation Driven University. Amity, being a very young University, has already undertaken over 145 Research Projects funded by leading government and non-government organizations and international funding agencies such as MoEF, DST, CSIR, DBT, ICMR, ICAR, DRDO, Bill and Melinda Gates Foundation, Leverhulme Trust of UK and USAID. State-of-the-art research infrastructures have been created, both through Government funded and S&T department as well as through Amity resources which included instruments such as FACS Accuri and FACS sorter, Confocal Microscope, Scanning Electron Microscope, FT-IR and HPLC.

Amity has been credited with filing over 1530 Patents and has consistently retained its position as the single largest patent-filing private Indian Institution next to all IITs together. Some of the technologies which have been transferred recently to Industries include Rootonics: a plant root fungus which has close to "magical" benefits for crops, Biodegradable Plastic, Milk Adulteration Kit, Photocatalytic Wastewater Treatment, Textile and Dyeing Industry, Nanomaterials for Dye Removal in Water, Iodine Based Fingerprint powder for developing Latent Fingerprint, LPG Sensor, Herbal Mosquito Repellent, Rechargeable Fly Ash Battery and Herbal Colors.

The Scientists and researchers of Amity University pursue research activities tirelessly, and publish their work in reputed, peer-reviewed refereed journals with high impact factors. Over 25000 research articles have been published by the researchers and scientists of the Amity University as books/chapters or in journals such as Cell, Lancet, ACS, Oxford Journals, Nature Publishing group, RSC Advances, Taylor & Francis, Springer, Elsevier and Wiley etc. Amity faculty have developed over 3500 case studies which have been referred to by leading institutions such as Harvard, MIT, Oxford and Stanford. The University has also made concerted efforts to strengthen Research and Innovation activities as evidenced by the Amity Innovation Incubator supported by DST, DST-Technology Enabling Center (TEC), NRDC-Amity Innovation Facilitation Centre and Amity Center for Entrepreneurship Development.

Amity follows Outcome-Based Education concept, and it is the first University in Asia that makes every day in the life of a student, a new learning. To fulfill the mission of outcome-based education system, Amity University offers different programs in Engineering, Biosciences, Architecture, Telecommunications, Arts, Journalism, Management, Humanities, Social Sciences, Hospitality, Law, Health, Insurance and many more. All the courses are designed to provide holistic development of the individuals with a Choice-Based Credit System. Beyond academics, Amity imparts attention to personality development and personal grooming. Special classes are held on leadership, teamwork and analytical skills to instill confidence among the students. Practical and a contemporary curriculum updated by Industry Advisory Board emphasizes industry-oriented teaching.

The education framework at Amity is rooted upon certain key values and ethics. It stresses on principles such as patriotism, courage and respect for elders. It is this sense of purpose that has led to the creation of the "101 Attributes of an Amitian". A center for corporate induction and a team with several corporate associations runs a Corporate Resource Centre to ensure continuous interaction with industry leaders, including CEOs, who are regularly invited to share their experiences of business and instill best practices within the university's student body, lending them an edge in the graduate employment market. Amity also offers an Incubation Centre, which supports budding industries and future entrepreneurs. For these reasons, Government of India has recognized Amity as a "Scientific and Industrial Research Organization" (SIRO).

Amity University is India's first University to receive the prestigious Institution of Engineering and Technology (IET), UK, Accreditation for its B.Tech. degrees. This stupendous achievement reflects the world-class quality of education provided by Amity Group. Amity University Uttar Pradesh has been accredited by National Assessment and Accreditation Council (NAAC) with "**A+**" Grade. Amity University Uttar Pradesh ranked 25th in the National Institutional Ranking Framework (NIRF) India Rankings, by Ministry of Education, Government of India, in 2022. Apart from this, there are several other accreditations, those convey the quality education Amity provides in different sectors. Amity University Uttar Pradesh has also been accredited by the Accreditation Council for Business Schools and Programmes (ACBSP, USA) for its management programs; Royal Institute of Charted Surveyors (RICS,

UK) for real estate and construction programs; European Foundation for Management Development—Technology Enhancement Learning (EFMD-CEL), Belgium; UNWTO.TeDQual, Andorra, for Travel and Tourism programs.

Amity University's innovative research has been identified as a particular strength, which has led to the inclusion of Amity Institutes among the top rankings and prestigious surveys conducted by leading publications in India.

About Amity School of Engineering and Technology

Programs Offered: B.Tech. (Electronics and Communication Engineering.), B.Tech. (Electronics and Communication Engineering)-3Continent, B.Tech. (Electronics and Telecommunication), B.Tech. (Robotics), B.Tech. Electronics Engineering (VLSI Design and Technology), B.Tech. (Computer Science and Engineering), B.Tech. (Civil Engineering), B.Tech. (Automobile Engineering), B.Tech. (Electrical and Electronics Engineering), B.Tech. (Information Technology), B.Tech. (Mechanical Engineering), etc.

About the Department of Electronics and Communication Engineering

Electronics and Communication Engineering department is the premium department of Amity School of Engineering and Technology (ASET), Noida, Uttar Pradesh, India. ASET Noida is accredited by IET, UK, and has been Ranked No. 3 among Top Private Engineering Schools in Times Engineering Institute Ranking survey and Ranked 25 in NIRF ranking by Government of India. ECE Department offers various UG Programs like B.Tech. (ECE), B.Tech. (ECE)-3C, B.Tech. (E&T), B.Tech. (Robotics), B.Tech. Electronics Engineering (VLSI Design and Technology) and PG Programs like M.Tech. (ECE), M.Tech. (VLSI), M.Tech. (Embedded System Technology), M.Tech. (Wireless Communication) and B.Tech. (ECE) + MBA (Integrated) and Ph.D. Programs. The department does cutting-edge research in areas such as signal processing, semiconductor chip design, telecommunication, data communication networks, satellite and radar link setup, RF and microwave, embedded systems, robotics and artificial intelligence. The department provides the students strong academic basics and ample opportunities for innovation and research through its 52 well-equipped labs, in-house training, industry training, and outcome based minor and major project works. As a result, the highly competent students of ECE under the guidance of experience and qualified faculty have published more than 800 research papers and filed more than 80 patents. Many students of ECE have been placed in Top organizations and MNCs including Google, Microsoft, Cisco, British Telecom, L&T, Videocon, Nokia, Thales, Tech. Mahindra, Samsung, Orange,

Workspace Dubai, Accenture, Wipro Technologies, TCS, HCL Technologies, Indigo, Indian Navy, Indian Air Force, Amazon, HFCL, etc. The department offers students a rich international exposure through guest lectures by internationally acclaimed scientists and academicians, and Student Exchange Programme and Study Abroad Programme (SAP). Department of ECE has been successfully hosting a prestigious International Conference on Signal Processing and Integrated Networks (SPIN) annually since 2014.

SPIN 2022 Conference at a Glance

The conference "*Signal Processing and Integrated Networks (SPIN)*" is devoted to the recent advancements in Signal Processing and Integrated Networks and its related areas. Over the past few years, SPIN has emerged as a well-recognized conference in the country and overseas because of its high-quality peer-reviewed research articles, invited talks from renowned industry and academic leaders showcasing their latest innovations and networking opportunities. The conference has proven to be a successive catalyst in fostering novel work, sharing views and getting innovative ideas in the field of Signal Processing and Integrated Networks. In succession of the International Conferences SPIN-2014, SPIN-2015, SPIN-2016, SPIN 2017, SPIN 2018, SPIN 2019, SPIN 2020 and SPIN 2021, **9th International Conference on Signal Processing and Integrated Networks (SPIN 2022)** is organized by Department of Electronics and Communication Engineering, Amity School of Engineering and Technology, Amity University Uttar Pradesh, Noida, UP, India, during **August 25–26, 2022**. The conference aimed to bring together scientists, academicians and industrialists working on the innovative advancements and development in the domain of Signal Processing, Image Processing, IoT, Wireless Communication, Robotics and Embedded Technology, Computational Electromagnetics, RF and Microwave; VLSI and Micro-electronics, etc. to discuss new concepts and endorse research work. The LNEE book series contains papers presented at the **9th International Conference on Signal Processing and Integrated Networks (SPIN 2022)**.

The call for papers received an overwhelming response. Each contributed paper was refereed before being accepted for publication in this book series. The papers were accepted for publication based on the theme, relevance, innovation and application to the field of signal processing and integrated networks. Over 500 scientific participants including students, academicians, government and industry personnel from India and abroad have participated in SPIN 2022.

This year, SPIN 2022 showcases over 35 Keynote/Invited Talks by eminent keynote speakers from across the world on a variety of topics.

We thank all the authors who have submitted and presented their research findings in SPIN 2022. Special thanks go to our **Technical Co-Sponsor Springer** for all the publication support, IEEE Council on RFID, and IEEE Antenna and Propagation

Society for the financial support. We thank our Academic Partners **Electromagnetics Laboratory, Chuo University, Tokyo, Japan**, and **Department of Electronics, Mathematics and Natural Sciences University of Gavle, Gavle, Sweden**.

We are grateful to the plethora of the eminent and distinguished speakers who have shown interest in SPIN 2022:

Dr. Vivek Lall, Chief Executive, General Atomics Global Corporation, San Diego County, California, USA.

Dr. Muriel Medard, Cecil H. and Ida Green Professor of EECS, Massachusetts Institute of Technology (MIT), Cambridge, Massachusetts, USA.

Dr. Jon Jenkins, Computer Scientist, NASA Ames Research Center, NASA, California, USA.

Dr. Jeff Grover, Research Scientist, Research Laboratory of Electronics, Massachusetts Institute of Technology (MIT), Massachusetts, USA.

Dr. Peter Stone, Professor, David Bruton, Jr. Centennial Professor, Director (Texas Robotics), Associate Chair, Department of Computer Science, The University of Texas, Austin, USA.

Dr. Ahmed Abdelgawad, Professor of Computer Engineering, College of Science and Engineering, Central Michigan University, Mount Pleasant, USA.

Dr. Annalisa Bruno, Principal Scientist, Energy Research Institute, Nanyang Technical University, Singapore.

Dr. Hien Quoc Ngo, Associate Professor, UKRI Future Leaders Fellow, Institute of Electronics, Communications and Information Technology (ECIT), Queen's University Belfast, UK.

Dr. Vijayan Sugumaran, Professor and Chair, Department of Decision and Information Sciences, Co-Director, Center for Data Science and Big Data Analytics, Oakland University, USA.

Dr. Lotte N. S. Andreasen Struijk, Associate Professor, Department of Health Science and Technology, The Faculty of Medicine, Neurorehabilitation Robotics and Engineering, Head of Center for Rehabilitation Robotics, Aalborg University, Denmark.

Dr. Peter Han Joo Chong, Professor and Associate Head of School (Research), School of Engineering, Computer and Mathematical Sciences, Faculty of Design and Creative Technologies, Auckland University of Technology, New Zealand.

Dr. Sunil Vadera, Professor of Computer Science, University of Salford, UK.

Dr. Mehmet Emir Koksalal, Professor, University of Twente, Department of Applied Mathematics, The Netherlands.

Dr. Carlos M. Travieso-González, Professor and Head of Signals and Communications Department, Institute for Technological Development, and Innovation in Communications (IDeTIC), University of Las Palmas de Gran Canaria (ULPGC), Spain.

Dr. Boris Novikov, Professor, Department of Informatics HSE University, St. Petersburg, Russia.

Dr. Muhammad Ijaz, Head of Laser and Optics Communication (LOC) lab, Department of Engineering, Faculty of Science and Engineering, Manchester Metropolitan University, Manchester, UK.

Dr. Miguel López-Benítez, Associate Professor, Department of Electrical Engineering and Electronics, University of Liverpool, UK.
Dr. Rowel Atienza, Professor and Scientist, University of the Philippines, Philippines.
Dr. Justin Dauwels, Associate Professor, Department of Microelectronics, TU Delft, The Netherlands.
Dr. Alex Casson, Associate Professor, Department of Electrical and Electronic Engineering, The University of Manchester, Manchester, England, United Kingdom.
Dr. Ajay Pandey, Professor, School of Electrical Engineering and Robotics, Queensland University of Technology, Brisbane, Australia.
Dr. Sarath Kodagoda, Professor of Robotics, President of Australian Robotics and Automation Association (ARAA), Director UTS Robotics Institute, Hornsby, New South Wales, Australia.
Dr. Zeljko Zilic, Professor, McGill University, Montreal, Quebec, Canada.
Dr. Gaurav Sharma, Professor, Department of Electrical and Computer Engineering, University of Rochester, USA.
Dr. Yury Shestopalov, Professor, University of Gävle, Gävle, Sweden.
Dr. Kazuya Kobayashi, Professor, Department of Electrical, Electronic, and Communication Engineering, Chuo University, Japan.
Dr. Nasimuddin, Institute for Infocomm Research, A-STAR, Singapore.
Dr. C. J. Reddy, Vice President of Business Development-Electromagnetics, Americas Altair Engineering, Inc, USA.
Dr. Shuai Zhang, Associate Professor, Aalborg University, Denmark.
Dr. Zhijiao Chen, School of Electronic Engineering, Beijing University of Posts and Telecommunications (BUPT), China.
Dr. Muhammad Ali Babar Abbasi, ECIT Institute, Queens University Belfast, UK.
Dr. Afaq Ahmad, Professor, Department of Electrical and Computer Engineering, Sultan Qaboos University, Sultanate of Oman, UAE.
Dr. Ankit Agrawal, Research Professor, Department of ECE, Northwestern University, USA.
Dr. Thuy T. Le, Professor and Department Chair, Electrical Engineering Department, San José State University, San Jose, California, USA.
Dr. Nandana Rajatheva, Professor, Digital Transmission Technologies, Centre for Wireless Communications, University of Oulu, Finland.

It would be unfair to not acknowledge the constant unwavering support and experience of outstanding researchers and leaders. The organizing chairs take this opportunity to thank Patrons-in-Chief (SPIN 2022) **Dr. Ashok K. Chauhan** (Hon'ble Founder President Ritnand Balved Education Foundation (RBEF), India), **Dr. Atul Chauhan** (Hon'ble Chancellor, Amity University Uttar Pradesh, Noida, India, and President RBEF, India), Chief Patron (SPIN 2022), **Prof. (Dr.) Balvinder Shukla** (Hon'ble Vice Chancellor, Amity University Uttar Pradesh, Noida, India), General Chair (SPIN 2022) **Prof. (Dr.) Manoj Kumar Pandey** (Joint Head, Amity School of Engineering and Technology, Amity University Uttar Pradesh, Noida, India), and Conference Chair (SPIN 2022) **Prof. (Dr.) J. K. Rai** (HoD, ECE, ASET, Amity University Uttar Pradesh, Noida, India).

On behalf of the Organizing Committee, we would like to thank Springer for publishing the proceedings of SPIN 2022.

We would also like to thank all our committee members and reviewers for dedicating their time and effort in technical evaluation of the conference papers and organizing this conference. We are very much grateful to the participants for their excellent papers, inspiring presentations and for sharing their knowledge with the researchers.

Finally, we thank our student volunteers for their enthusiastic organizing and managing efforts, which has made this conference SPIN 2022 a grand success.

Organizing Chairs (SPIN 2022)
Prof. (Dr.) Pradeep Kumar
Prof. (Dr.) Ashwani Kumar Dubey
Amity University Uttar Pradesh, Noida, India

Keynote/Invited Talks

Keynote 1

Dr. Muriel Medard

Professor of EECS, Massachusetts Institute of Technology, Cambridge, Massachusetts, United States

Dr. Muriel Médard is the Cecil H. and Ida Green Professor in the Electrical Engineering and Computer Science (EECS) Department at MIT, where she leads the Network Coding and Reliable Communications Group in the Research Laboratory for Electronics at MIT. She obtained three Bachelor's degrees (EECS 1989, Mathematics 1989 and Humanities 1991), as well as her M.S. (1991) and Sc.D (1995), all from MIT. She is a Member of the US National Academy of Engineering (elected 2020), a Fellow of the US National Academy of Inventors (elected 2018), American Academy of Arts and Sciences (elected 2021) and a Fellow of the Institute of Electrical and Electronics Engineers (elected 2008). She holds an Honorary Doctorate from the Technical University of Munich (2020).

She was co-winner of the MIT 2004 Harold E. Egerton Faculty Achievement Award and was named a Gilbreth Lecturer by the US National Academy of Engineering in 2007. She received the 2017 IEEE Communications Society Edwin Howard Armstrong Achievement Award and the 2016 IEEE Vehicular Technology James Evans Avant Garde Award. She received the 2019 Best Paper award for IEEE Transactions on Network Science and Engineering, the 2018 ACM SIGCOMM Test of Time Paper Award, the 2009 IEEE Communication Society and Information Theory Society Joint Paper Award, the 2009 William R. Bennett Prize in the Field of Communications Networking, the 2002 IEEE Leon K. Kirchmayer Prize Paper Award, as well as eight conference paper awards. Most of her prize papers are co-authored with students from her group.

She has served as technical program committee co-chair of ISIT (twice), CoNext, WiOpt, WCNC and of many workshops. She has chaired the IEEE Medals committee, and served as member and chair of many committees, including as inaugural chair of the Millie Dresselhaus Medal. She was Editor-in-Chief of the IEEE Journal on Selected Areas in Communications and has served as editor or guest editor of many IEEE publications, including the IEEE Transactions on Information Theory, the IEEE Journal of Lightwave Technology and the IEEE Transactions on Information Forensics and Security. She was a member of the inaugural steering committees for the IEEE Transactions on Network Science and for the IEEE Journal on Selected Areas in Information Theory. She currently serves as the Editor-in-Chief of the IEEE Transactions on Information Theory. Muriel was elected president of the IEEE Information Theory Society in 2012, and serves on its board of governors, having previously served for 11 years.

Muriel received the inaugural 2013 MIT EECS Graduate Student Association Mentor Award, voted by the students. She set up the Women in the Information Theory Society (WithITS) and Information Theory Society Mentoring Program, for which she was recognized with the 2017 Aaron Wyner Distinguished Service Award. She served as undergraduate Faculty in Residence for 7 years in two MIT dormitories (2002–2007). She was elected by the faculty and served as member and later chair of the MIT Faculty Committee on Student Life and as inaugural chair of the MIT Faculty Committee on Campus Planning. She was chair of the Institute Committee on Student Life. She was recognized as a Siemens Outstanding Mentor (2004) for her work with High School students. She serves on the Board of Trustees since 2015 of the International School of Boston, for which she is treasurer.

She has over 50 US and international patents awarded, the vast majority of which have been licensed or acquired. For technology transfer, she has co-founded CodeOn, for which she consults, and Steinwurf, for which she is Chief Scientist. Muriel has supervised over 40 master's students, over 20 doctoral students and over 25 postdoctoral fellows.

Guessing Random Additive Noise Decoding (GRAND) or How to Stop Worrying About Error-Correcting Code Design

Abstract. To maintain data integrity in the face of network unreliability, systems rely on error-correcting codes. System standardization, such as this, has been occurring for 5G, is predicated on co-designing these error-correcting codes and, most importantly, their generally complex decoders, into efficient, dedicated and customized chips. In this talk, we show that this assumption is not necessary and has been leading to significant performance loss. We describe "Guessing Random Additive Noise Decoding", or GRAND, by Duffy, Médard and their research groups, which renders universal, optimal, code-agnostic decoding possible for low to moderate redundancy settings. Moreover, recent work with Yazicigil and her group has demonstrated that such decoding can be implemented with extremely low latency in silicon. GRAND enables a new exploration of codes, in and of themselves, independently of tailored decoders, over a rich family of code designs, including random ones. Surprisingly, even the simplest code constructions, such as those used merely for error checking, match or markedly outperform state-of-the-art codes when optimally decoded with GRAND. Without the need for highly tailored codes and bespoke decoders, we can envisage using GRAND to avoid the issue of limited and sub-optimal code choices that 5G encountered, and instead have an open platform for coding and decoding.

Keynote 2

Dr. Jon Jenkins

Computer Scientist, NASA Ames Research Center, California, United States

Dr. Jon M. Jenkins is Co-Investigator for Data Processing for both the Kepler Mission and the upcoming TESS Mission. He conducts research at NASA Ames Research Center on data processing and detection algorithms for discovering transiting extrasolar planets. Dr. Jenkins joined the Kepler team in 1995 to help develop the technology for Kepler before it was selected for flight in 2002. He led the design and development of the science processing pipeline for Kepler, which takes the data from raw pixels to the detection and initial characterization of transiting planet candidates. In February 2014, Dr. Jenkins joined NASA Ames Research Center to lead the effort to design and build a science processing center for NASA's newly selected TESS Mission which will perform an all-sky transit survey to identify the closest and best Earth-size and super-Earth-size planets for follow-up and characterization. Dr. Jenkins received NASA's Exceptional Technology Achievement Medal as well as NASA's prestigious Software of the Year Award in 2010 for his work on Kepler. He holds a Ph.D. in Electrical Engineering from Georgia Tech Institute of Technology and has co-authored over 120 scientific papers in planetary science, astrophysics and technology.

Stochastic Signal Processing and Machine Learning Approaches for Exoplanet Transit Surveys

Abstract. Since 1995, over 5000 exoplanets orbiting other stars in our galaxy have been discovered. By far, the most productive detection method has been transit photometry which has detected ~3900. In transit surveys, repeated small dips in brightness of a star can reveal the presence of an exoplanet crossing the face of its host star as viewed from our solar system. The change in brightness caused by a planetary transit varies from less than 80 ppm to ~2% for small terrestrial planets like Earth to large, inflated hot Jupiters. These brightness changes can be difficult to detect as they are buried in intrinsic stellar brightness variations such as star spots and instrumental effects. In this talk, I present an adaptive wavelet-based matched filter approach to detecting these weak planetary signatures that has resulted in the discovery of ~2900 of the current exoplanet catalog from observations by NASA's Kepler and Transiting Exoplanet Survey Satellite (TESS) missions. This talk also discusses the application of machine learning techniques for vetting and validation of transiting exoplanet candidates.

Keynote 3

Dr. Justin Dauwels

Associate Professor, Department of Microelectronics, TU Delft, The Netherlands

Dr. Justin Dauwels is an Associate Professor at the TU Delft (Circuits and Systems, Department of Microelectronics). He was an associate professor of the School of Electrical and Electronic Engineering at the Nanyang Technological University (NTU) in Singapore till the end of 2020. He was the Deputy Director of the ST Engineering—NTU corporate lab, which comprises 100+ Ph.D. students, research staff and engineers, developing novel autonomous systems for airport operations and transportation.

His research interests are in data analytics with applications to intelligent transportation systems, autonomous systems, and analysis of human behavior and physiology. He obtained his Ph.D. degree in electrical engineering at the Swiss Polytechnical Institute of Technology (ETH) in Zurich in December 2005. Moreover, he was a postdoctoral fellow at the RIKEN Brain Science Institute (2006–2007) and a research scientist at the Massachusetts Institute of Technology (2008–2010). He has been a JSPS postdoctoral fellow (2007), a BAEF fellow (2008), a Henri-Benedictus Fellow of the King Baudouin Foundation (2008) and a JSPS invited fellow (2010, 2011).

He served as Chairman of the IEEE CIS Chapter in Singapore from 2018 to 2020, and serves as Associate Editor of the IEEE Transactions on Signal Processing (since 2018), Associate Editor of the Elsevier journal Signal Processing (since 2021), member of the Editorial Advisory Board of the International Journal of Neural Systems, and organizer of IEEE conferences and special sessions. He is also Elected Member of the IEEE Signal Processing Theory and Methods Technical Committee and IEEE Biomedical Signal Processing Technical Committee, both since 2018. His research team has won several best paper awards at international conferences and journals.

His research on intelligent transportation systems has been featured by the BBC, Straits Times, Lianhe Zaobao, Channel 5 and numerous technology websites. Besides his academic efforts, the team of Dr. Justin Dauwels also collaborates intensely with local start-ups, SMEs and agencies, in addition to MNCs, in the field of data-driven transportation, logistics and medical data analytics. His academic lab has spawned four start-ups across a range of industries, ranging from AI for health care to autonomous vehicles.

Artificial Intelligence for Applications in Neurology

Abstract. Many tasks in medicine still involve substantial manual work. In many cases, there is strong potential for intelligent automation by Artificial Intelligence (AI), leading possibly to a reduction in costs and man-hours, while increasing the quality of clinical service. In this talk, we will consider applications of AI in the domain of neurology.

We are developing a low-cost validated system to automatically interpret EEG via remote access. Diagnosis and management of neurological disorders rely on visual review of EEG data by specialized physicians. As the duration of EEG recordings ranges from 30 minutes to several weeks, the visual review is time-consuming, and accounts for approximately 80% of total cost associated with EEG reading. Our system has the potential to reduce expenses associated with EEG testing and allows physicians to devote more quality time to their patients. One of the applications that we have explored so far is diagnosis of epilepsy of EEG. In this talk, we will show numerical results on large EEG data sets of epilepsy patients and healthy control subjects for multiple centers.

Keynote 4

Dr. Rowel Atienza

Professor and Scientist, University of the Philippines, Philippines

Dr. Rowel Atienza is a Professor at the Electrical and Electronics Engineering Institute of the University of the Philippines, Diliman. He holds the Dado and Maria

Banatao Institute Professorial Chair in Artificial Intelligence. Dr. Rowel has been fascinated with intelligent robots since he graduated from the University of the Philippines. He received his M.Eng. from the National University of Singapore for his work on an AI-enhanced four-legged robot. He finished his Ph.D. at The Australian National University for his contribution on the field of active gaze tracking for human-robot interaction. Dr. Rowel's current research work focuses on AI, computer vision, speech and language. He dreams of building useful machines that can perceive, understand and reason. To help make his dreams become real, Rowel has been supported by grants from the Department of Science and Technology (DOST), Samsung Research Philippines and Commission on Higher Education-Philippine California Advanced Research Institutes (CHED-PCARI).

Building Efficient Deep Neural Networks

Abstract. For untethered machines like mobile robots that rely on their own source of power, efficient execution of tasks is crucial in increasing their overall usefulness. Although deep learning has been pervasive in autonomous systems, there is a lack of complete understanding of how and where to extract efficiency in the design and deployment of models. In this talk, we examine design philosophies, algorithmic implementations, number representations, operator fusion and model packaging formats as mechanisms for achieving efficient deep learning models. We quantify the contribution of each optimization procedure to the overall goal of efficiency.

Keynote 5

Dr. Sunil Vadera

Professor of Computer Science, University of Salford, UK

Dr. Sunil Vadera is a Professor of Computer Science at the University of Salford. He is a Fellow of the British Computer Society, a Chartered Engineer (CEng) and Chartered IT Professional (CITP). He has held many leadership roles including as Dean of the School of Computing, Science and Engineering, Head of Computer Science, Associate Dean of Research and Director of Informatics Research Institute. He was Chair and Vice Chair of British Computer Society (BCS) Accreditations Committee from 2007 to 2010 which has responsibility for accreditation of the UK degrees in Computing. Dr. Vadera was awarded the UK BDO best Indian Scientist and Engineer in 2014 and the Amity Research Award for AI and Neural Networks in 2018.

Dr. Vadera gained a first class B.Sc. (Hons) in Computer Science and Mathematics from the University of Salford in 1982, receiving three best student prizes. He holds a Ph.D. from the University of Manchester in the area of Formal Methods of Software Development which was awarded in 1992. Following graduation, he began his career as a Research Assistant and progressed to a Lectureship in Computer Science in 1984. He was promoted to a Senior Lecturer in 1997 and to a Chair in Computer Science in 2000.

Dr. Vadera has served as an external examiner for several institutions including Loughborough University, Liverpool University, Leeds Metropolitan University, Northumbria University, London Southbank University and Middlesex University. He has also carried out reviews of Computer Science departments in Universities in Jordan, Algeria, China, India and Sri Lanka.

Methods for Pruning Deep Neural Networks

Abstract. Convolutional neural networks, that can take images as input, learn to identify key features and perform classification which are the heart of many of the proposed applications in medical diagnosis such as detecting breast cancer, predicting Alzheimer's disease and grading brain tumors. These neural networks can, however, be very large, taking up memory and requiring significant computational resources.

This talk presents a walk-through of research on methods for pruning deep neural networks. It begins by categorizing over 150 studies based on the underlying approach used and then focuses on three categories: methods that use magnitude-based pruning, methods that utilize clustering to identify redundancy and methods that use sensitivity analysis to assess the effect of pruning. Some of the key influencing studies within these categories are presented to highlight the underlying approaches and results achieved. Most studies present results which are distributed in the literature as new architectures, algorithms and data sets that have developed with time, making comparison across different studies difficult. The paper therefore provides a resource for the community that can be used to quickly compare the results from many different methods on a variety of data sets, and a range of architectures, including AlexNet, ResNet, DenseNet and VGG. The resource is illustrated by comparing the results published for pruning AlexNet and ResNet50 on ImageNet and ResNet56 and VGG16 on the CIFAR10 data to reveal which pruning methods work well in terms of retaining accuracy while achieving good compression rates.

The presentation concludes by identifying some research gaps and promising directions for future research.

Keynote 6

Prof. Afaq Ahmad

Professor, Department of Electrical and Computer Engineering, Sultan Qaboos University, Sultanate of Oman

Dr. Afaq Ahmad is a Professor at the Electrical and Computer Engineering department, Sultan Qaboos University, Oman. From 1980 until 1992, he was with the College of Engineering, Aligarh Muslim University, India. He received his Ph.D. in Computer Engineering in 1990 from Indian Institute of Technology, Roorkee, India. He also earned his M.Sc. and B.Sc. degrees in Electrical, Electronics and Communication Engineering and a post-graduate diploma in industrial management. Ahmad, a recipient of various scholarships, award and recognitions, has authored more than 200 scientific papers, books, book chapters, numerous technical reports and manuals. He received best scientific paper awards. He serves as editor, associate editor, member of international advisory boards and reviewer for many worlds' reputed journals. He is honored to serve as program chairs, technical chairs and tutorial chairs for many international conferences. He chaired many technical sessions, meetings and panel discussions of international conferences, symposia and meetings. He conducted many workshops and short courses. He has delivered many invited talks and keynote addresses on current issues in various areas of importance. His research interests include VLSI testing, Fault-Tolerant Computing, data coding and security, Mathematical Morphology and education. He has over 40 years of professional experience with universities and industries. He has credit for developing and assessing the curricula and programs for educational institutions. He is Fellow member of IETE, senior member of IEEE, life member of SSI, senior member of IACSIT, and member of IAENG and ISIAM societies.

Bit Swapped Test Technique with Statistical Signal Properties for EVS Applications

Abstract. The intent of this talk is to introduce the idea of a new scheme devised to reduce test power, with employment of a low-transition test pattern generator. Various techniques have been proposed to reduce the test power such as Bit Insertion and Bit swapping. In the proposed scheme, the Bit Swapping technique is integrated with pseudorandom signal analyzer to make further reduction in the test power. A hardware complexity estimation for testing Embedded Vision Systems (EVS) will also be discussed.

Keynote 7

Dr. Alex Casson

Associate Professor, Department of Electrical and Electronic Engineering, The University of Manchester, Manchester, England, United Kingdom

Dr. Alex Casson is a Reader in the Materials, Devices and Systems division of the Department of Electrical and Electronic Engineering at the University of Manchester; Visiting Reader in the School of Medicine at the University of Leeds; Bioelectronics technology platform lead for the EPSRC Henry Royce Institute for advanced materials; and a Fellow of the Alan Turing Institute for data science and artificial intelligence.

His research focuses on non-invasive bioelectronic interfaces: the design and application of wearable sensors, and skin-conformal flexible sensors, for human body monitoring and data analysis from highly artifact-prone naturalistic situations. This work is highly multi-disciplinary, and he has research expertise in Ultra low power microelectronic circuit design at the discrete and fully custom microchip levels; Sensor signal processing and machine learning for power and time-constrained motion artifact-rich environments; and Manufacturing using 3D printing, screen printing and inkjet printing. He has particular interests in closed loop systems: those which are tailored to the individual by personalized manufacturing via printing; and tailored to the individual by adjusting non-invasive stimulation (light, sound, electrical current) using data-driven responses/outputs from real-time signal processing. Dr. Casson's ultra low power sensors work is mainly for health and wellness applications, with a strong background in brain interfacing (EEG and transcranial current stimulation) and heart monitoring. Applications focus on both mental health situations including chronic pain, sleep disorders and autism, and physical health/rehabilitation applications including diabetic foot ulceration and chronic kidney disease.

Dr. Casson gained his undergraduate degree from the University of Oxford in 2006 where he read Engineering Science specializing in Electronic Engineering (M.Eng.).

He completed his Ph.D. from Imperial College London in 2010, winning the prize for best doctoral thesis in electrical and electronic engineering. Dr. Casson worked as a research associate and research fellow at Imperial College until 2013 when he joined the faculty at the University of Manchester. He is an ambassador for the Manchester Integrating Medicine and Innovative Technology (MIMIT) scheme for systematically connecting clinicians and engineers to address unmet clinical needs.

Dr. Casson is currently a Senior Member of the IEEE, Fellow of the Higher Education Academy and chair of the Institution of Engineering and Technology's healthcare technologies network.

Edge Signal Processing for Wearables: In-wearable Processing and Energy Harvesting

Abstract. Wearable devices are having a profound impact on health and social care delivery, with remote out-of-the-clinic measurement of physiological parameters now possible. This is leading to a new era of personalized medicine as longitudinal data can be collected about each person, and "Big Data" medicine as data and insights can be easily collected from a large population. Nevertheless, current wearables have a number of limitations, from the quality of the data collected to limited battery life, to questions on the privacy considerations of the data collected. Signal processing at the edge, that is, operating on the wearable device itself, is an important part of next generation wearables which aim to overcome these issues. This presentation will explore technologies being created at the University of Manchester for next generation wearables, focusing on signal processing and Big Data analyses. It will give an example of how multi-modal wearables can provide data fusion for improving data quality, and edge machine learning for human activity recognition. It will conclude by discussing energy harvesting and the potential for future self-powered wearable devices using insights obtained from the UK Biobank wearable data collected from 67,000 participants.

Keynote 8

Dr. Ahmed Abdelgawad

College of Science and Engineering, Central Michigan University, Mount Pleasant, USA

Dr. Ahmed Abdelgawad received his M.S. and a Ph.D. degree in Computer Engineering from University of Louisiana at Lafayette in 2007 and 2011 and subsequently joined IBM as a Design Aids and Automation Engineering Professional at Semiconductor Research and Development Center. In Fall 2012, he joined Central Michigan University as a Computer Engineering Assistant Professor. In Fall 2017, Dr. Abdelgawad was early promoted as a Computer Engineering Associate Professor. He is a senior member of IEEE. His area of expertise is distributed computing for Wireless Sensor Network (WSN), Internet of Things (IoT), Structural Health Monitoring (SHM), data fusion techniques for WSN, low power embedded system, video processing, digital signal processing, Robotics, RFID, Localization, VLSI and FPGA design. He has published two books and more than 88 articles in related journals and conferences. Dr. Abdelgawad served as a reviewer for several conferences and journals, including IEEE WF-IoT, IEEE ISCAS, IEEE SAS, IEEE IoT Journal, IEEE Communications Magazine, Springer, Elsevier, IEEE Transactions on VLSI, and IEEE Transactions on I&M. He served on the technical committees of IEEE ISCAS 2007/8 and IEEE ICIP 2009 conferences. He served on the administration committee of IEEE SiPS 2011. He also served on the organizing committee of ICECS2013 and 2015. Dr. Abdelgawad was the publicity chair in North America of the IEEE WF-IoT 2016/18/19 conferences. He was the finance chair of the IEEE ICASSP 2017. He is the TPC Co-Chair of I3C'17, the TPC Co-Chair of GIoTS 2017 and the technical program chair of IEEE MWSCAS 2018. He is the technical program chair of IEEE WF-IoT 2020. He delivered many tutorials at international conferences including IEEE SOCC, IEEE MWSCAS, IEEE SiPS and APCCAS. In addition, he taught many short IoT courses in different countries. He was the keynote speaker for many international conferences and conducted many webinars. He is currently the IEEE Northeast Michigan section chair and IEEE SPS Internet of Things (IoT) SIG Member. In addition, Dr. Abdelgawad served as a PI and Co-PI for several funded grants from NSF.

Using Internet of Things (IoT) to Fight Covid-19

Abstract. The Internet of Things (IoT) has opened up a world of opportunities and numerous applications in health care, from smart sensors to remote monitoring and smart medical devices integration that can collect invaluable additional data. IoT has the potential to not only keep patients safe and healthy but also give extra insight into symptoms, improve how physicians deliver care, enable remote care and generally give patients more control over their treatment. IoT is one of the latest technologies that will change our lifestyle in the coming years. This talk aims to give a comprehensive introduction to IoT and how to use IoT to fight Covid-19.

Keynote 9

Dr. Jeff Grover

Research Scientist, Research Laboratory of Electronics, Massachusetts Institute of Technology, Massachusetts, USA

Dr. Jeff Grover is a Research Scientist in the EQuS Group and PI in the Center for Quantum Engineering, both at MIT. He focuses on software infrastructure and

building extensible calibration, control and metrology protocols for multi-qubit systems. Prior to joining EQuS, he spent 5 years in industry—first as a Senior Data Scientist at Athenahealth applying machine learning to health care and later as a Senior Physicist at Northrop Grumman working on quantum annealing. Dr. Grover received his Ph.D. in Physics from the Joint Quantum Institute at the University of Maryland in 2015. He graduated from Amherst College in 2009, majoring in Physics and Mathematics.

Waveguide QED with Artificial Atoms

Abstract. The strong-coupling regime of atom-photon interactions, in which a single atom and single photon can coherently exchange an excitation, is a common target for quantum optics and quantum information platforms. This is now readily achieved in cavity Quantum Electrodynamics (QED) by confining real or artificial atoms within the modes of cavities. There is a complementary paradigm, dubbed waveguide QED, where instead atoms couple to itinerant photons propagating in waveguides. The small mode volume of the waveguide can enable the strong-coupling regime even without cavity enhancement. In this talk, we discuss waveguide QED experiments with superconducting artificial atoms coupled to microwave coplanar waveguides, in which we realize substantial atom-photon coupling. This allows us to engineer so-called "giant atoms" that break the dipole approximation, and to deterministically emit single photons into a chosen direction. This architecture is a potential building block for remote entanglement or quantum communication protocols.

Keynote 10

Professor David Bruton, Jr. Centennial

University Distinguished Teaching Professor, Director, Texas Robotics, Associate Chair, Department of Computer Science, The University of Texas, USA

Dr. Peter Stone is the David Bruton, Jr. Centennial Professor and Associate Chair of Computer Science, as well as Director of Texas Robotics, at the University of Texas at Austin. He received his Ph.D. in 1998 and his M.S. in 1995 from Carnegie Mellon University, both in Computer Science. He received his B.S. in Mathematics from the University of Chicago in 1993. From 1999 to 2002, he was a Senior Technical Staff Member in the Artificial Intelligence Principles Research Department at AT&T Labs—Research.

Professor Stone's research interests in Artificial Intelligence include planning, machine learning, multiagent systems and robotics. Application domains include robot soccer, autonomous traffic management and human-interactive robots. His doctoral thesis research contributed a flexible multiagent team structure and multiagent machine learning techniques for teams operating in real-time noisy environments in the presence of both teammates and adversaries. He has developed teams of robot soccer agents that have won fifteen robot soccer tournaments (RoboCup) in both simulation and with real robots. He has also developed agents that have won ten auction trading agent competitions (TAC). Professor Stone is the author of "Layered Learning in Multiagent Systems: A Winning Approach to Robotic Soccer" (MIT Press, 2000) as well as a co-author of "Autonomous Bidding Agents: Strategies and Lessons from the Trading Agent Competition" (MIT Press, 2007).

Professor Stone is an Alfred P. Sloan Research Fellow, Guggenheim Fellow, AAAI Fellow, IEEE Fellow, AAAS Fellow, ACM Fellow, Fulbright Scholar and 2004 ONR Young Investigator. In 2013, he was awarded the University of Texas System Regents' Outstanding Teaching Award and in 2014 he was inducted into the UT Austin Academy of Distinguished Teachers, earning him the title of University Distinguished Teaching Professor. In 2003, he won an NSF CAREER award for his proposed long-term research on learning agents in dynamic, collaborative and adversarial multiagent environments, in 2007 he received the prestigious IJCAI Computers and Thought Award, given biannually to the top AI researcher under the age of 35, and in 2016 he was awarded the ACM/SIGAI Autonomous Agents Research Award.

Professor Stone co-founded Cogitai, Inc., a start-up company focused on continual learning, in 2015, and served as President and COO until 2019. He is currently Executive Director of Sony AI America.

Machine Learning for Robot Locomotion: Grounded Simulation Learning and Adaptive Planner Parameter Learning

Abstract. Robust locomotion is one of the most fundamental requirements for autonomous mobile robots. With the widespread deployment of robots in factories, warehouses and homes, it is tempting to think that locomotion is a solved problem. However, for certain robot morphologies (e.g. humanoids) and environmental conditions (e.g. narrow passages), significant challenges remain.

This talk begins by introducing Grounded Simulation Learning as a way to bridge the so-called reality gap between simulators and the real world in order to enable transfer learning from simulation to a real robot (sim-to-real). It then introduces Adaptive Planner Parameter Learning as a way of leveraging human input (learning from demonstration) toward making existing robot motion planners more robust, without losing their safety properties.

Grounded Simulation Learning has led to the fastest known stable walk on a widely used humanoid robot, and Adaptive Planner Parameter Learning has led to efficient learning of robust navigation policies in highly constrained spaces.

Keynote 11

Professor Peter Han Joo Chong

School of Engineering, Computer and Mathematical Sciences, Faculty of Design and Creative Technologies, Auckland University of Technology

Professor Peter Han Joo Chong is the Associate Head of School (Research) at the School of Engineering, Computer and Mathematical Sciences, Auckland University of Technology, New Zealand. Between 2016 and 2021, he was the Head of Department of Electrical and Electronic Engineering at AUT. He received his B.Eng. (with distinction) in Electrical Engineering from the Technical University of Nova Scotia, Canada, in 1993, and the M.A.Sc. and Ph.D. degrees in Electrical and Computer Engineering from the University of British Columbia, Canada, in 1996 and 2000, respectively. He has visited Tohoku University, Japan, as a Visiting Scientist in 2010 and Chinese University of Hong Kong (CUHK), Hong Kong, between 2011 and 2012. He is currently an Adjunct Professor at the Department of Information Engineering, CUHK. He is a Fellow of the Institution of Engineering and Technology (FIET), UK.

Before joining AUT in 2016, Prof. Chong was an Associate Professor (tenured) from July 2009 to April 2016 and Assistant Professor from May 2002 to June 2009 at the School of Electrical and Electronic Engineering, Nanyang Technological University (NTU), Singapore. Between 2011 and 2013, he was the Assistant Head of Division of Communication Engineering. Between 2013 and 2016, he was the Director of Infinitus, Centre for Infocomm Technology. He was the recipient of "EEE Teaching Excellence Award" and "Nanyang Award Excellence in Teaching" in 2010, and "Nanyang Education Award (College)" in 2015. In 2015, he became a Fellow of the Teaching Excellence Academy in NTU. From February 2001 to May 2002, he was with the Radio Communications Laboratory at Nokia Research Center, Finland. Between July 2000 and January 2001, he worked in the Advanced Networks Division at Agilent Technologies Canada Inc., Canada. He co-founded P2 Wireless Technology in Hong Kong in 2009 and Zyetric Technologies in Hong Kong, New Zealand and the US in 2017.

His current research projects focus on machine learning techniques applied to software-defined vehicular networks. He has been developing techniques of deep reinforcement learning (DRL)-based resource management for future 5G-V2X networks. His research interests are in the areas of wireless/mobile communication systems including radio resource management, multiple access, MANETs/VANETs, green radio networks and 5G-V2X networks. He has published over 200 journal and conference papers, 1 edited book and 9 book chapters in the relevant areas.

MAC Layer Protocols in Vehicular Networks: State of the Art and Future Directions

Abstract. Traffic density around the globe is increasing on a day-to-day basis, resulting in more accidents, congestion and pollution. Timely delivery of safety and non-safety messages is necessary for traffic congestion control and for avoiding road mishaps. For efficient resource sharing and optimized channel utilization, the media access control (MAC) protocol plays a vital role.

This talk gives a review of the applications, characteristics and challenges faced in the design of MAC protocols. A classification of the MAC protocol is presented based on contention mechanisms and channel access. The classification based on contention is oriented as contention-based, contention-free and hybrid, whereas the classification based on channel access is categorized as distributed, centralized, cluster-based, cooperative, token-based and random access. An analysis of the objectives, mechanisms, advantages/disadvantages and simulators used in specified protocols is presented. In the end, the conclusion with a discussion on the future scope and open challenges for improving the MAC protocol design will be given.

Keynote 12

Dr. Hien Quoc Ngo

Institute of Electronics, Communications and Information Technology (ECIT), Queen's University Belfast, UK.

Dr. Hien Quoc Ngo is currently a Reader (Associate Professor) at Queen's University Belfast, and a UKRI Future Leaders Fellow. My specific research interests include massive (large-scale) MIMO, cell-free massive MIMO and cooperative communications. I have received three prestigious prizes: the IEEE ComSoc Stephen O. Rice Prize in 2015, the IEEE ComSoc Leonard G. Abraham Prize in 2017 and the Best Ph.D. award in 2018 by the European Association for Signal Processing (EURASIP).

Professional Activities: Editor of IEEE Transactions on Wireless Communications, Editor of IEEE Wireless Communications Letters, Editor of Elsevier Physical Communication (PHYCOM), Executive Editor of Transactions on Emerging Telecom. Technologies (Wiley), Editor of Digital Signal Processing, Editor of IEICE Transactions on Fundamentals of Electronics, Communications and Computer Sciences, Editor of REV Journal on Electronics and Communications, Guest Editor of IET Communications, special issue on "Recent Advances on 5G Communications", 2017, Guest Editor of IEEE Access, special issue on "Modelling, Analysis, and Design of 5G Ultra-Dense Networks", 2017, Technical Program Chair of the International Conference on Recent Advances in Signal Processing, Telecommunications and Computing (Jan. 2018), Technical Program Committee (TPC): IEEE Wireless Communications and Networking Conference (WCNC); International Conference on Wireless Communications and Signal Processing (WCS); The International Conference on Advanced Technologies for Communications (ATC); The International Symposium on Wireless Communication Systems (ISWCS); IEEE International Conference on Communications (ICC); The International Conference on Coastal Engineering (ICCE); The Global Communications Conference (GLOBECOM); The Vehicular Technology Conference (VTC).

Cell-Free Massive MIMO for 6G: Fundamentals, Myths, and Challenges

Abstract. Cell-free massive multiple-input multiple-output (MIMO) is a system where many (hundreds or thousands) access points or base stations coherently serve many (tens or hundreds) users. Different from the current cellular (mobile) networks where the coverage area is divided into cells, in cell-free massive MIMO, there are no cells or cell boundaries. Cell-free massive MIMO is expected to overcome the boundary effect—the inherent limitation of the current cellular networks which persists over the last 50 years. It is expected to ensure everything and everywhere get connected, and, hence, fulfills the key requirements of next generation wireless communication systems (beyond 5G and toward 6G).

In this talk, we first provide some basics of cell-free massive MIMO systems. We then discuss some key myths and explain why they are not true. Finally, we conclude with a range of important topics and future directions.

Keynote 13

Prof. Yury Shestopalov

University of Gävle, Gävle Sweden

Professor Yury Shestopalov received his M.Sc. in applied mathematics in 1975 and Ph.D. and Doctor of Science degrees in mathematics and physics in 1978 and 1988, respectively, from Moscow State University. In 1989, he was a post-doc at Royal Institute of Technology and Stockholm University, Sweden.

In 1978–2000, he was assistant, associate and full professor and department head at the Faculty of Computational Mathematics and Cybernetics at Kolmogorov School of Moscow State University. Since 2000, he is docent and since 2008 professor in applied mathematics at Karlstad University and since 2013 at University of Gävle.

Professor Shestopalov has been active in several scientific fields, such as mathematical methods for electromagnetics, partial differential and integral equations, numerical methods, inverse problems, nonlinear analysis and various applications in electromagnetic field theory connected with wave scattering, diffraction and propagation in linear and nonlinear media, investigations of materials, optics, contact mechanics, solution to large-scale problems with uncertain data and supercomputer technologies. Among his scientific interests are studies in didactics of applied mathematics. He is currently professor at University of Gävle.

On a Mathematical Theory of Resonance Scattering and Partial Invisibility

Abstract. We consider an introduction to the mathematical theory of resonance scattering, cloaking and invisibility. The developed theoretical and numerical techniques end up with explicit formulas that enable one to calculate resonance and cloaking parameters for a wide family of layered cylindrical structures having circular symmetry, like a homogeneous dielectric rod or Goubau line.

Keynote 14

Prof. Kazuya Kobayashi

Department of Electrical, Electronic, and Communication Engineering, Chuo University, Japan

Professor Kazuya Kobayashi received his B.S., M.S. and Ph.D. degrees, all in electrical engineering, from Waseda University, Tokyo, Japan, in 1977, 1979 and 1982, respectively. In 1982, he joined the Department of Electrical, Electronic, and Communication Engineering, Chuo University, Tokyo, Japan, where he has been Professor since 1995. He held central management positions at Chuo University including Vice President (2006–2008), Director of International Center (2003–2006) and Secretary to President (2001–2003). Dr. Kobayashi held Visiting Professor and

Visiting Scientist positions at various institutions including Macquarie University, Sydney, Australia (2014, 2016, 2017, 2019); Karpenko Physico-Mechanical Institute of the National Academy of Sciences of Ukraine, Lviv, Ukraine (2001, 2018); Physical Research Laboratory, Ahmedabad, India (2015); Institute of Radiophysics and Electronics of the National Academy of Sciences of Ukraine, Kharkov, Ukraine (2001); University of Wisconsin-Madison, Madison, Wisconsin, USA (1987–1988). He has been Adjunct Professor at The Electromagnetics Academy at Zhejiang University, Hangzhou, China (since 2004), and Honorary Professor, Amity University, Noida, India (since 2019). Dr. Kobayashi received a number of distinguished awards including The President's Award (2020) from URSI (International Union of Radio Science) for his leadership and untiring efforts in initiating, organizing and establishing AP-RASC as one of the URSI Flagship Meetings; The Governor's Award (2018) from Toyama Prefectural Government, Japan, for outstanding contribution to the promotion of tourism in Toyama Prefecture and raising international recognition through the holding of large-scale international conferences; JNTO Best International Convention Award (2016, 2018) from Japan National Tourism Organization (JNTO), The Government of Japan for outstanding contribution to the invitation of large-scale international congresses to Japan; M. A. Khizhnyak Award (2016) at "16th International Conference on Mathematical Methods in Electromagnetic Theory" (MMET*2016) for contribution to electromagnetic theory. Dr. Kobayashi is a Member of Science Council of Japan, a Fellow of The Electromagnetics Academy and a Fellow of URSI (International Union of Radio Science). He also serves as MICE Ambassador, JNTO since 2016. He has held various important positions in the international radio science and electromagnetics/optics communities including Vice-President of URSI (since 2021); Vice-Chair (2014–2017) and Chair (2017–2021) of URSI Commission B; URSI Assistant Secretary-General AP-RASC (since 2015); President (2008–2018) and Vice-President (since 2021) of the Japan National Committee of URSI; Chair of the PIERS Awards Committee, The Electromagnetics Academy (since 2018); Series Editor of the book series "Springer Series in Optical Sciences", Springer Nature (since 2020); Editor of the journal "Radio Science" (since 2019). Dr. Kobayashi has contributed significantly to organization of numerous international conferences including PIERS (Photonics and Electromagnetics Research Symposium) and URSI conferences as General Chair, General Co-Chair and Chair/Co-Chair of the Technical Program Committee. He is currently involved in organizing "XXXV URSI General Assembly and Scientific Symposium" (URSI GASS 2023; August 2023, Sapporo, Japan) as General Chair and Associate Scientific Program Coordinator. His research interests are in the areas including analytical regularization methods; Wiener-Hopf methods; high-frequency methods; electromagnetic theory; canonical problems; scattering and diffraction; radar cross section.

Radar Cross Section Analysis of a Finite Parallel-Plate Waveguide with Material Loading: A Rigorous Wiener-Hopf Approach

Abstract. The analysis of electromagnetic scattering by open-ended metallic waveguide cavities is an important subject in the prediction and reduction of the radar cross section (RCS) of a target. This problem serves as a simple model of duct structures such as jet engine intakes of aircraft and cracks occurring on surfaces of general complicated bodies. Some of the diffraction problems involving two- and three-dimensional cavities have been analyzed thus far based on high-frequency techniques and numerical methods. It appears, however, that the solutions due to these approaches are not uniformly valid for arbitrary dimensions of the cavity. Therefore, it is desirable to overcome the drawbacks of the previous works to obtain solutions which are uniformly valid in arbitrary cavity dimensions. The Wiener-Hopf technique is known as a powerful, rigorous approach for analyzing scattering and diffraction problems involving canonical geometries. In this talk, we shall consider a finite parallel-plate waveguide with four-layer material loading as a geometry that can form cavities, and analyze the plane wave diffraction rigorously using the Wiener-Hopf technique. Both E and H polarizations are considered. Introducing the Fourier transform of the scattered field and applying boundary conditions in the transform domain, the problem is formulated in terms of the simultaneous Wiener-Hopf equations. The Wiener-Hopf equations are solved via the factorization and decomposition procedure leading to an exact solution. However, this solution is formal since infinite series with unknown coefficients and infinite branch-cut integrals with unknown integrands are involved. For the infinite series with unknown coefficients, we shall derive approximate expressions by taking into account the edge condition. For the branch-cut integrals with unknown integrands, we assume that the waveguide length is large compared with the wavelength and apply a rigorous asymptotics. This procedure yields high-frequency asymptotic expressions of the branch-cut integrals. Based on these results, an approximate solution of the Wiener-Hopf equations, efficient for numerical computation, is explicitly derived, which involves a numerical solution of appropriate matrix equations. The scattered field in the real space is evaluated by taking the inverse Fourier transform and applying the saddle point method. Representative numerical examples of the RCS are shown for various physical parameters, and the far field scattering characteristics of the waveguide are discussed in detail. The results presented here are valid over a broad frequency range and can be used as a reference solution for validating other analysis methods such as high-frequency techniques and numerical methods.

Keynote 15

Dr. C. J. Reddy

Vice President of Business Development-Electromagnetics, Americas Altair Engineering, Inc, USA

Dr. C. J. Reddy is the Vice President, Business Development Electromagnetics for Americas at Altair Engineering, Inc. (www.altair.com). Dr. Reddy was a research fellow at the Natural Sciences and Engineering Research Council (NSERC) of Canada during 1991–1993 and was awarded the US National Research Council (NRC) Resident Research Associateship in 1993 to work at NASA Langley Research Center in Hampton, Virginia. While conducting research at NASA Langley, he developed various computational codes for electromagnetics and received a Certificate of Recognition from NASA for development of a hybrid Finite Element Method/Method of Moments/Geometrical Theory of Diffraction code for cavity backed aperture antenna analysis. Dr. Reddy is a Fellow of IEEE, Fellow of Applied Computational Electromagnetics Society (ACES) and a Fellow of Antenna Measurement Techniques Association (AMTA). Dr. Reddy served on ACES Board of Directors from 2006 to 2012 and again from 2015 to 2018. Dr. Reddy was awarded Distinguished Alumni Professional Achievement Award by his alma mater, National Institute of Technology (NIT), Warangal, India, in 2015. He published 37 journal papers, 77 conference papers and 18 NASA Technical Reports to date. Dr. Reddy is a co-author of the book, “Antenna Analysis and Design Using FEKO Electromagnetic Simulation Software”, published in June 2014 by SciTech Publishing (now part of IET). Dr. Reddy was the General Chair of ACES 2011 Conference held in Williamsburg, VA, during March 27–31, 2011, and also ACES 2013 conference, Monterey, CA (March 24-28, 2013), as well as the General Chair of ACES 2015 conference held in Williamsburg, Virginia, during March 22-26, 2015. He was the Co-General Chair of 2014 IEEE International Symposium on Antennas and Propagation and USNC-URSI Radio Science Meeting held during July 6-11, 2014, in Memphis, TN. Dr. Reddy was the General Chair for AMTA 2018 conference held in Williamsburg, Virginia, during November 3-8, 2018. Dr. Reddy is elected as a member of AMTA Board of Directors for a three-year term starting January 2020 and will be the Technical Coordinator for AMTA 2020 Conference. Dr. Reddy is also serving as an Associated Editor for the newly introduced, IEEE Open Journal of Antennas of Propagation.

Pushing the Boundaries of Computational Electromagnetics—Application to Antenna Designs and Placement of Antennas

Abstract. Advances in computational electromagnetic tools have made possible antenna design and integration of antennas on various ground, sea, air and space platforms. Now numerical simulations can be performed to evaluate the effects of antenna design, placement, radiation hazard, EMC/EMI, etc. for wide-ranging industry applications. Numerical approaches that include full-wave techniques such as Method of Moments (MoM), Multilevel Fast Multipole Method (MLFMM) and asymptotic techniques such as Physical Optics (PO) and Uniform Theory of Diffraction (UTD) are being utilized to solve many challenging problems that were not possible in the past. For many practical applications, sometimes it is necessary to study the electromagnetic behavior of a specific structure over a broad frequency band. In this talk, an overview of various advanced numerical techniques that are useful for antenna designs and placement studies will be presented.

Keynote 16

Dr. Mehmet Emir Koksal

Ondokuz Mayis University, Turkey

Dr. Mehmet Emir Koksal is a Visiting Professor in the Department of Applied Mathematics at the University of Twente, and he is also a Professor of Applied Mathematics in the Department of Mathematics at the Ondokuz Mayis University. He received his Ph.D. in Mathematics from Gebze Institute of Technology, Turkey, in 2009. After receiving his Ph.D., he studied as a Postdoctoral Researcher at the Department of Mathematical Sciences, Florida Institute of Technology, Melbourne, during 2009–2010. In 2014, he passed the exam held by The Presidency of the Inter-University Council of Turkey and received the title of Associate Professor in Mathematics.

His research interests are in systems and control theory, differential equations and numerical analysis. The main focus is on the development and analysis of commutativity conditions of time-varying systems, feedback systems, decomposition and transitivity properties of commutativity, and applications of commutativity in physical systems using differential and difference equations. He also studies ordinary and partial differential equations, their numerical solution methods, and mathematical modeling and analysis of various engineering problems using differential equations. His research has been published in top international journals. He has received numerous grants and awards. He has reviewed over 200 manuscripts for 60 different international peer-reviewed journals and 40 different international conferences. His research has been mainly supported by the Scientific and Technological Research Council of Turkey. He has led two national projects, one of which was awarded a project performance award. He has been a keynote speaker, invited speaker and a member of the organizing and scientific committees of many well-known international conferences.

Some Properties of Time-Varying Linear Systems

Abstract. In this talk, commutativity, decomposition and transitivity properties of continuous and discrete time-varying linear systems are described. Necessary and sufficient conditions for these properties are presented. Some applications of these properties are discussed.

Keynote 17

Dr. Shuai Zhang

Aalborg University, Denmark

Dr. Shuai Zhang (SM'18) received his B.E. degree from the University of Electronic Science and Technology of China, Chengdu, China, in 2007 and his Ph.D. degree in electromagnetic engineering from the Royal Institute of Technology (KTH), Stockholm, Sweden, in 2013. After his Ph.D. studies, he was a Research Fellow at KTH. In April 2014, he joined Aalborg University, Denmark, where he currently works as Associate Professor. He is also the group leader of antennas in the department of Electronic Systems at Aalborg University with 12 staff. In 2010 and 2011, he was a Visiting Researcher at Lund University, Sweden, and at Sony Mobile Communications AB, Sweden, respectively. He was also an external antenna specialist at Bang and Olufsen, Denmark, in 2016–2017. He has supervised/co-supervised 7 Postdocs and 18 Ph.D. students. He has co-authored over 100 articles in well-reputed international journals (where 50 are published in various IEEE Trans.) and over 16 US or WO patents (where all the patents are sold to industrial partners). His citations in Scopus are over 3000 with H index of 28. His current research interests include millimeter-wave antennas for cellular communications, biological effects, Meta surfaces, CubeSat antennas, Massive MIMO antennas, wireless sensors and RFID antennas.

He is an IEEE Senior member. He is the Associate Editor for 3 journals: IEEE Antennas and Wireless Propagation Letters (impact factor 3.834); Sensors (impact factor 3.576 and BFI level 2); IET Microwaves, Antennas and Propagation (impact factor 1.972). He was the Lead Guest Editor in IEEE Antennas and Wireless Propagation Letters (impact factor 3.834) for the special cluster of "5G/6G Enabling Antenna Systems and Associated Testing Technologies". He is the Guest Editor in Frontiers in Materials (impact factor 3.515) for the special cluster of "Transformation optics and its frontier branches". He is also a reviewer for all the top IEEE and IET journals in antenna areas, where he got the prize of "Top Reviewers in IEEE Transactions on Antennas and Propagation 2019–2020 and 2020–2021". He is the General Co-Chair for iWAT2023 at Aalborg, Denmark, the Super Reviewer (previously known as Super TPC or Vice Chair) for IEEE APS 2020 and 2021 and the TPC for several top IEEE conferences. He has also been intensively invited to international conferences and industry to give keynote/plenary speech and presentations. He was invited to serve as a reviewer for Icelandic Research Fund in 2019 and 2020. From 2019–2023, He was the Management Committee for EU COST Action CA18223 of SyMat, which mainly focuses on high symmetrical periodic structures or metamaterials.

Isolation Enhancement of Massive MIMO Antenna Arrays for Base Stations

Abstract. Massive multiple-input multiple-output (MIMO) technology has served as an indispensable technology for fifth-generation (5G) communications. In 5G beyond cellular systems, millimeter-wave technology will be applied to massive MIMO arrays. Mutual coupling in massive MIMO or large-scale phased arrays will deteriorate the performance such as channel capacity, active impedance matching and total radiation efficiency. Furthermore, in most massive MIMO arrays, there are no isolators between antenna elements and power amplifiers. The efficiency of power amplifiers is very sensitive to the mutual coupling between array elements. Massive MIMO antenna arrays are typically dual-polarized as well. It is very challenging and essential to suppress the mutual coupling among adjacent and non-adjacent antenna elements for dual-polarized Massive MIMO arrays. This presentation will focus on introducing the challenges and recent progress in isolation enhancement of massive MIMO antenna arrays in base stations at both sub-6GHz and millimeter-wave frequencies. The future research directions of massive MIMO will also be addressed.

Keynote 18

Dr. Zeljko Zilic

McGill University, Montreal, Quebec, Canada

Dr. Zeljko Zilic received his Ph.D. and M.Sc. from the University of Toronto, and his B.Eng. from University of Zagreb, all in Electrical and Computer Engineering. From 1996 till 1997, he worked for Lucent Microelectronics, where as a Member of Technical Staff he was involved in the design, test and verification of Orca FPGAs. He joined McGill University in 1998, where he is now an Associate Professor. Dr. Zilic has used a sabbatical leave in 2004/2005 to work with ST Microelectronics in Ottawa on assertion-based verification of multiprocessors on chip, conducted in cooperation with the IBM research lab in Haifa.

Dr. Zilic conducts research on various aspects of the design and test of Microsystems, including programmable logic cores. He has published over 200 publications, for which he received several awards, including Myril B. Reed Best Paper award from Midwest Symposium on Circuits and Systems, and Best Paper Award from Design and Verification Conference and Exposition. He has also received several Honorary Mention Awards, such as from IEEE Microelectronics Systems Education Conference, ETAN and CMC Symposia. Dr. Zilic has led or participated in program committees for the High-level Design Validation and Test, International Symposium on Field Programmable Gate Arrays, International Test Conference, Silicon Debug and Diagnosis, Midwest Symposium on Circuit and Systems, NEWCAS and AQTR conferences, Electronic Circuits and Systems Conference and Workshop on Open-Source Test Technology. Dr. Zilic has supervised over 70 post-graduate students, who have received numerous awards for their theses, and the graduating students have moved on to leading industrial and academic institutions upon their graduation.

Dr. Zilic has been granted the Chercheur Strategique research chair from the Province of Quebec. He has also been awarded the Wighton Fellowship for laboratory course teaching, by Sandford Fleming Foundation and the National Council of Deans of Engineering and Applied Science (NCDEAS). He is the Senior Member of IEEE and ACM.

From Internet of Things to Blockchain to Healthcare

Abstract. In this talk, an evolution of the Internet of Things (IoT) support was described for health care, starting with numerous sensing devices and Body Area Network (BAN), and ending up in a full-scale Blockchain support for health care. The focus is on the emerging issues and the subtle blockchain operation aspects. Regarding the blockchain implementation, the alternatives among the public and private blockchain are described, as well as the first performance evaluation criteria established directly for the healthcare blockchain.

Keynote 19

Dr. Sarath Kodagoda

Professor of Robotics, President of Australian Robotics and Automation Association (ARAA), Director UTS Robotics Institute, Hornsby, New South Wales, Australia

Dr. Sarath Kodagoda is an award-winning and influential researcher in the field of robotics and mechatronics, whose passion for teaching has resulted in a number of significant education innovations. Dr. Sarath was instrumental in the development of the mechatronics courses at UTS and designed the University's innovative Robotic Remote Lab teaching facilities. He has received three teaching awards, including the prestigious UTS Medal for teaching and research integration, and earned a 2015 Citation for Outstanding Contributions to Student Learning from the Australian Government's Office for Learning and Teaching. With over 20 years of experience in robotics, Dr. Sarath's broad aim is to develop robotic solutions that allow people to carry out dirty, dull and dangerous tasks safely and efficiently. At UTS, he is Acting Director of the internationally acclaimed UTS Robotics Institute, and his research interests include sensor fusion, sensor networks, data processing and decision-making through machine learning. He is the founder of the iPipes lab, which hosts state-of-the-art research and development in robotics for wastewater infrastructure. He was a co-chair of the first version of A Robotics Road Map for Australia, which was launched in 2018, as well as currently contributing to the second version. The robotic roadmap is a guide to how Australia can exploit the benefits of a new robotic economy. Dr. Sarath has attracted many research and development grants, is an editorial board member, associated editor and reviewer on a number of robotics journals and conferences, and is regularly invited to give keynote addresses or to chair conferences. He has served the research community as President of the Australian Robotics and Automation Association (ARAA), and he is an ambassador for the NSW Smart Sensing Network. Dr. Sarath has published more than 170 papers, including a paper in the prestigious Scientific Reports, Nature, attracted over $6 million in industry grants and won eight research awards, including state, national and international awards. He has supervised 12 Ph.D. students for successful completions, who are currently doing well at Amazon, Google and ABB. In 2018, he was a Visiting Professor at Stanford University in the US. He is a regular media commentator about his specialist areas.

Robotics for Underground Pipe Inspections

Abstract. Underground systems are an important national infrastructure for any country, and they are realized through a web of underground pipes and tunnels—some of them could be 100m below ground. These infrastructures consist of pipes and tunnels of varying sizes, shapes and effluents, making a unique monitoring solution very challenging. Majority of those infrastructures are old, and they undergo a significant strain due to urbanization and industrialization. Yet they cannot fail needing comprehensive inspections and timely intervention. Robotic technologies have shown tremendous advancements in the past decade and penetrated into many application areas to make our life safer, easier and efficient. This talk focuses on the need, role and developments of the robotic technologies in relation to underground pipe condition monitoring.

Keynote 20

Dr. Boris Novikov

Professor, Department of Informatics HSE university, St. Petersburg, Russia

Dr. Boris Novikov is a Professor, the Department of Informatics at National Research University Higher School of Economics in Saint Petersburg. He graduated from Leningrad University (school of mathematics and mechanics) in 1972 and joined Operations Research laboratory at the same university (Saint-Petersburg University since 1991). He has been a professor since 2000, and he was the department chair in 2013–2018 at Saint Petersburg University. Dr. Novikov joined National Research University "Higher School of Economics" in January 2019. He visited CRAI (Italy) and Aalto University (Finland), and was a leading researcher for several projects funded by INTS and Russian agencies, as well as industrial partners. He is a member of journal editorial boards of "Programming and Computer software" and "Computer Science and Information Systems", and served as a program committee chair for several international conferences. His research interests are in a broad area of information management and include several aspects of design, development and tuning databases, applications and database management systems, as well as distributed scalable systems for stream processing and analytics.

Adaptive Query Execution in Distributed Stream Processing

Abstract. Big data comes through networks and require online processes. Several problems depending on network data processing require efficient online analytics. An incomplete list includes a discovery of concept drift, detection of anomalies such as intrusion detection and much more. The high expressiveness of declarative languages provides for adequate specifications of required processing, while the power of query optimizers enables the generation of efficient executable code. However, changes in the statistical properties of data streams may result in sub-optimal execution. Adaptive optimization and execution techniques capturing thing dynamics of incoming data streams are discussed.

Keynote 21

Dr. Carlos M. Travieso-González

Professor and Head of Signals and Communications Department, Institute for Technological, Development and Innovation in Communications (IDeTIC), University of Las Palmas de Gran Canaria (ULPGC), Spain

Dr. Carlos M. Travieso-González received his M.Sc. degree in 1997 in Telecommunication Engineering at Polytechnic University of Catalonia (UPC), Spain; and Ph.D. degree in 2002 at University of Las Palmas de Gran Canaria (ULPGC-Spain). He is Full Professor and Head of Signals and Communications Department at ULPGC. He belongs to ULPGC from 2001, teaching subjects on signal processing, pattern recognition and learning theory. His research lines are biometrics, biomedical signals and images, data mining, classification system, signal and image processing, machine learning and environmental intelligence. He has conducted research in more than 50 International and Spanish Research Projects, in some of them as head researcher. He is co-author of 4 books, co-editor of 25 Proceedings Books, Guest Editor for eight JCR-ISI international journals and up to 24 book chapters. He has over 460 papers published in international journals and conferences (83 of them indexed on JCR—ISI—Web of Science). He has published 7 patents in Spanish Patent and

Trademark Office. He has been supervisor on 9 Ph.D. Thesis (11 more are under supervision) and 130 Master Thesis. He is founder of The IEEE IWOBI conference series and President of its Steering Committee, of The InnoEducaTIC conference series; and of The APPIS conference series. He is an evaluator of project proposals for European Union (H2020 and Horizon Europe), Medical Research Council (MRC—UK), Spanish Government (ANECA), Research National Agency (ANR—France), DAAD (Germany), Argentinian Government and Colombian Institutions. He has been a reviewer in different indexed international journals (<70) and conferences (<240) since 2001. He has been a member of IASTED Technical Committee on Image Processing from 2007 and member of IASTED Technical Committee on Artificial Intelligence and Expert Systems from 2011. He was the General Chair, APPIS 2020, IEEE-IWOBI 2020, APPIS 2019 General Chair and IEEE-IWOBI 2019, IEEE-IWOBI 2018 General Chair, APPIS 2018 General Chair, InnoEducaTIC 2017 General Chair, IEEE-IWOBI 2017 General Chair, IEEE-IWOBI 2015 General Chair, InnoEducaTIC 2014 General Chair, IEEE-IWOBI 2014 General Chair, IEEE-INES 2013 General Chair, NoLISP 2011 General Chair, JRBP 2012 General Chair and IEEE-ICCST 2005 Co-Chair. He is an Associate Editor in Computational Intelligence and Neuroscience journal (Hindawi—Q1 JCR-ISI), Sensors (MDPI—Q1 JCR-ISI) and Entropy (MDPI—Q2 JCR-ISI). He was the Vice-Dean from 2004 to 2010 in Higher Technical School of Telecommunication Engineers in ULPGC; and Vice-Dean of Graduate and Postgraduate Studies from March 2013 to November 2017.

Intelligent Information Processing on Neurodegenerative Diseases

Abstract. The use of image processing methods is a useful tool in order to extract information from persons for different application. In particular, soft biometrics application can be applied to detect race, age, gender, expression, etc. In this case, this information is used to extract the grade of emotions. It can be an important and interesting indicator for medical doctors, and to have more diagnostic evidence about a decision or criterion in neurodegenerative diseases. There is that to difference the concept of expression and the grade of expression or arousal. For this kind of disease, the important concept is the grade of expression because it says if the person loses that grade or not, with independence of the valence, positive or negative. The neurodegenerative disease is present in the arousal. This kind of study can represent new paradigms for the medicine and in great input from the technology field, reaching precision medicine. The added value of this kind of proposal is its low cost and easy use.

Keynote 22

Dr. Miguel López-Benítez

Department of Electrical Engineering and Electronics, University of Liverpool, United Kingdom

Dr. Miguel López-Benítez received his B.Sc. and M.Sc. degrees (Hons.) in Telecommunication Engineering from Miguel Hernández University, Elche, Spain, in 2003 and 2006, respectively, and his Ph.D. degree (summa cum laude) in Telecommunication Engineering from the Technical University of Catalonia, Barcelona, Spain, in 2011. From 2011 to 2013, he was a Research Fellow with the Centre for Communication Systems Research, University of Surrey, Guildford, UK. In 2013, he became a Lecturer (Assistant Professor) with the Department of Electrical Engineering and Electronics, University of Liverpool, UK., where he has been a Senior Lecturer (Associate Professor) since 2018. Since 2018, he has also been an Affiliate Senior Research Associate with the ARIES Research Centre, Antonio de Nebrija University, Madrid, Spain. His research interests include the field of wireless communications and networking, with special emphasis on mobile communications, dynamic spectrum access and sharing, and the Internet of Things. He is an Associate Editor of IEEE Access, IET Communications, and Wireless Communications and Mobile Computing, and has served as Guest Editor for other journals' special issues as well as a member of the organizing committee for several IEEE workshops.

Ultra-Reliable Low Latency Communications (URLLC) for 5G Networks and Beyond

Abstract. The 5th Generation of cellular mobile communication systems introduced the concept of Ultra-Reliable Low-Latency Communication (URLLC) as a new use case to support innovative application scenarios such as Industry 4.0, intelligent transportation systems, tactile Internet, remote health care, mission-critical services and ad-hoc disaster/emergency relief, among others. The demanding requirements set for URLLC make the wireless access design very challenging in terms of protocols and associated transmission techniques. This talk will provide a brief overview of the new 5G use cases with a special emphasis on URLLC and will introduce a novel hybrid transmission technique specifically devised to improve the link reliability for 5G URLLC with efficient use of resources

Keynote 23

Dr. Gaurav Sharma

Professor, Department of Electrical and Computer Engineering, Department of Computer Science, University of Rochester, USA

Dr. Gaurav Sharma is a Professor in the Departments of Electrical and Computer Engineering, Computer Science, and Biostatistics and Computational Biology, and a Distinguished Researcher in Center of Excellence in Data Science (CoE) at the Goergen Institute for Data Science at the University of Rochester. He received his

Ph.D. degree in Electrical and Computer engineering from North Carolina State University, Raleigh, in 1996. From 1993 through 2003, he was with the Xerox Innovation group in Webster, NY, most recently in the position of Principal Scientist and Project Leader. His research interests include data analytics, cyber physical systems, signal and image processing, computer vision and media security, areas in which he has 54 patents and has authored over 220 journal and conference publications. He served as the Editor-in-Chief for the IEEE Transactions on Image Processing from 2018 through 2020, and for the Journal of Electronic Imaging from 2011 through 2015. He is a member of the IEEE Publications, Products, and Services Board (PSPB) and chaired the IEEE Conference Publications Committee in 2017–2018. He is the editor of the Digital Color Imaging Handbook published by CRC press in 2003. Dr. Sharma is a fellow of the IEEE, a fellow of SPIE, a fellow of the Society for Imaging Science and Technology (IS&T) and has been elected to Sigma Xi, Phi Kappa Phi and Pi Mu Epsilon. In recognition of his research contributions, he received an IEEE Region I technical innovation award in 2008 and the IS&T Bowman award in 2021. Dr. Sharma served as a 2020–2021 Distinguished Lecturer for the IEEE Signal Processing Society.

Wearable Sensor Signal Processing and Data Analytics for Health Applications

Abstract. Advances in nano-fabrication and MEMS devices have led to radical improvements in sensing technologies in recent years. These improvements are most visible to all of us in our Smartphones that already feature a panoply of miniaturized sensors. Many of the same sensors are also positively impacting several other application domains. In this talk, it is highlighted how smart lightweight body-worn sensors are set to revolutionize health care and the practice of medicine by providing technologies for assessing biomarkers for physiological and physical attributes related to disease condition, treatment effectiveness and longitudinal progression. In contrast with the subjective, sporadic in-clinic assessments that are in common use today, body-worn sensors can provide objective and repeatable measurements and based on extended periods of continuous monitoring. Examples from recent and ongoing research that features lightweight, low-power sensors that can be affixed to the body like adhesive temporary tattoos, in a diverse set of health monitoring applications including quantification of movement disorders for Parkinson's and Huntington's diseases, stroke rehabilitation and cardiac monitoring are presented. Examples of signal processing and data analytics for these applications that effectively exploit the sensor measurements are presented. Finally, the ongoing and emerging directions for research and development are highlighted.

Keynote 24

Dr. Vijayan Sugumaran

Prof. & Chair, Department of Decision and Information Sciences, Co-Director, Center for Data Science and Big Data Analytics, Oakland University, USA

Dr. Vijayan Sugumaran is a Distinguished University Professor of Management Information Systems and Chair of the Decision and Information Sciences department at Oakland University. He is also the Co-Director of the Center for Data Science and Big Data Analytics at Oakland University. He received his Ph.D. in Information Technology from George Mason University, Fairfax, Virginia, USA. Over the years, he has taught courses at the Graduate and Undergraduate levels in Object-Oriented Systems Development, C++, Java, JavaScript, Database Management Systems and Data Warehouses, Advanced Databases and Big Data Management, Deep Learning and Text Analytics, Systems Analysis and Design, Electronic Commerce, and Introduction to MIS. His research interests are in the areas of Big Data Analytics, Business Intelligence, Ontologies and Semantic Web, Intelligent Agent and Multi-Agent Systems, Component Based Software Development, Knowledge-Based Systems, and Data and Information Modeling. He has published over 250 peer-reviewed articles in Journals, Conferences and Books. He has edited twenty books and serves on the Editorial Board of eight journals. His recent publications have appeared in Information Systems Research, ACM Transactions on Database Systems, IEEE Transactions on Engineering Management, IEEE Transactions on Big Data, IEEE Transactions on Education, IEEE software, Communications of the ACM, Healthcare Management Science, Data and Knowledge Engineering, The DATABASE for Advances in Information Systems, Information Systems Journal, and Journal of Information Systems and E-Business Management.

Dr. Sugumaran is the Editor-in-Chief of the International Journal of Intelligent Information Technologies and also serves on the editorial board of eight other journals. He was the Program Co-Chair for the International Conference on Applications of Natural Language to Information Systems (NLDB 2008, NLDB 2013, NLDB 2016 and NLDB 2019). He was also the Chair of Intelligent Information Systems track for the Information Resources Management Association International Conference (IRMA 2001, 2002, 2005–2007) and the Intelligent Agent and Multi-Agent Systems in Business mini-track for Americas Conference on Information Systems (AMCIS 1999–2021). He served as Program Co-Chair for the 14th Workshop on E-Business (WeB2015), 29th Australasian Conference on Information Systems (ACIS 2018), 14th Annual Conference of Midwest Association for Information Systems (MWAIS 2019), and the 5th IEEE International Conference on Big Data Service and Applications (BDS 2019). He also served as Chair of the E-Commerce track for Decision Science Institute's Annual Conference, 2004. He was the Information Technology Coordinator for the Decision Sciences Institute from 2007 to 2010. He has presented his research results in various national and international conferences, including the International Conference on Information Systems and Americas Conference on Information Systems. He also regularly serves as a program committee member for numerous national and international conferences.

Data Management and Analytics for Wearable Technologies in Healthcare

Abstract. With the advent of Internet of Things and wearable devices, a huge digital footprint is left with every step taken in our daily lives. Though these devices positively impact health and lifestyle of users, on the dark side violations of health information privacy have become ubiquitous, diverting academic attention toward these issues. This study focuses on understanding the coping behavior of fitness tracker users when confronted with data privacy breach fear appeal.

Keynote 25

Dr. Lotte N. S. Andreasen Struijk

Department of Health Science and Technology, The Faculty of Medicine, Neurorehabilitation Robotics and Engineering, Center for Rehabilitation Robotics, Aalborg University, Denmark

Dr. Lotte N. S. Andreasen Struijk is presently Head of Center for Rehabilitation Robotics and Associate Professor in the Department of Health Science and Technology and in the Faculty of Medicine, Neurorehabilitation Robotics and Engineering, Aalborg University, North Jutland, Denmark. She has researched rehabilitation for more than 20 years. Her focus of the research is to increase the quality of life for individuals living with severe disabilities through new knowledge and technologies within rehabilitation engineering and robotics.

Dr. Lotte is an Associate Editor of IEEE Transactions on Neural Systems and Rehabilitation Engineering (SCI, Impact Factor: 3.34). She has published more than 112 research papers in International Journals and Conferences of repute.

Dr. Lotte has been awarded "Vanførefondens Forskerpris Researcher Award" in May 2019 by the Danish Foundation of Physically Disabled Persons. Her research outcome, "iTongue" was awarded the Clinica Award at the MedTech Nordic Investing and Partnering event held on September 3, 2013, in Helsinki, Finland.

She is Member of the following Boards: "Health, Disease Prevention, and Health Technology—the Danish Academy of Technical Sciences" and "The Think-thank on welfare technology of IDA—the Danish Union of Engineers".

Dr. Lotte has the following projects to her credits: Center for Rehabilitation Robotics, Intelligent Hybrid Light Weight Tendon Based Exoskeleton for Severely Disabled Individuals, HyperHand: Intelligent hybrid control of an exoskeleton hand based on non-invasive tongue interfacing combined with electrical muscle stimulation for restoration of hand function in paralyzed individuals GripX: A telerehabilitation platform for training with balls, H2020 Eurostar Project—iTooBot—Tongue Control of an Assistive Robotic Manipulator, H2020 Eurostar project Robert-SAS, Hybrid FES-EMG robotic rehabilitation, MultiRob: Multimodal control of assistive robotic arms for severely disabled individuals, EXOTIC: Assistive personal robotics platform with an exoskeleton using the tongue for intelligent control as a use case (EXOTIC) and Using iTongue for tongue control of an assistive robotic arm. She will start working on the nerve cuff electrode and cutaneous sensory feedback and has completed her work on tongue control of a multi-functional arm prosthesis and prostheses and Implants.

Her industry projects include Grant from the Danish Ministry of Science for an international Ph.D. on Efficient Tongue-Computer Interfacing for People with Upper-Limb Impairments Tongue control system.

Exoskeletons and Biorobotics in Rehabilitation

Abstract. The demographic changes result in a growing need for robotics in therapeutic and assistive rehabilitation. The combination of technologies, users with disabilities and their use in society poses requirements for such robotics systems that go beyond the traditional basic technical requirements. The talk gives examples of exoskeletons and bio-robotics controlled with tongue and brains signals that seek address such broader requirements.

Keynote 26

Dr. Ajay Pandey

School of Electrical Engineering and Robotics, Queensland University of Technology, Brisbane, Australia

Dr. Ajay Pandey was awarded a Ph.D. in Physics from the University of Angers (France) with a thesis titled "Physics and Models of Plastic Photovoltaic Devices". Currently, he is an Associate Professor at the School of Electrical Engineering and Robotics. He held prestigious fellowships including QUT-Vice Chancellor's Senior Research Fellowship in Medical Robotics, and Australian Renewable Energy Agency (ARENA) Research Fellowship in Next generation Photovoltaic Technology. His research interests have the interdisciplinary mix of Condensed Matter Physics, Molecular Electronics, Materials Robotics and Quantum Sensing. He leads a number of major research projects, including development and implementation of bio-inspired advanced optoelectronic devices for applications in Neuroscience, Bionics, Medical Robotics, Energy Up-Conversion and Robotic Vision. He is a Program Leader within QUT's Centre for A Waste Free World and also serves as an Editorial Board Member of Scientific Reports—a Springer Nature Group open access journal. He has graduated 4 Ph.D.s and 1 M.Phil. student and currently has the research supervision of 12 Ph.D. students. He is one of the unit coordinators of Master of Professional Engineering curriculum and teaches Electronics to undergraduate engineering students. He has published >85 peer-reviewed research publications, and his works are consistently cited by top R&D labs across the globe.

Intelligent Robotic Imaging Systems for Minimally Invasive Surgeries

Abstract. Minimally invasive surgeries are complex to perform, and surgeons often have to work within a limited field of view. Current state-of-the-art robot-assisted surgeries are mostly limited to open surgeries, and so far, there is a lack of reliable vision system that can identify and track soft tissue and interfaces between different tissues intra-operatively. Advancing the medical imaging technology will greatly enhance surgeons' clinical abilities and drastically lower the physical footprint and the cost of surgical procedures.

In this keynote, robotic vision approaches to overcome this limitation and share latest developments in applying intelligent sensors and machine learning to medical imaging for achieving enhanced situational awareness in medical robotics and bionics were introduced.

Keynote 27

Dr. Annalisa Bruno

Principal Scientist, Energy Research Institute, Nanyang Technical University, Singapore

Dr. Annalisa received her B.S., M.S. and Ph.D. degrees in Physics and Applied Physics from the University of Naples Federico II, Italy, where she also worked as a post-doc in the Physics and Chemical Engineering Departments. She then joined the Chemistry Department of Imperial College London, first as a Post-Doctoral Research Associate, and then as Long-Term Visiting Staff studying organic and hybrid materials for optoelectronic applications. During her scientific career, she has also been a visiting researcher at Lund University, Lawrence Berkeley National Laboratory, and the Strathclyde University, UK. In 2011 she became a tenured Senior Staff Scientist at Italian National Agency for New Technologies, Energy, and Sustainable Economic Development (ENEA). In 2014, she also joined the Energy Research Institute at Nanyang Technological University (ERI@N) as a Senior Scientist. Since 2017, she has been coordinating the Thermally Evaporated and Tandem Solar Cells team in ERI@N. They aim at developing highly efficient large-area solar cells.

Intrinsically-stable and Versatile n.i.p and p.i.n Co-Evaporated Halide Perovskites Solar Cells

Abstract. Metal-halide perovskites made a breakthrough in photovoltaic and light-emitting technologies in the last ten years. MHPs are one of the most promising low-cost materials, due to their excellent optoelectronic properties and fabrication versatility. Since the advent of the first perovskite solar cells (PSCs) in 2009, their power conversion efficiency (PCE) has now reached 25.6% [1], for active areas smaller than 1 cm2 and their operational stability is constantly improving [2-4]. The interest in transferring the existing technology into large-area perovskite modules using industrial-compatible techniques is exploding.

In this talk, it has been shown why thermal evaporation is a promising perovskite fabrication technique to bring this technology closer to reliable and extended production, by relying on excellent size scalability, promising stability, fine composition control and surface adaptability [5]. The co-evaporated perovskite thin films are uniform over large areas with low surface roughness and a long carrier lifetime. The presentation was about the highly efficient, large area, PSCs where the MAPbI3 perovskite is deposited by thermal co-evaporation. Developing optimization strategies customized for n.i.p [6-8] and p.i.n [9] architectures, the PSCs achieved PCEs above 20% in both configurations. Moreover, the co-evaporated MAPbI3 is formed intrinsically strain-free and the PSCs showed remarkable structural robustness and impressive thermal maintaining over ≈80% of their initial PCE after 3600 under continuous thermal aging at 85 °C without encapsulation.

Extrapolating the optimization strategies over large areas, the co-evaporated mini-modules achieved record PCEs up to 18.7% for active areas larger than 10 cm2 [5] [10]. Looking toward building-integrated photovoltaics, semi-transparent PSCs and mini-modules with a wide range of colors have also been developed.

Keynote 28

Dr. Ankit Agrawal

Research Professor, Department of ECE, Northwestern University, USA

Dr. Ankit Agrawal is a Research Professor in the Department of Electrical and Computer Engineering at Northwestern University, USA. He specializes in interdisciplinary artificial intelligence (AI) and big data analytics via high-performance data mining, based on a coherent integration of high-performance computing and data mining to develop customized AI solutions for big data problems with real-world impact. His research has contributed to large-scale data-driven discoveries in various scientific and engineering disciplines, such as materials science, health care, social media and bioinformatics. He has co-authored 200+ peer-reviewed publications, co-developed and released 20+ software, delivered 50+ invited/keynote talks at major conferences, universities and companies all over the world, been on program/organizing committees of 50+ conferences/workshops and served as a PI/Co-PI on 20+ projects funded by various US federal agencies (e.g. NSF, DOE, AFOSR, NIST, DARPA and DLA) as well as industry (e.g. Toyota Motor Corporation Japan). He is one of the few computer scientists who are actively introducing AI and advanced data science techniques in the field of materials science and has successfully led several large-scale materials informatics projects. As an example, he is co-leading the AI group at the Center for Hierarchical Materials Design (CHiMaD), which is a $60 million NIST-sponsored center of excellence. He is also serving as the Editor-in-Chief of Computers, Materials and Continua.

Artificial Intelligence and High-Performance Data Mining for Accelerating Scientific Discovery

Abstract. The increasing availability of data from the first three paradigms of science (experiments, theory and simulations), along with advances in artificial intelligence and machine learning (AI/ML) techniques, has offered unprecedented opportunities for data-driven science and discovery, which is the fourth paradigm of science. Within the arena of AI/ML, deep learning (DL) has emerged as a game-changing technique in recent years with its ability to effectively work on raw big data, bypassing the (otherwise crucial) manual feature engineering step traditionally required for building accurate ML models, thus enabling numerous real-world applications, such as autonomous driving. In this talk, the ongoing research in AI and high-performance data mining, along with illustrative real-world scientific applications were presented. In particular, approaches to gainfully apply DL on big data (by accelerating DL and enabling deeper learning) as well as on small data (deep transfer learning) in the context of materials science were discussed. Some of the software tools developed were demonstrated.

Keynote 29

Dr. Muhammad Ijaz

Manchester Metropolitan University, UK.

Dr. Muhammad Ijaz is Senior Lecturer, the head of Laser and Optics communication lab, and the lead International at the Department of Engineering, Manchester Metropolitan University, UK.

Dr. Ijaz has received his M.Sc. in Optoelectronic and Communication Engineering degree from Northumbria University at Newcastle upon Tyne, UK, in 2009. After the successful completion of his M.Sc., he has been awarded Northumbria University studentship to pursue his Ph.D. He was awarded his Ph.D. degree in 2013 at the same Northumbria University, UK, for his research work in Free-Space Optical (FSO) Communications. He also worked as Research Fellow in Visible Light Communication funded by prestigious EPSRC (£4.6 million research grant) UP-VLC project, where he worked in modeling and designing a large-scale multiple-input multiple-output (MIMO)-based visible light communication (VLC) system using spatial modulation and for novel methods of energy harvesting ideas in the field of wireless communications. He worked in one of the prestigious EPSRC funded (£4.6 million research grant) UP-VLC project. Dr. Ijaz is also leading two Knowledge exchange partnership (KTP) projects of £500K with local industrial partners funded by Innovate UK. He has successfully supervised a number of Ph.D. and M.Sc. students, and KTP associates on various internally and externally funded projects. His specialized areas of research and interests are optical wireless/fibre communications, Visible light communication, Free space optics (FSOO), Visible light positioning (VLP), LiFi, Advanced modulation schemes including OFDM, Powerline communications (PLC), photonics networks for 5G networks, IoTs and optical sensors design.

He has published over 80 peer-reviewed research papers including conferences and journals in the areas of optical wireless communications. He is also Guest Editor of various journals including MDPI Information 2020–2021 and MDPI Electronics

2021–2022 and program committee member on a number of conferences including ICFNDS 2017. He is a member of IET and a member of IEEE.

Optical Wireless Communication and Its Applications for Industry 4.0

Abstract. There is a forecast for exceptional digital data traffic growth due to the digitization of industrial applications using the Internet of Things. As a result, a great need for high bandwidth and faster transmission data rates for future wireless networks has emerged. One of the considered communication technologies that can assist in satisfying this demand is optical wireless communication (OWC) or LiFi. OWC is an emerging technology that uses the visible light spectrum by mainly utilizing light emitting diodes (LEDs) for simultaneous indoor lighting and high bandwidth wireless communication. Some of the applications of VLC are to provide high data rate Internet in homes, offices, campuses, hospitals and several other areas. One of these promising areas of application is industrial wireless communications. The research talk will provide a review of OWC applications intended for industrial applications with an emphasis on visible light positioning (VLP).

Keynote 30

Dr. Zhijiao Chen

School of Electronic Engineering, Beijing University of Posts and Telecommunications (BUPT), China

Dr. Zhijiao Chen is an Associate Professor in the School of Electronic Engineering, Beijing University of Posts and Telecommunications (BUPT). She received her B.Sc. from Beijing University of Posts and Telecommunications (BUPT) in 2010, his Ph.D. degree in electronic engineering from Queen Mary University of London (QMUL) in 2014, and joined the BUPT as a Lecturer in 2014. Dr. Chen has authored/co-authored more than 30 journal articles, one English book and 19 authorized patents. She leads

12 research projects funded by both academic institutions and industries. She was the recepient of the Best Paper Award in iWAT 2013 (Karlsruhe, Germany), TICRA Travel Grant in EuCAP 2014 (Hague, the Netherland) and Best Student Paper in APS/URSI 2013 (Orlando, FL). She serves as the Associate Editor for Microwave and Optical Technology Letters. Her research interests include but are not limited to dielectric resonator antennas, millimeter-wave antenna array, base station antennas and antennas for radio astronomy.

Low-cost Millimeter Wave Antenna Array for Wireless Future

Abstract. With the adoption of millimeter-wave (mmWave) technologies in 5G and B5G, we face challenges on high performance and low-cost antenna implementations. To address the issues in mmwave communications such as high propagation loss, antennas integrated into base station (BS) and user equipment (UE) should be verified with wideband, high gain and high efficiency. On the other hand, conventional large-volume high-cost antenna solutions should be replaced with low-cost planar ones because of the high amount of antenna adoptions and the limited space of the mmWave devices. In this case, many proposals have been made to design the low-cost planar mmWave antennas. This talk analyzes challenges for mmWave communications in wireless future and contrasts advantages and limitations of the existing mmWave antenna technologies. The challenges and opportunities of designing mmWave antennas to satisfy the specifications of the 5G and B5G will be summarized. Her recent research on mmWave antenna array designs will be discussed with a highlight of their design considerations. Finally, further works on mmWave antenna design for wireless future will be presented.

Keynote 31

Dr. Muhammad Ali Babar Abbasi

ECIT Institute, Queens University Belfast, Queens Road, Belfast, United Kingdom

Dr. Muhammad Ali Babar Abbasi is Assistant Professor with the Centre of Wireless Innovation (CWI) at the Queen's University Belfast since 2019. He is an IEEE-APS Young-Professional Ambassador since 2022 and IET-young-professional since 2017. Dr. Abbasi is the Grand Prize winner of the GSMA Mobile World Scholar challenge 2019, for his innovation of a dual Rotman lens-based 28 GHz beamformer (since then, it evolved into a lens-based mmWave beamformer for mmWave in 26–40GHz).

Beyond 5G Beamformers: Lean Toward Smart Radio Environment

Abstract. It is surprising to see what an array of co-located beamforming antennas can do to make sure that wireless devices stay connected to the Internet. The role of antenna beamformers in 5G and beyond wireless networks is to review the latest trends and applications, and forecast future beamformers design for our hyper-connected world are discussed.

Keynote 32

Dr. Nasimuddin

Institute for Infocomm Research, A-STAR, Singapore

Dr. Nasimuddin is a scientist at the Institute for Infocomm Research, Singapore. His research interests include multilayered microstrip-based structures; millimeter waves; radio-frequency (RF) identification readers; GPS/Global Navigation Satellite System satellites; TV white space; RF energy-harvesting systems; and ultrawideband, beam-scanning/beamforming, metamaterial, circularly polarized microstrip antennas. He is a Senior Member of the IEEE.

Overview of Beam Steering Antenna Techniques

Abstract. Reconfigurable/millimeter-wave (mmW) technology is playing a significant role in next generation wireless communication and connectivity systems. The k/Ka-band and even higher frequency bands provide wide spectrum availability for high-speed data transfer with large data and connectivity. The beam steering antenna is the key element for mmW wireless communication systems. To ensure highly reliable wireless communication at the mmW, the beamforming/beam steering antenna techniques must be used to compensate the high path loss and improve the signal-to-interference/noise ratio (SINR). In this talk, an overview of various beam steering/beamforming antenna technologies will be discussed based on some figures of merit and their applications. The frequency scanning/switched beam antenna designs will be discussed in detail with applications. These antenna techniques are useful for radar, tracking, navigation, imaging, satellite, mobile wireless and 5G/connectivity system designs.

Keynote 33

Dr. Thuy T. Le

San Jose State University, United States of America

Thuy T. Le received his Ph.D. (1990), M.S. (1987) and B.S. (1985) degrees all from the University of California at Berkeley. Presently, he is a Professor and Department Chair of Electrical Engineering Department at San Jose State University. At San Jose State University, he is responsible for teaching graduate and undergraduate courses in digital system design, computer and microprocessor architecture, computer interfacing, system-on-chip design, numerical methods, probability and random processes, and linear system theory. He also serves as research advisor in the areas of digital system and logic design, ASIC, SOC, hardware accelerators, computational electronics, microcontroller and microprocessor. Professor Le has also been working on several projects with local companies; these projects include high-performance system architectures, parallel algorithms and applications, digital arithmetic algorithms and circuit design, and System-on-Chip verification and validation. He also served as an Adjunct Professor in the Mathematical and Engineering Department at the University of South Carolina. At Savannah River Laboratory, Dr. Le participated in the design of five new nuclear reactors for Tritium production. In this capacity, he co-authored several computational nuclear reactor physics and radiation shielding codes, which include the three-dimensional vectorized reactor design code (GRIMHX3), the improved collision probability assembly resonance treatment code (MARJORI), the three-dimensional radiation shielding analysis code (ROOM-DOSE) and the two-dimensional generalized geometry discrete ordinates code. Dr. Le was also a member of the Quality Assurance team who is responsible for the verification and validation of the computer codes in use at Savannah River Site. These codes include the nuclear fuel cycle and spent fuel analysis codes, reactor safety analysis systems, and the reactor charge design codes.

From 1988 to 1990, Dr. Le was a Physics Instructor at U.C. Berkeley and College of Alameda, and was also a Research Assistant in the Physics Department at Lawrence Berkeley National Laboratory. While there he participated in a large DOE funded research project in hydrogenated amorphous silicon, which includes the studies of hydrogenated amorphous silicon structure, very low noise preamplifier, radiation effects on semiconductor, memory, and electronic systems, and high-energy electromagnetic shower simulation. During the period of 1986 and 1990, Dr. Le also performed another research project in High-performance Computational Nuclear Reactor Physics in the Nuclear Engineering Department at University of California Berkeley. In this project, he developed a Mathematical Nodal model that can speed up the solutions of three-dimensional neutron diffusion problems in nuclear reactor cores. He successfully prototyped and benchmarked his work by implementing three-dimensional neutron diffusion with thermal-hydraulic feedback code running on 64KB IBM PC-XT.

During the periods of 1993 to 1996 and from 1989 to 1990, Dr. Le was an independent contractor of Sierra Nuclear Corporation, California. In this capacity, he served as an independent technical reviewer to review and give comments on documents and reports in the areas of nuclear fuel storage calculations, cask design and criticality analysis for the transportation of nuclear spent fuels. During the period of 1985 to 1988, he worked as a Reactor Operator and Nuclear Health Physicist for the Triga Mark III nuclear research reactor at the University of California, Berkeley. Professor Le has served as keynote speaker, general chair, technical program chair, session chair, reviewer and committee member of a number of technical international conferences and symposiums. His publications include technical papers, seminars and reports in broad areas of parallel and distributed computing and algorithms, computer architectures and computer networks, hardware accelerators for complex algorithms, and computational reactor physics. Professor Le's current research interests include topics in SoC and Embedded System Design, Hardware Accelerator, Microprocessor and Microcontroller, Quantum Information and Computing, implementation of Probability theory and Monte Carlo simulation, radiation effects on electronic devices and systems. Professor Le is also a Co-Founder and Advisor of the Vietnamese Strategic Ventures Network (now Strategic Alliance Vietnamese Ventures International) and Chairman of the Board of the United States—Vietnam Foundation.

Quantum Technologies and the Future Workforce

Abstract. Quantum technologies have become a reality and are expected to be commercialized soon. The United States and several other countries worldwide have initiated quantum programs to accelerate the commercialization of quantum technologies. Besides the significant technical challenges in quantum research and development, the shortage of a diverse, inclusive and sustainable quantum workforce plays an important factor. In this presentation, the speaker will outline some current technology challenges in quantum technologies, which include quantum computing, sensing, networking and discuss possible pathways of quantum workforce development. The speaker will also share the recent development of Quantum Computing specialization in the Master of Science in Electrical Engineering program at San Jose State University.

Contents

Artificial Intelligence, Machine Learning and Data Sciences

Alzheimer's Disease Classification Using Ensemble Methods 3
Prashant Upadhyay and Pradeep Tomar

Difference-Attribute-Based Clustering for Ordinal Survey Data 17
Bhupendera Kumar and Rajeev Kumar

Machine Learning Algorithms for Binary Classification of Breast Cancer 29
Preeti Katiyar and Krishna Singh

Applications, Issues, and Deep Learning Analysis of Electronic Health Records 43
Shahid Ahmad Wani, S. M. K. Quadri, and Gowhar Mohiudin

Design and Implementation of Multiple-User Variable Rate Channel Assignment Method in Cognitive Radio System 59
Gowda C. P. Mallikarjuna, Kuber Vijaykumar Bellad, P. Prajwal, V. Hima, and Niharika Suresh

Wireless Communication, WSN and Adhoc Networks

Design of Frequency Beam Scanning Antenna Using SIW for mm-wave Applications 73
Arman Singh, Arun Kumar, Arun Singh, Ayush Jaiswal, and Ankit Sharma

QUICLORIS: A Slow Denial-of-Service Attack on the QUIC Protocol 85
Anirudh Srinivas Balaji, V. Anil Kumar, P. P. Amritha, and M. Sethumadhavan

Preventing Data Leakage by Trojans in Commercial and ASIC Applications Using TDM and DES Encryption and Decryption 95
Shashank Kumar Pathak and M. Nirmala Devi

Design of Helmet-Mounted Dual-Band Conformal Antenna for Military Applications 111
Govind Mishra, Shuchismita Pani, and M. S. Prasad

An Experimental Study of IEEE 802.11n Devices for Vehicular Networks with Various Propagation Loss Models 125
Preeti Rani and Rohit Sharma

Cloud Computing, 5G and IoT

Performance Evaluation of Genetic Algorithm and Flower Pollination Algorithm for Scheduling Tasks in Cloud Computing 139
Pardeep Singh, Gourav Bathla, Deepak Panwar, Alankrita Aggarwal, and Shivani Gaba

Malicious Node Detection in Heterogeneous Internet of Things 155
Supriya Goel, Somya, Swapnil Sutar, and Priyanka Mekala

Performance Evaluation of HSRP and GLBP Over OSPF and RIP Routing Protocols 173
Aayush Agarwal, Shilpi Sharma, and Evin Tom Xavier

Comparative Analysis of Load Balancing Algorithm in Cloud Computing 187
Prakash Choubey and B. Mohapatra

Robotics and Embedded Systems

Securely Importing of Back-Up File into Cryptowallet 203
P. Gnanendra Prasad, P. P. Amritha, and M. Sethumadhavan

Trajectory Optimization of a Satellite 215
Mahika Sachdeva, M. S. Prasad, and Shivani Verma

Artificial Intelligence-Aided Trajectory Planning for UAV 227
Sanjoy Debnath, Achanta Sai Mani Teja, and Miriyala Naveen

Sensors and Sensor Technology

9T1R nvSRAM Cell with Improved Read Delay and Margin 245
Damyanti Singh, Neeta Pandey, and Kirti Gupta

Fractional-Order Capacitor Realization Based upon Active Inductor ... 259
Garima Varshney, Neeta Pandey, and Rajeshwari Pandey

Modeling of Martian Ascent Launch Vehicle 271
Mahima Soota and Sudhir Nadda

About the Editors

Ashwani Kumar Dubey received his M.Tech. degree in Instrumentation and Control Engineering from Maharshi Dayanand University, India, in 2007, and a Ph.D. from the Faculty of Engineering and Technology, Jamia Millia Islamia, India, in 2014. Currently, he is working as a professor in the department of Electronics and Communication Engineering, ASET at Amity University Uttar Pradesh, India. He has over twenty years of experience in different positions. Dr. Dubey has filed 23 patents and has regularly contributed to over 90 research papers in various reputed international and national journals and conferences of repute. He is a corporate member of The Institution of Engineers (India). He has developed many visual inspection systems for various industrial and agricultural applications to improve and monitor the quality and reliability of the products/systems. He has guided five Ph.D. research scholars and is presently guiding five Ph.D. research scholars. He has been awarded the 'SSPS Award of Excellence 2022' by The Strategic School of Professional Skills, SK Edu Trust, Delhi, India. His research interest includes wireless sensor networks, IoT, IIoT, digital and medical image processing, food, and fruit quality inspections through computer vision, MEMS/NEMS-based smart sensors and systems, etc.

Vijayan Sugumaran is a Distinguished University Professor of Management Information Systems and Chair of the Decision and Information Sciences department at Oakland University, USA. He is also Co-Director of the Center for Data Science and Big Data Analytics at Oakland University. He received his Ph.D. in Information Technology from George Mason University, USA. He has taught courses such as object-oriented systems development, C++, Java, Javascript, database management systems and data warehouses, advanced databases and Big Data management, Deep Learning and text analytics, systems analysis and design, electronic commerce, and introduction to MIS. His research interests are in the areas of Big Data analytics, business intelligence, ontologies and semantic web, intelligent agent and multi-agent systems, component-based software development, knowledge-based systems, and Data and Information Modeling. He has published over 250 peer-reviewed articles in journals, conferences, and books. He has edited twenty books and serves on the editorial board of eight journals.

Peter Han Joo Chong is a Full Professor and an Associate Head of the School (of Research), at the School of Engineering, Computer and Mathematical Sciences, at Auckland University of Technology, New Zealand. Previously, he was a Head of the Department of Electrical and Electronic Engineering at AUT. Dr. Chong received a B.Eng. (with distinction) in Electrical Engineering from the Technical University of Nova Scotia, Canada, in 1993, and the MA.Sc. and Ph.D. degrees in Electrical and Computer Engineering from the University of British Columbia, Canada, in 1996 and 2000, respectively. He visited Tohoku University, Japan, as a Visiting Scientist in 2010 and the Chinese University of Hong Kong, Hong Kong, between 2011 and 2012. He is currently an Adjunct Professor at the Department of Information Engineering, CUHK. He is currently an Editorial Board Member of several international journals. His research interests are mobile communications systems including radio resource management, multiple access, MANETs/VANETs, multihop cellular networks, and Internet of Things/Vehicles and artificial intelligence for wireless networks. He has published over 300 research articles. He is an Honorary Professor at Amity University, India. He is a Fellow of the Institution of Engineering and Technology (FIET), UK. Prof. Chong is listed in the World's Top 2% Scientists published by Stanford University in 2022.

Artificial Intelligence, Machine Learning and Data Sciences

Alzheimer's Disease Classification Using Ensemble Methods

Prashant Upadhyay and Pradeep Tomar

Abstract Alzheimer's disease (AD) is a "neurodegenerative disorder" that leads the brain cells to deteriorate and die. It is a chronic illness. It is an incurable condition that causes memory loss as it proceeds. This research focuses on a detailed investigation of effective and consistent learning methodologies for AD classification. The main machine learning approaches examined in this study are convolutional neural networks, support vector machine, and ensembled approach. This study finds that the accuracy level of ensembled approach (SVM + CNN) surpasses both the other approaches in diagnosing neurological illnesses. Accuracy levels examined in this study are SVM 86.5, CNN 96.3, and hybrid 97.1%

Keywords Ensembled methods · Convolutional neural network (CNN) · Magnetic resonance imaging (MRI) · Alzheimer's disease (AD)

1 Introduction

Alzheimer's disease (AD) is one of the most typical causes of dementia in today's culture. According to the World Alzheimer Report 2018 [1] the disease affected around 50 million people in 2018, with the number expected to double by 2050. Machine learning (ML) has been successfully used in biological data mining [2, 3], image analysis [4, 5], financial forecasting [6, 7], anomaly detection [8, 9], and natural language processing [10, 11] throughout the previous decade. Although AD is now incurable, it can be stopped in its tracks if caught early [12]. In 2006, 26.6 million people were affected with AD. By 2050, AD is expected to impact 1 in 85 persons worldwide, with around 43% of cases requiring high-level care [13].

P. Upadhyay (✉) · P. Tomar
Gautam Buddha University, Greater Noida, India
e-mail: prashanttheace@gmail.com

P. Tomar
e-mail: pradeep.tomar@gbu.ac.in

A. K. Dubey et al. (eds.), *Advanced IoT Sensors, Networks and Systems*,
Lecture Notes in Electrical Engineering 1027,
https://doi.org/10.1007/978-981-99-1312-1_1

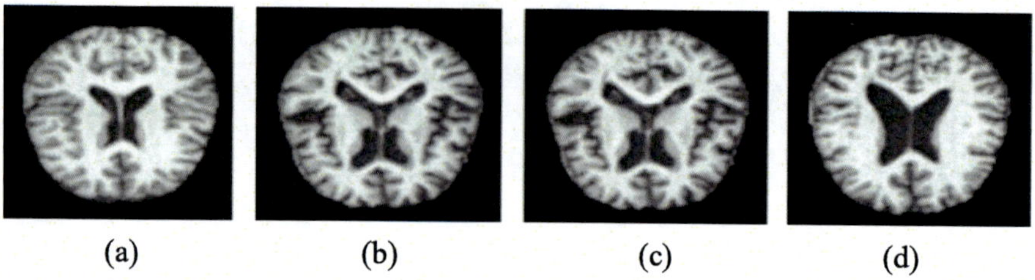

Fig. 1 Different brain MRI images presenting different AD stage: **a** mild dementia; **b** moderate dementia; **c** very mild dementia; **d** Non-demented

Dementia is a syndrome that primarily affects older people. It has an impact on brain function, daily activities, and communication [14, 15]. AD is the most common type of adult-onset dementia. Early dementia diagnosis is important for commencing therapies and predicting disease outcomes, according to certain research [16–18]; however, there are no acceptable procedures for early AD diagnosis. Figure 1 depicts various brain MRI images depicting various stages of AD. The objective is to determine which biomarkers and factors can be used to detect AD, which datasets are available, which pre-processing techniques are required to deal with biomarkers (particularly in neuroimaging), how to extract single features from 3D brain scans, which deep models can capture disease-related patterns of AD, and how to manage multi-modal data.

2 Literature Survey

This section is divided into three subsection:

Support Vector Machine: The findings of all SVM articles analyzed in this study are summarised in Table 1. The use of SVM for AD on numerous factors is examined in detail in the subsections that follow:

Artificial Neural Networks (ANN): This machine learning model is capable of modeling highly nonlinear data patterns. In this section, we'll examine how ANNs can be used to classify AD. The studies that have employed ANN for Alzheimer's are included in Table 2.

Deep Learning and Ensemble Methods: The DL is included, as well as more complicated networks such as convolutional neural networks (CNN) are incorporated. Finally, we look at the strategies for classifying AD using ensemble methods. Table 3 summarizes studies that used DL and ensemble techniques to classify AD.

Table 1 Alzheimer's data classification using SVM

SN	Refs.	Spec (%)	Acc (%)	Sens (%)	Dataset	Machine learning	Validation
1	Khedher [19]	76	92	86	818 (229 CN, 200AD, 389 MCI)	SVM (RBF)	k-fold
2	Khedher [20]	74.49	77.62	80.27	818 (229 CN, 389 MCI, 200 AD)	SVM (Linear)	Twofold
3	Klöppel [21]	91.2	85.6	75.8	90 (57 CN, 33 AD)	SVM (Linear)	LOOCV
4	Kulkarni [22]	–	96	–	100 (50 CN, 50 AD)	SVM	LOOCV
5	Lahmiri [23]	96.07	97.08	98.09	33 (11 CN, 11 MCI, 11AD)	SVM	
6	Lahmiri [24]	100	100	100	70 (35 CN, 35 AD)	SVM (Polynomial)	Tenfold
7	Lee [25]	100	100	100	84 (39 CN, 45 MCI)	SVM (RBF)	Tenfold
8	Long [26]	94.82	97.1	87.37	427 (135 CN,132 sMCI, 95pMCI, 65 AD)	SVM (Linear)	Tenfold
9	Lu [27]	93.33	90.53	90.63	272 (152 CN,120 MCI)	RF-RSVM	Threefold
10	Magnin [28]	96.6	94.5	91.5	38 (22 CN, 16 AD)	SVM (RBF)	LOOCV

3 Materials and Methods

Material and Methods section is divided into two subsections:

Dataset Used: The dataset for this study is taken from https://www.kaggle.com/datasets/tourist55/alzheimers-dataset-4-class-of-images. [49] MRI scans make up the dataset. Dataset is further divided into 80:20: Train and Test, which contains approximately 6400 photos. Four classes of AD are as follows: Mild Demented (MDe), Moderate Demented (MoDe), Non-Demented (NDe), and Very Mild Demented (VMDe).

Table 2 Alzheimer's data classification using ANN

SN	Refs.	Acc (%)	Sens (%)	Spec (%)	Machine learning	Dataset	Validation
1	Veen [29]	–	91.5	91	LGMLVQ	304 (146 PD, 82 CN, 76 AD)	–
2	Yang [30]	100	100	100	ANN	481 (408 AD, 73 CN)	Fivefold
3	Zhang [31]	92.17	79.9	89	PCA + BPNN	52 (24 AD, 28 CN)	LOOCV
4	Zheng [32]	94	94	94	MlapRls	202 (99 MCI, 52 CN, 51 AD)	10-fold
5	Zhou [33]	88.52 ± 8.61	84.60 ± 15.24	92.20 ± 12.61	SVM + RBM	103 (52 CN, 51 AD)	10-fold
6	Jie [34]	93.75	87.5	100	TrAdaBoost	543 (245 MCI, 103 CN,195 AD)	–
7	Joshi [35]	91.9	100	88	SVM	37 (25 CN, 12 aMCI)	LOOCV
8	Kar [36]	97.137	–	–	RBF NN	890 (578 AD, 312 PD)	–
9	Khazaee [37]	100	100	100	ANN	20(9 AD,11 CN)	–
10	Liu [38]	95.09 ± 2.28	92	98	SVM	168 (89MCI, 45 CN and 34 AD)	–

Table 3 Alzheimer's data classification using deep learning

SN	Refs.	Acc (%)	Sens (%)	Spec (%)	Machine learning	Dataset	Validation
1	Suk [39]	–	86	93	LR	413 (86 AD, 190MCI, 97 MCIc, 93MCInc, 137 CN)	LOOCV
2	Lebedev [40]	92.87	–	–	CNN	780 (148 AD,403MCI, 167 MCIc,236 MCInc, 229 CN)	RP
3	Li [41]	–	88.6	92	RF	575 (185 AD, 225CN, 35 MCIc, 130MCInc)	
4	Lillemark [42]	74.7	66.7	81.4	RF	293 (161 MCIc, 132MCInc)	Fivefold
5	Korolev [43]	87.7	–	–	LDA	524 (170 CN, 114AD, 240 MCI)	–
6	Suk [44]	80	83.4	76.4	pMKL	259 (120 MCIn,139 pMCI)	Tenfold
7	Zhang [45]	91.02	92.72	89.94	SAE	805 (186 AD, 393MCI (167 MCIc,226 MCIs), 226 CN)	–
8	Hor [46]	81	83	78	Adaboost	810 (194 AD, 228CN, 246 MCInc,142 MCIc)	–
9	Clark [47]	79.5	73.7	89.7	RF	608 (108 AD, 322MCI, (96 MCIc,126 MCInc), 178 CN)	–
10	MRI Images for Alzheimer Detection [48]	83.2	62.5	89.2	Ensemble classifier	158 (51 CN, 24MCIc, 83 MCInc)	–

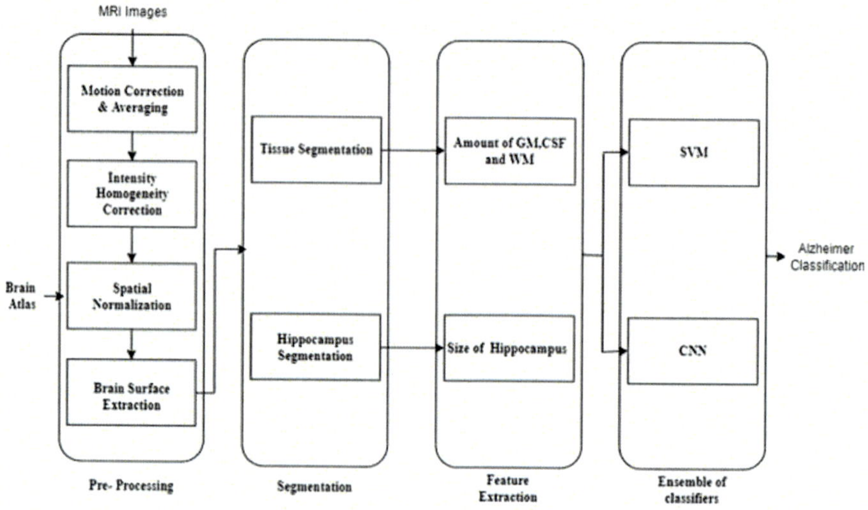

Fig. 2 Flow diagram of the methodology

Methodology: The input picture (MRI) is initially prepared in the first phase for further analysis. The second step divides the image into several segments. The third phase involves extracting features. The photograph is classified as if it is belonging to a normal subject or an AD patient in the terminal phase. Figure 2 shows the proposed methodology in detail.

4 Result and Discussions

There are two subsections in this section. The experimental setup and the performance evaluation metrics are discussed in this section.

Experimental Setup: The proposed model was built with Python library [50] on TensorFlow software library [51], Google Collaborative [52], and Kaggle, where the dataset was downloaded, and the Scikit library [53] for all statistical performance evaluation.

Performance Evaluation: This study looked at three machine learning methods for predicting AD. In cross-validation, a total of 6400 MRI samples were used for testing and training. Confusion matrix was analyzed using three machine learning approaches: CNN, SVM, and ensemble method, as shown in Tables 4, 5, and 6.

Table 4 Confusion matrix for CNN

			Predicted								
			0				1				
			MDe	MoDe	NDe	VMDe	MDe	MoDe	NDe	VMDe	Σ
Actual	0	MDe	121	0	17	29	9	0	2	1	179
		MoDe	1	8	1	1	0	1	0	0	12
		NDe	3	0	530	46	1	0	54	6	640
		VMDe	11	0	57	338	0	0	3	39	448
	1	MDe	6	0	1	1	569	2	60	78	717
		MoDe	0	1	0	0	2	45	1	3	52
		NDe	1	0	21	5	35	0	2304	194	2560
		VMDe	1	0	5	20	58	1	246	1461	1792
		Σ	144	9	632	440	674	49	2670	1782	

Table 5 Confusion matrix for SVM

			Predicted								
			0				1				
			MDe	MoDe	NDe	VMDe	MDe	MoDe	NDe	VMDe	Σ
Actual	0	MDe	106	0	10	52	7	0	1	3	179
		MoDe	4	0	1	6	0	0	1	0	12
		NDe	15	0	390	194	2	0	37	2	640
		VMDe	6	0	61	359	0	0	7	15	448
	1	MDe	8	0	0	4	488	0	97	120	717
		MoDe	0	0	0	2	11	24	7	8	52
		NDe	0	0	40	17	334	0	1470	699	2560
		VMDe	3	0	9	34	430	0	438	878	1792
		Σ	142	0	511	668	1272	24	2058	1725	

The confusion matrices in Tables 4, 5, and 6 clearly show that CNN and SVM are the worst performers in trials (TN). The hybrid strategy outperformed two other machine learning methods, with maximum classification accuracy. Table 7 shows the comparison of SVM, CNN, and ensemble systems.

The Calibration plot is a probability-based curve. As illustrated in Figs. 3 and 4, an ensemble approach beats all other methods in predicting the presence and absence of AD.

Table 6 Confusion matrix for proposed (ensemble)

			Predicted								
			0				1				
			MDe	MoDe	NDe	VMDe	MDe	MoDe	NDe	VMDe	∑
Actual	0	MDe	130	0	17	24	6	0	2	0	179
		MoDe	1	88	1	1	0	1	0	0	92
		NDe	5	0	555	38	1	0	35	6	640
		VMDe	13	0	55	350	0	0	2	28	448
	1	MDe	10	0	3	1	559	2	57	85	717
		MoDe	0	1	0	0	2	40	0	9	52
		NDe	1	1	34	7	44	0	2287	186	2560
		VMDe	1	0	7	25	55	2	243	1459	1792
		∑	161	90	672	446	667	45	2626	1773	

Table 7 Comparison of algorithms

Model	Area under the curve	F measure	Recall	Precision
SVM	0.866	0.586	0.58	0.614
CNN	0.963	0.839	0.84	0.839
(Ensemble) proposed	0.971	0.841	0.842	0.841

5 Conclusion

By predicting Alzheimer's disease efficiently, we might be able to save lives and shift the course of treatment. In this paper, it is shown that ensemble methods can predict MDe, MoDe, NDe, and VMDe Alzheimer's Disease. The study's classification systems assess more effective and coherent learning strategies for Alzheimer's disease categorization which are based on a number of precise criteria.

The main machine learning approaches examined in this study are support vector machine, convolutional neural networks, and ensemble approach. This study finds that the accuracy level of ensembled approach (SVM + CNN) surpasses both the other approaches in diagnosing neurological illnesses. Accuracy level examined in this study is SVM 86.5, CNN 96.3, and Hybrid 97.1%.

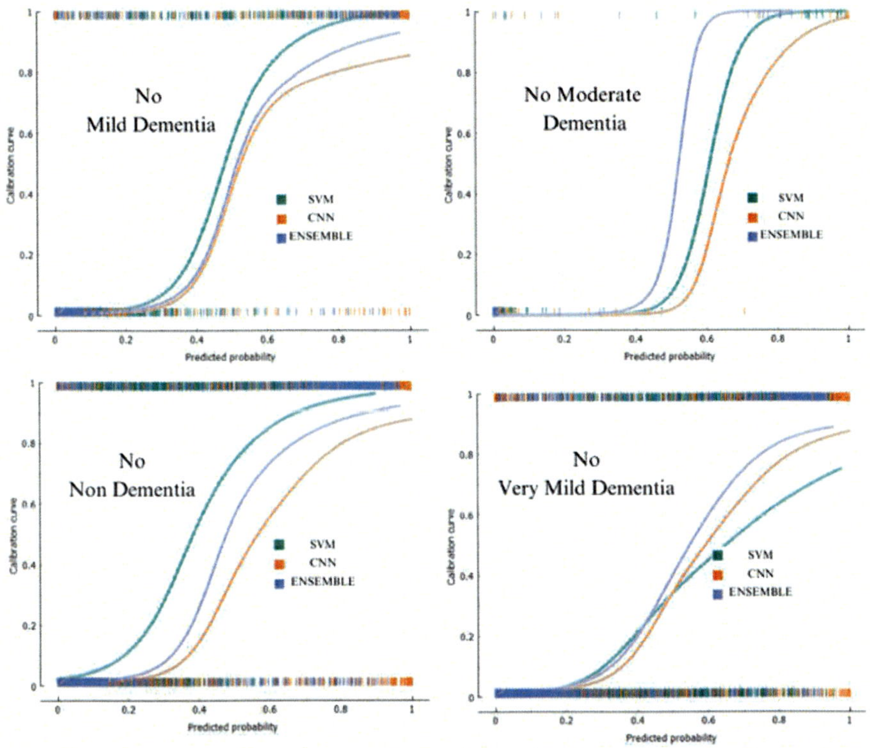

Fig. 3 Probabilistic prediction in the absence of Alzheimer's disease

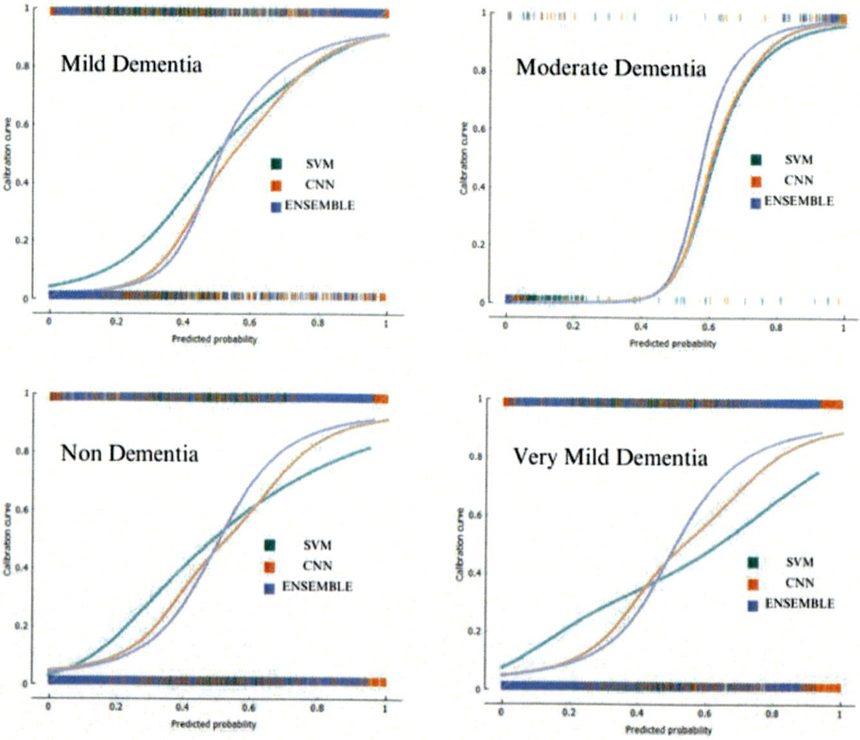

Fig. 4 Probabilistic prediction in the presence of Alzheimer's disease

References

1. Patterson C (2018) The state of the art of dementia research: new frontiers. World Alzheimer's Report.
2. Mahmud M, Kaiser MS, Hussain A, Vassanelli S (2018) Applications of deep learning and reinforcement learning to biological data. IEEE Trans Neural Netw Learn Syst 29(6):2063–2079. https://doi.org/10.1109/TNNLS.2018.2790388
3. Mahmud M, Kaiser MS, Hussain A (2020) Deep learning in mining biological data. CoRR arXiv:abs/2003.00108, 1–36. https://arxiv.org/abs/2003.00108.
4. Ali HM, Kaiser MS, Mahmud M (2019) Application of convolutional neural network in segmenting brain regions from mri data. In: International Conference on Brain Informatics. Springer, pp 136–146. https://doi.org/10.1007/978-3-030-37078-7_14.
5. Orojo O, Tepper J, McGinnity TM, Mahmud M (2019) A multi-recurrent network for crude oil price prediction. In: Proceedings of the 2019 IEEE Symposium Series on Computational Intelligence (SSCI), pp 2953–2958. https://doi.org/10.1109/SSCI44817.2019.9002841.
6. Yahaya SW, Lotf A, Mahmud M (2019) A consensus novelty detection ensemble approach for anomaly detection in activities of daily living. Appl Soft Comput 83:105613. https://doi.org/10.1016/j.asoc.2019.105613
7. Fabietti M, Mahmud M, Lotf A, Averna A, Guggenmo D, Nudo R, Chiappalone M (2020) Neural network-based artifact detection in local feld potentials recorded from chronically implanted neural probes. In: Proceedings of 2020 International Joint Conference on Neural Networks (IJCNN). pp 1–8.

8. Noor MBT, Zenia NZ, Kaiser MS, Mahmud M, Al Mamun S (2019) Detecting neurodegenerative disease from mri: a brief review on a deep learning perspective. In: International Conference on Brain Informatics. pp 115–125. https://doi.org/10.1007/978-3-030-37078-7_12.
9. Miah Y, Prima CNE, Seema SJ, Mahmud M, Kaiser MS (2020) Performance comparison of machine learning techniques in identifying dementia from open access clinical datasets. In: Proceedings of ICACIn. pp 69–78.
10. Rabby G, Azad S, Mahmud M, Zamli KZ, Rahman MM, Teket: a tree-based unsupervised keyphrase extraction technique. Cogn Comput. https://doi.org/10.1007/s12559-019-09706-3
11. Watkins J, Fabietti M, Mahmud M (2020) Sense: a student performance quantifer using sentiment analysis. In: Proceedings of 2020 International Joint Conference on Neural Networks IJCNN. pp 1–6.
12. Lodha P, Talele A, Degaonkar K (2018) "Diagnosis of Alzheimer's disease using machine learning". In: Fourth International Conference on Computing Communication Control and Automation (ICCUBEA). Pune, India, pp 1–4. https://doi.org/10.1109/ICCUBEA.2018.8697386.
13. Brookmeyer R, Johnson E, Ziegler-Graham K, Arrighi H (2007) "Forecasting the global burden of Alzheimer's disease". Alzheimer's & Dement 3(3): 186–191. Available:https://doi.org/10.1016/j.jalz.2007.04.381.
14. Harvey RJ, Skelton-Robinson M, Rossor MN (2003) The prevalence and causes of dementia in people under the age of 65 years. J Neurol, Neurosurg & Psychiatry 74: 1206–1209. [CrossRef] [PubMed].
15. McMurtray A, Clark DG, Christine D, Mendez MF (2006) Early-onset dementia: frequency and causes compared to late-onset dementia. Dement Geriatr Cogn Disord 21: 59–64. [CrossRef] [PubMed].
16. Battineni G, Chintalapudi N, Amenta F (2019) Machine learning in medicine: performance calculation of dementia prediction by support vector machines (SVM). Inform Med Unlocked [CrossRef].
17. Chen,R, Herskovits EH (2010) Machine-learning techniques for building a diagnostic model for very mild dementia. Neuroimage 52: 234–244. [CrossRef].
18. Alam MAU, Roy N, Holmes S, Gangopadhyay A, Galik E (2016) Automated functional and behavioral health assessment of older adults with dementia. In: Proceedings of the IEEE 1st International Conference on Connected Health: Applications, Systems and Engineering Technologies (CHASE 2016), Washington, DC, USA, 27–29 June 2016. pp 140–149. [CrossRef].
19. Khedher L, Illán IA, Górriz JM, Ramírez J, Brahim A, Meyer-Baese A (2017) Independent component analysis-support vector machine-based computer-aided diagnosis system for Alzheimer's with visual support. Int J Neural Syst.
20. Khedher L, Ramírez J, Górriz JM, Brahim A, Illán IA (2015) Independent compo-nent analysis-based classification of Alzheimer's disease from segmented MRI data. In: Proceedings of the International Work-Conference on the Interplay between Natural and Artificial Computation. Springer, pp 78–87.
21. Klöppel S, Stonnington CM, Chu C, Draganski B, Scahill RI, Rohrer JD, Fox NC, Jack Jr CR, Ashburner J, Frackowiak RSJ (2008) Automatic classification of MR scans in Alzheimer's disease. Brain.
22. Kulkarni NN, Bairagi VK (2017) Extracting salient features for EEG-based diagnosis of Alzheimer's disease using support vector machine classifier. IETE J Res.
23. Lahmiri S, Boukadoum M (2014) New approach for automatic classification of Alzheimer's disease, mild cognitive impairment and healthy brain magnetic resonance images. Healthcare Technol Lett.
24. Lahmiri S, Shmuel A (2019) Performance of machine learning methods applied to structural MRI and ADAS cognitive scores in diagnosing Alzheimer's disease. Biomed Signal Process Control 52 (2019).
25. Lee W, Park B, Han K (2013) Classification of diffusion tensor images for the early detec-tion of Alzheimer's disease. Comput Biol Med 43: 10.

26. Long X, Chen L, Jiang C, Zhang L, Alzheimer's Disease Neuroimaging Initiative, et al (2017) Prediction and classification of Alzheimer's disease based on quantification of MRI.
27. Lu S, Xia Y, Cai W, Fulham M, Feng DD, Alzheimer's Disease Neuroimaging Initiative, et al (2017) Early identification of mild cognitive impairment using incomplete random forest-robust support vector machine and FDG-PET imaging. Comput Med Imag Graph. 35–41.
28. Magnin B, Mesrob L, Kinkingnéhun S Pélégrini-Issac M, Colliot O, Sarazin M, Dubois B, Lehéricy S, Benali H (2009) Support vector machine-based classification of Alzheimer's disease from whole-brain anatomical MRI. Neuroradiol. 73–83.
29. van Veen R, Talavera Martinez L, Kogan RV, Meles SK, Mudali D, Roerdink JBTM, Massa F, Wang S, Zhang Y, Dong Z, Du S, Ji G, Yan J, Yang J, Wang Q, Feng C, Phillips P (2015) Feed-forward neural network optimized by hybridization of PSO and ABC for abnormal brain detection. Int J Imag Syst Technol. 153–164.
30. Yang ST, Lee JD, Huang CH, Wang JJ, Hsu WC, Wai YY (2010) Computer-aided diagnosis of alzheimer's disease using multiple features with artificial neural network. In: Proceed-ings of the Pacific Rim International Conference on Artificial Intelligence. Springer, pp 699–705.
31. Zhang D, Shen D (2011) Semi-supervised multimodal classification of alzheimer's disease. In: Proceedings of the IEEE International Symposium on Biomedical Imaging: From Nano to Macro. IEEE, pp 1628–1631.
32. Zheng X, Shi J, Zhang Q, Ying S, Li Y (2017) Improving MRI-based diagnosis of Alzheimer's disease via an ensemble privileged information learning algorithm. In: Proceedings of the IEEE 14th International Symposium on Biomedical Imaging (ISBI'17). IEEE, pp 456–459.
33. Zhou K, He W, Xu Y, Xiong G, Cai J (2018) Feature selection and transfer learning for Alzheimer's disease clinical diagnosis. Appl Sci.
34. Jie B, Zhang D, Wee CY, Shen D (2014) Topological graph kernel on multiple thresh-olded functional connectivity networks for mild cognitive impairment classification. Hum Brain Map. 2876–2897.
35. Joshi S, Shenoy D, Rrashmi PL, Venugopal KR, Patnaik LM (2010) Classification of Alzheimer's disease and parkinson's disease by using machine learning and neural network methods. In: Proceedings of the 2nd International Conference on Machine Learning and Computing. IEEE, pp 218–222.
36. Kar S, Majumder DD (2019) A novel approach of diffusion tensor visualization-based neuro fuzzy classification system for early detection of Alzheimer's disease. J Alzheimer's Dis Rep. 1–18.
37. Khazaee A, Ebrahimzadeh A, Babajani-Feremi A (2016) Application of advanced machine learning meth-ods on resting-state fMRI network for identification of mild cognitive impairment and Alzheimer's disease. Brain Imaging Behav. 799–817.
38. Liu X, Tosun D, Weiner MW, Schuff N, Alzheimer's Disease Neuroimaging Initiative (2013) Locally linear embedding (LLE) for MRI-based Alzheimer's disease classification. Neuroimage. 148–157.
39. Suk Hl, Lee SW, Shen D, Alzheimer's Disease Neuroimaging Initiative (2014) Hierarchical feature representation and multimodal fusion with deep learning for AD/MCI diagnosis. NeuroImage. 569–582.
40. Lebedev AV, Westman E, Van Westen GJP, Kramberger MG, Lundervold A, Aarsland D, Soininen H, Kłoszewska I, Mecocci P, Tsolaki M et al (2014) Random forest ensembles for detection and prediction of Alzheimer's disease with a good between-cohort robustness. NeuroImage: Clin. 115–125.
41. Li H, Liu Y, Gong P, Zhang C, Ye J, Alzheimer's Disease Neuroimaging Initiative (2014) Hierarchical interactions model for predicting mild cognitive impairment (MCI) to Alzheimer's disease (AD) conversion. PloS One.
42. Lillemark L, Sørensen L, Pai A, Dam EB, Nielsen M (2014) Brain region's relative proximity as marker for Alzheimer's disease based on structural MRI. BMC Med Imag.
43. Korolev IO, Symonds LL, Bozoki AC, Alzheimer's disease neuroimaging initiative (2016) Predicting progression from mild cognitive impairment to Alzheimer's dementia using clinical, MRI, and plasma biomarkers via probabilistic pattern classification. PloS One.

44. Suk HI, Shen D (2016) Deep ensemble sparse regression network for Alzheimer's disease diagnosis. In: Proceedings of the International Workshop on Machine Learning in Medical Imaging. Springer, pp 113–121.
45. Zhang J, Stonnington C, Li Q, Shi J, Bauer RJ, Gutman BA, Chen K, Reiman EM, Thompson PM, Ye J et al (2016) Applying sparse coding to surface multivariate tensor-based morphometry to predict future cognitive decline. In: Proceedings of the IEEE 13th International Symposium on Biomedical Imaging (ISBI'16). IEEE, pp 646–650.
46. Hor S, Moradi M (2016) Learning in data-limited multimodal scenarios: scandent decision forests and tree-based features. Med Image Anal. 30–41.
47. Clark DG, McLaughlin PM, Woo E, Hwang K, Hurtz S, Ramirez L, Eastman J, Dukes RM, Kapur P, DeRamus TP et al (2016) Novel verbal fluency scores and structural brain imaging for prediction of cognitive outcome in mild cognitive impairment. Alzheimer's Dementia: Diagn Assess Dis Monitor. 113–122.
48. MRI Images for Alzheimer Detection. Available Online: https://www.kaggle.com/datasets/tourist55/alzheimers-dataset-4-class-of-images.
49. Chollet (2021) Keras Library. [Online]. Available:https://keras.io/.
50. Google Brain Team (2021) TensorFlow Library. [Online]. Available: https://www.tensorflow.org/.
51. Google (2021) Google Colab Notebook., [Online]. Available: https://colab.research.google.com/.
52. Cournapeau D (2021) Scikit-Learn Library. [Online]. Available: https://scikit-learn.org/.
53. Rossum GV (2021) Python Library. [Online]. Available: https://www.python.org/.

Difference-Attribute-Based Clustering for Ordinal Survey Data

Bhupendera Kumar and Rajeev Kumar

Abstract When survey data are collected based on a questionnaire for understanding issues, most responses may vary in some context, like two or more having the same intention. Still, the choosing ordinal values of the Likert scale may be different. Some of them use different lower and upper bounds. There may be a relation among the attributes. For example, one responds to the questions starting with 2–4, and another gives 1–3. They both have the same intention concerning questions. So basically, these two should belong to the same category and lie in the same cluster. In this paper, we used the concept of differences in attributes. The relationship can be identified based on such a phenomenon. The patterns can be identified which have a strong relationship to such values of attributes. This work is to find out such relationship in patterns for further research, which can have a significant impact on decision-making based on survey data.

Keywords Relationship · Likert scale · Pattern · Survey data · Ordinal data · Clustering

1 Introduction

Today's age is heavily dependent on surveys. Data collected from a survey is extensive in attributes and instances. Data may be of any type like categorical, ordinal, numerical, etc. In this paper, we used ordinal data which is used in most surveys. As we have said, the data collected by these procedures are cumbersome. Therefore, mining these databases has attracted a lot of researchers' interests toward finding a wide range of applications for the improvement of decision-making strategies.

Survey data collection methods can be different for different problems. Specifically, it may be of questionnaire method or a physical appearance method. In today's

B. Kumar (✉) · R. Kumar
Data to Knowledge (D2K) Lab, School of Computer and Systems Sciences, Jawaharlal Nehru University, New Delhi 110 067, India
e-mail: bkchauhan86@gmail.com

A. K. Dubey et al. (eds.), *Advanced IoT Sensors, Networks and Systems*, Lecture Notes in Electrical Engineering 1027,
https://doi.org/10.1007/978-981-99-1312-1_2

age, survey data collection is done online or the computer-generated responses. It is an enormous challenge for anyone to find meaningful inferences from such data. One has to drill and see the behavior of every respondent for a problem for a specific objective, or finding many statistical solutions to the problem. Insufficient efforts and careless responding also play a crucial role in objective findings. Not all respondents, but some immature and unmotivated participants may respond carelessly. These insufficient efforts in responses may have a relatively good or adverse effect on the survey output. There are lots of methods used to detect and fix these careless responses [1]. There are methods which play a great role when using supervised and unsupervised learning on a large dataset. Yet, this remains largely an open challenge.

Clustering is the process of grouping similar types of data. Data clustering, which comes under unsupervised learning, is a well-known technique that can find meaningful information from a large dataset. This proved to be very useful for study in many research communities, e.g., statistical pattern recognition, machine learning, information retrieval, and data mining. Most decisions are taken on the Likert scale survey in today's era. Unsupervised learning will play a tactical role in this scenario. K-means algorithm provides the benchmark for clustering. In the following paper, we have used different clustering methods for analyzing the survey results. Based on the fact that two or more respondents can have a similar type of thinking, we try to figure out the patterns to put them in the same cluster. While giving the answers to the questionnaire, there may be a connection among the questions. These questions become attributes later. So in this paper, we use the difference-attributes approach for finding the directional vectors.

The rest of the paper is organized as follows. Section 2 discusses the characteristics of survey data with their categories. Section 3 discusses some related work to pattern-based clustering. Section 4 identifies some research questions and defines datasets to find the answer. Section 5 proposed the methodology of clustering. Section 6 has extensive performance results followed by discussion. Section 7 concludes the paper.

2 Characteristics of Survey Data

Basically, we have to ensure that survey properties can be categorized into the following parts: (i) The data required for the statistical analysis, (ii) Analyze the relationship among the variables of the population, and (iii) Finally, the selected part of the population from which the findings can be generalized back to the population.

Generally, the dataset of surveys is collected and divided into two categories:

1. **Categorical data**: Most of the data collected in different surveys are categorical means in which no numerical values are involved. Further, these can be categorized as (i) Nominal Data (there is no order, e.g., yes or no), and (ii) Ordinal Data (there is an order).
2. **Quantitative data**: In this type of survey data, we have some numerical values that are related by comparative relationship. Further these can be categorized: (i)

Discrete data (Which has the values, e.g., 0, 1, 2, etc.,) and (ii) Continuous data (Some real values) [2].

To analyze ordinal survey data, we have to collect the data in some number points, which is why the Likert scale is used. It may be of many types, e.g., (i) Agreement (strongly agree, Agree, Neither agree nor disagree, Disagree, Strongly disagree); (ii) Quality (Very poor, Poor, Fair, Good, Excellent); (iii) Likelihood (Not at all likely, somewhat likely, neutral, extremely likely); (iv) Experience (very negative, somewhat negative, neutral, somewhat positive, very positive); (v) Frequency (Always, often, sometimes, rarely, never).

Generally, we use the five-point Likert scale in ordinal survey data. This is widely used in survey data. The above types of values of Likert scale are of the range 1 to 5; anyone of the above can be used in the survey data. For example, take the following response pattern on a set of ten items, assuming 1 to 5 on the Likert scale from strongly disagree to strongly agree.

3 Related Work

Many categorical data methods were developed in early literature, such as logistic regression and log-linear model for finding inferences. The cumulative probability models become famous when the popular ordinal response was being considered. Liu and Agresti [3] discussed the impact of inference from the ordinal responses data using a mathematical approach such as probability distribution and maximum likelihood techniques.

However, all the statistical approaches gave relevant results, but the large data size may create a problem in finding the result precisely. For avoiding the curse of dimensionality in data, respondents should be in cluster form. The respondent's behavior may be clusterized to find out more precise inference. Kamishima and Akaho [4] suggest a more appropriate method for clustering the ordinal data named $k - o$ means, which outperformed the hierarchical clustering.

The traditional clustering method is based on the fact that a similar type of object should be in the same group. Most clustering approaches are based on a single pattern model based on respondents' specific pattern, but similar patterns in a subspace can also be identified in pattern-based clustering. Multiple patterns can also be found based on shifting, scaling, and others based on selecting features simultaneously [5]. Suppose the ordinal data are treated as a discrete variable. In that case, these variables come into the latent Gaussian variables, the heterogeneous population may view the response items differently [6].

McParland and Gormley [7] developed a clustMD model for latent Gaussian variables, which generates ordinal data into a mixture of different kinds of data. It may be ordinal or nominal, continuous, or discrete simultaneously. Over the web-based surveys, one participant may enroll more than one time. Hierarchical clustering with Euclidean distance matrix generated by respondents identifies the group of stakeholder respondents by separating the non-independent cluster from the mixed population [8].

Further, the information among the categories of the attributes for the clustering is to benefit the researchers for taking the valuable decision. Zhang et al. [9] proposed a unified entropy-based distance metric which was applied to ordinal and nominal data. Many commands were introduced such as ineqord [10] and ordinalClust [11] for the analysis of ordinal data clustering.

Dinh et al. create an algorithm called Clustering by combining the imputation and clustering phases into single step-missing values in a mixture of categorical and numeric data k-CMM. There are three phases to the algorithm. The input dataset is divided into two categories during the initialization phase: attribute types and missing values. The impudence phase finds the collection of associated data objects using a decision tree-based approach. The clustering phase creates cluster centers at numerical and categorical characteristics using the mean and kernel-based strategies. The program also calculates the separations between items and the cluster center using the squared Euclidean and information-theoretic based dissimilarity measure [12].

As a result, Zhang et al. [13] examine the inherent distinction and relationship between nominal and ordinal attribute values from a perspective similar to the graph. To quantify the intra-attribute distances of nominal and ordinal attributes in a unified manner while maintaining the order connection among ordinal values, they proposed a unique distance metric. This clustering technique combines the learning of intra-attribute distance weights and partitions of data objects into a single learning paradigm rather than two distinct phases.

For finding effective results from the ordinal data, pattern-based clustering is well defined over the survey data because of the closeness and similarity in data which shows the behavior of the respondents. The response of the respondents can be seen as a directional pattern. Sadh and Kumar [14] used the concept of directional patterns for clustering and validated the stakeholder theory over the K-means algorithm.

4 Datasets and Preprocessing

In this work, we have used two datasets. One is a real dataset with 2533 instances & eleven attributes. The second is a synthetic dataset with 1913 instances & eleven attributes. This real dataset was previously used in PCSD3 [14]. The real dataset was collected from six categories. These categories are Faculty, Parents, Postgraduates (PG), Professional, Researcher, and Undergraduates (UG). All these six categories have their responses to eleven attributes on the quality parameters. Like the real dataset, we have taken the six categories in the synthetic dataset and compared the results on stakeholders. In these instances, we have the following Table 1 which gives a summary of all respondents.

We have already seen that most popular K-means and density-based clustering algorithms are based on numerical mathematical formulas; they did not provide good results. A new approach is used in PCSD3, a directional vector called pattern-

Table 1 Number of respondents for real and synthetic data

Categories	Real dataset	Synthetic dataset
UG	438	333
Researchers	447	342
Professionals	389	289
PG	463	363
Parents	395	295
Faculty	401	291

based clustering. All these methods give relevant results on a particular dataset. We have applied different clustering methods such as K-means clustering, density-based clustering, and PCSD3 on these two datasets.

We take different datasets to answer the following two research questions:

RQ1 Can two respondents give responses with the same frame of mind, which differ in choosing a minimum and maximum range of scale?

RQ2 What will happen if two or more reference vectors are the same as in the case of PCSD3 method?

5 The Proposed Methodology

The proposed method works on the concept of directional vectors. This method basically finds a new pattern in the form of vectors. It measures the ordinal data responses and then converts them into vector forms. Every respondent may have the same mentality to give responses on a different scale. Some start giving from "1" as the minimum whereas others start from "2". So every time, we have to find the correlation vector. It helps to find out the relation among attributes. Our work is planned according to the following steps:

1. Convert the ordinal data to a vector form,
   (a) Find minimum and maximum for each row,
   (b) Find average of minimum and maximum values,
   (c) Subtract average from each attributes of every row, and
   (d) assign −1, 0, 1 accordingly for each value.
2. Find category-wise representative vector as a cluster by some probabilistic measure, and
3. Compare and cluster each instance vector with representative vectors as generated in Step 2 above.

6 Results and Discussion

Pattern-based clustering plays an important role in today's surveying techniques for finding the correlation between attributes. On behalf of this technique, many crucial results are formed. Here in this paper, we took the concept of directional vectors. Before going to the result, we would like to brief the proposed method.

Every respondent has a different approach to giving the responses with minimum and maximum range. That's why we calculate the minimum and maximum of every respondent and find the average of these two. After that, this average is subtracted from each response. After calculating real values by this process, we convert them into a vector form by applying the rule: If the value is less than 0, then it is -1. If greater than 0, then 1; otherwise, 0. We have taken two datasets to compare our results. In the following figure, we take a small example to show how the initial vector is created. The difference-attributes approach is used for finding the initial vector. Later this approach is used for the complete dataset. (see Table 2.)

6.1 Synthetic Dataset

We took synthetic data of 1913 respondents with the same eleven attributes and six categories in this case study. But the instances are different in each category. Again we perform all the clustering methods on stakeholders. K-means and Density-Based (DB) were performed as these were on a real dataset. The PCSD3 does not perform well on this dataset because out of the six reference vectors, four vectors are the same during clustering. This means only two clusters are distinct, which violates the stakeholder's theory. Again, this is a challenging issue as this method is based on pattern-based clustering with directional vectors. We improvise this by applying the concept of collectively correlated vectors, and the results are quite distinct accordingly.

Table 3 gives a summary of performance measures TP (True Positive), TN (True Negative), FP (False Positive), FN (False Negative) with all six respondents on syn-

Table 2 Creation of sample response vector

Respondent	Category	v1	v2	v3	v4	Respondent	v1	v2	v3	v4
Respondent1	A	3	3	5	2	Respondent1	0	0	1	−1
Respondent2	B	1	2	3	4	Respondent2	−1	0	0	1
Respondent3	C	4	1	4	3	Respondent3	1	−1	1	0
Respondent4	A	3	2	5	3	Respondent4	0	−1	1	0
Respondent5	B	1	2	3	3	Respondent5	−1	0	1	1
Respondent6	A	5	2	3	4	Respondent6	1	−1	0	0
Respondent7	B	4	1	3	3	Respondent7	1	−1	0	0
Respondent8	C	3	2	4	5	Respondent8	0	−1	0	1
Respondent9	B	2	3	5	2	Respondent9	−1	0	1	−1
Respondent10	A	2	3	4	5	Respondent10	−1	0	0	1

Table 3 Performance measures for synthetic dataset with respondents

Categories	True Positive (TP) count			
	K-MEANS	Density based	PCSD3	Proposed
Faculty	54	44	140	87
Parents	55	44	0	46
PG	70	68	0	22
Professional	51	68	0	57
Researcher	68	73	0	30
UG	62	59	182	82
MEAN	60	59.33	54	54
Categories	True Negative (TN) count			
	K-MEANS	Density based	PCSD3	Proposed
Faculty	1387	1404	843	1110
Parents	1365	1378	1618	1375
PG	1310	1305	1550	1448
Professional	1360	1289	1624	1348
Researcher	1282	1285	1571	1440
UG	1308	1347	768	1255
MEAN	1335.33	1334.67	1329	1329.33
Categories	False Positive (FP) count			
	K-MEANS	Density based	PCSD3	Proposed
Faculty	235	218	151	204
Parents	253	240	295	249
PG	240	245	363	341
Professional	264	335	289	232
Researcher	289	286	342	312
UG	272	233	151	251
MEAN	258.83	259.50	265.17	264.83
Categories	False Negative (FN) count			
	K-MEANS	Density based	PCSD3	Proposed
Faculty	237	247	779	512
Parents	240	251	0	243
PG	293	295	0	102
Professional	238	221	0	276
Researcher	274	269	0	131
UG	271	274	812	325
MEAN	258.83	259.50	265.17	264.83

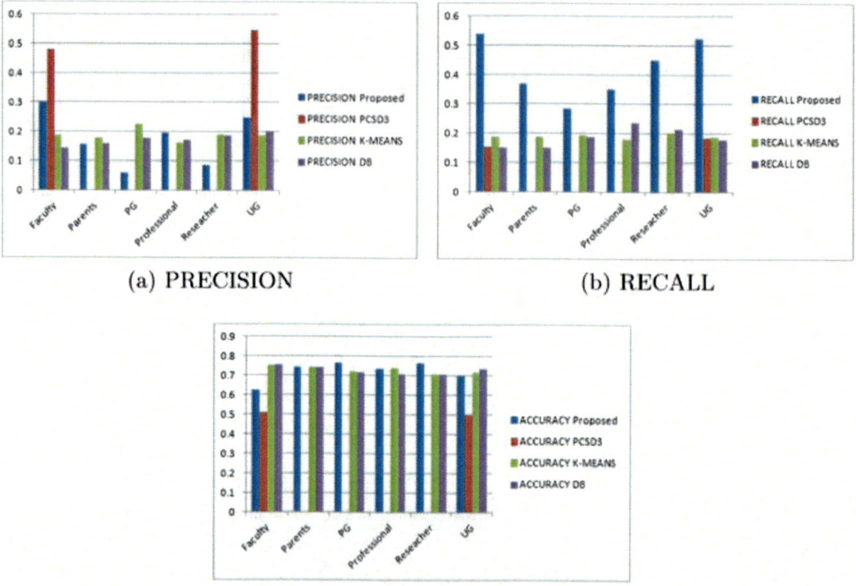

(a) PRECISION (b) RECALL

(c) ACCURACY

Fig. 1 Comparison of different clustering methods on stakeholder for synthetic dataset

thetic dataset. It shows the outcome of the proposed model correctly predicts the positive class for each category. The proposed method reduces the falsely clustered category.

The performance parameters precision, recall, and accuracy are shown comparatively for the six categories in following graph for synthetic dataset (see Fig. 1).

6.2 *Real Dataset*

In this case study, we have taken a real dataset of 2533 respondents and eleven attributes with six categories (UG, PG, Researchers, Faculty, Parents, and Professionals). We apply the different clustering techniques (K-means, Density-Based (DB), PCSD3, and the proposed one to stakeholders with six clusters. We have made exactly the same clusters in all the above clustering techniques.

Table 4 gives a summary of performance measures TP (True Positive), TN (True Negative), FP (False Positive), and FN (False Negative) with all respondents on the real dataset. The distribution of each category is shown here.

Thereafter, the comparison of the performance measures precision, recall, and accuracy is shown in the following graph for real dataset (see Fig. 2). The proposed method shows that the precision and recall for most of the categories are on the higher side. Higher recall and higher precision indicate that the proposed method provides the majority and relevance to the results. That's how this method improves accuracy.

Table 4 Performance Measures for real dataset with respondents

Categories	True Positive (TP) counts			
	K-MEANS	Density based	PCSD3	Proposed
Faculty	197	209	279	258
Parents	98	94	186	222
PG	130	75	141	105
Professional	198	197	191	171
Researcher	117	189	159	70
UG	191	193	215	234
MEAN	155.17	159.50	195.17	176.67
Categories	True Negative (TN) counts			
	K-MEANS	Density based	PCSD3	Proposed
Faculty	1923	1935	1921	1912
Parents	1676	1790	1870	1760
PG	1753	1798	1790	1807
Professional	1953	1973	1968	1829
Researcher	1878	1709	1936	2001
UG	1880	1884	1818	1883
MEAN	1843.83	1848.17	1883.83	1865.33
Categories	False Positive (FP) counts			
	K-MEANS	Density based	PCSD3	Proposed
Faculty	209	197	122	143
Parents	462	348	209	173
PG	317	272	322	358
Professional	191	171	198	218
Researcher	208	377	288	377
UG	215	211	223	204
MEAN	267	262.67	227	245.50
Categories	False Negative (FN) counts			
	K-MEANS	Density based	PCSD3	Proposed
Faculty	204	192	211	220
Parents	297	301	268	378
PG	333	388	280	263
Professional	191	192	176	315
Researcher	330	258	150	85
UG	247	245	277	212
MEAN	267	262.67	227	245.50

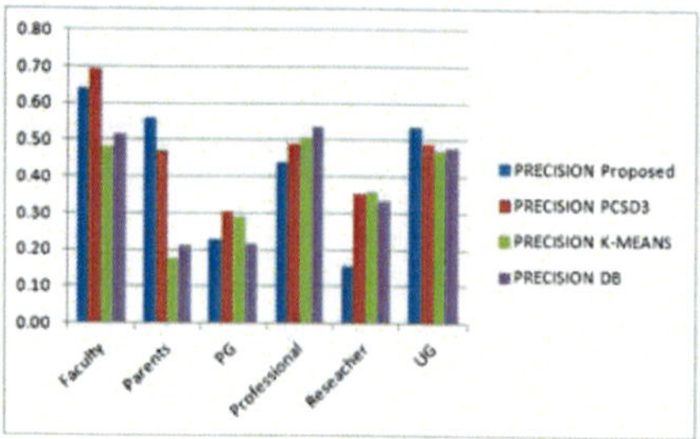

(a) PRECISION

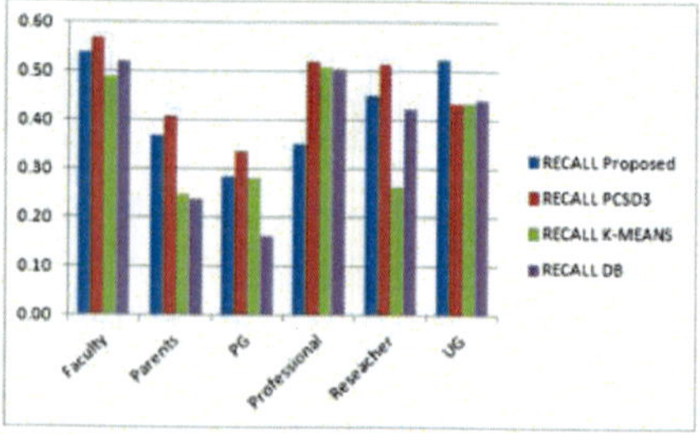

(b) RECALL

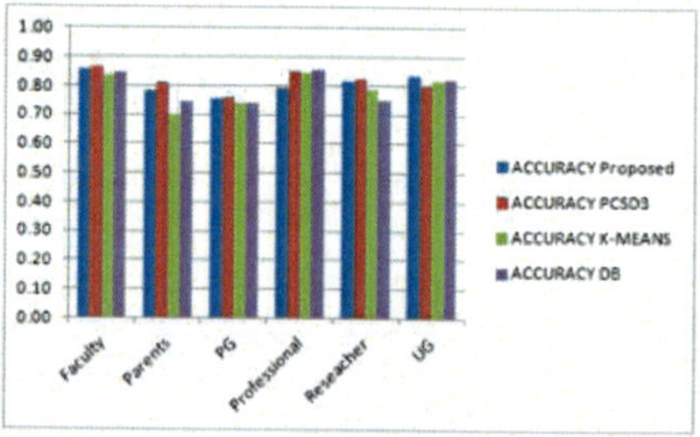

(c) ACCURACY

Fig. 2 Comparison of different clustering methods on stakeholder for real dataset

In all the three measures, we can easily detect that the proposed method performs better than K-means and Density-Based (DB) clustering and comparable to PCSD3.

6.3 Discussion

The performance measure results of the proposed method and the comparison with Density-Based, K-means, and PCSD3 suggest that the proposed method works satisfactorily on ordinal data. Wildly it improvises the previously defined techniques. It is basically based on the fact of a collective correlation of attributes, as we can say that the grouping of a set of instances should be more generalized. It means selecting instances into a group should have the same frame of thinking, not exactly the same type of responses.

7 Conclusion

This paper proposed an improvised clustering method on collectively correlated attributes of PCSD3. This is designed for ordinal data. The method proposed in this paper is based on differences among the attributes. The difference-attributes consider the fastening among attributes. This method also works on pattern similarity but performs better, as seen in the result section. This is shown to be a better method than K-means and Density-Based clustering. It basically generates a new set of patterns.

We compared the proposed method with PCSD3, K-means, and Density-based clustering on two different datasets. One is synthetic data described in case study 1, and another is the real dataset used in PCSD3, as described in case study 2. In both studies, we found that the proposed method performs better.

In this paper, we proposed a method that is limited to ordinal data. In future, we wish to focus on adapting the proposed methodology for nominal and mixed data.

References

1. Huang JL, Curran PG, Keeney J, Poposki EM, DeShon RP (2012) Detecting and deterring insufficient effort responding to surveys. J Bus Psychol 27(1):99–114
2. Gibson D, Kleinberg J, Raghavan P (2000) Clustering categorical data: an approach based on dynamical systems. The VLDB J 8(3):222–236
3. Liu I, Agresti A (2005) The analysis of ordered categorical data: an overview and a survey of recent developments. Test 14(1):1–73
4. Kamishima T, Akaho S (2009) Efficient clustering for orders. In: Mining complex data. Springer, pp 261–279
5. Ma Q, Guo J (2012) Mining multi-patterns in pattern-based clustering. Procedia Eng 29:3179–3183
6. McParland D, Gormley IC (2013) Clustering ordinal data via latent variable models. In: Algorithms from and for nature and life. Springer, pp 127–135
7. Damien McParland and Isobel Claire Gormley (2016) Model based clustering for mixed data: clustMD. Adv Data Anal Classification 10(2):155–169
8. Handorf EA, Heckman CJ, Darlow S, Slifker M, Ritterband L (2018) A hierarchical clustering approach to identify repeated enrollments in web survey data. PLoS One 13(9):e0204394
9. Zhang Y, Cheung YM, and Tan KC (2019) A unified entropy-based distance metric for ordinal-and-nominal-attribute data clustering. IEEE Trans Neural Netw Learn Syst 31(1):39–52
10. Jenkins SP (2020) Comparing distributions of ordinal data. The Stata J 20(3):505–531
11. Selosse M, Jacques J, Biernacki C (2021) ordinalclust: an r package to analyze ordinal data. The R J 12(2)
12. Dinh D-T, Huynh V-N, Sriboonchitta S (2021) Clustering mixed numerical and categorical data with missing values. Inf Sci 571:418–442
13. Zhang Y, Cheung YM (2021) Learnable weighting of intra-attribute distances for categorical data clustering with nominal and ordinal attributes. IEEE Trans Pattern Anal Mach Intell
14. Sadh R, Kumar R (2021) Directional pattern based clustering for quantitative survey data: method and application. Survey Res Methods 15(2):169–185

Machine Learning Algorithms for Binary Classification of Breast Cancer

Preeti Katiyar and Krishna Singh

Abstract Breast cancer is one of the most serious diseases in women, and it is the second largest cause of cancer mortality. Breast cancer occurs when malignant, cancerous tumors form from breast cells. Self-tests and routine clinical examinations aid in early detection and, as a result, dramatically improve survival chances. Breast cancer classification is a medical strategy that researchers and scientists face great challenges. Neural networks have acquired appeal as a technique for cancer data classification. This research begins by looking into the use of machine learning algorithms to classify breast masses for various diagnostic, predictive, or prognostic tasks in a number of imaging modalities, including MRI, ultrasound, digital breast tomosynthesis, and mammography. For evaluation, we proposed various machine learning algorithms such as support vector machine, random forest, Naive Bayes, logistic regression, K-nearest neighbor, artificial neural network, and deep neural network. We have used publicly accessible Breast Cancer Wisconsin (Diagnostic) datasets from the UCI Repository. As demonstrated in the results sections, the DNN classifier has a high level of accuracy (99.42%), showing that it outperforms traditional models.

Keywords Machine learning algorithms · Classification methods · Wisconsin breast cancer dataset · Deep neural network

1 Introduction

Deaths from breast cancer are increasing rapidly on a global scale. A statistical study conducted by the World Health Organization (WHO) revealed that 1.5 million women worldwide have carcinoma cancer. Due to inadequate awareness, this cancer has a

P. Katiyar (✉)
GGSIP University, New Delhi, India
e-mail: preetikatiyar27@gmail.com

K. Singh
G.B.Pant Institute of Engineering and Technology, New Delhi, India

A. K. Dubey et al. (eds.), *Advanced IoT Sensors, Networks and Systems*,
Lecture Notes in Electrical Engineering 1027,
https://doi.org/10.1007/978-981-99-1312-1_3

high fatality rate in India [1]. Breast cancer is a type of malignant tumor that begins in breast tissue, mainly in the internal lining of the breast lobules or milk ducts, and spreads to other regions of the body. Breast cancer is the world's second most common disease. As a result, it is critical to monitor the number of premature deaths due to breast cancer in successful therapy and the drop. Breast cancer research has increased during the previous decade [2]. Breast cancer medical imaging can be used to see within the human body as a non-invasive means of assisting doctors in diagnosing and treating patients [3, 4]. Any of the various imaging modalities can result in an early breast cancer diagnosis; however, it cannot be proved that these images are malignant on their own [5]. When a breast cancer tumor is at least 20 mm in size, it is frequently identifiable during screening. As a result, allowing for early identification of breast cancer is critical for facilitating early treatment [6]. Following detection via screening procedures such as clinical breast examination (CBE) and breast self-examination (BSE), early therapy may be advantageous. CBE is a routine medical exam performed by healthcare experts to detect breast lesions, whereas BSE is performed by an individual to notice physical changes and the look of the breasts. The practice of BSE encourages women to take charge of their health. As a result, the WHO recommends using BSE to raise awareness among at-risk women [7]. For image processing, several machine learning (ML), artificial intelligence (AI), and neural network approaches have recently been tested. The CAD system's main accomplishment is the creation of a genuine and dependable system capable of limiting experimental errors and assisting in the more accurate separation of benign and malignant tumors [8]. These technologies are used to improve image quality for human judgment as well as to automate image readability for improved comprehension and interpretation.

Computer science divides machine learning into three categories: reinforcement learning, unsupervised learning, and supervised learning. In supervised learning, labeled data is given, and the computer uses that labeled data to infer the label of the unlabeled input attributes. In contrast, all of the features are available in unsupervised learning even though there is no label or output class. In the case of reinforcement learning, it is the process of letting the models learn on their own. In this type of learning, after completing a task, a computer decides whether to penalize itself based on a set of established rules. The main goal is to improve the overall return [9, 10].

In order to identify the type of breast cancer in a suspected patient, this study employed the Wisconsin Breast Cancer Dataset (WBCD) of the FNA biopsy system and used various machine learning (ML) classifiers such as Support Vector Machine (SVM), random forest, Naive Bayes, logistic regression, K-nearest neighbor, Artificial Neural Network (ANN), and Deep Neural Network (DNN) classification models. The results are then evaluated in order to compare the algorithms and choose the model that is most effective at predicting breast cancer.

2 Literature Review

There are frequent researches on breast cancer diagnosis based on modern technologies such as machine learning, fuzzy system, and genetic algorithm.

Hijab et al. [11] used transfer learning to fine-tune the VGG16 network to diagnose breast cancers in a 1300-picture database in the United States. To avoid overfitting, the author employed image augmentation to expand the dataset and create a new one with 21,600 photos. The new dataset was then divided among 15,120 images for training and the remainder for testing. The fresh training set was used to fine-tune the weights on the VGG16 network's final convolutional layer and assess the system's effectiveness in classifying the images in the test set. The authors attained accuracy and AUC values of 0.97 and 0.98, respectively.

Zheng et al. [12] proposed using CNN based transfer learning to define breast masses for a variety of diagnostic, predictive, and prognostic purposes, as well as in a variety of imaging modalities, including MRI, ultrasound, digital breast tomosynthesis, and mammography. Several convolutional layers, LSTM, and max-pooling layers are included in the deep learning framework. A fully connected layer and a softmax layer have categorization and error estimation built in. This work focuses on finding the best way by integrating these machine learning methodologies with methods for choosing and extracting features, as well as evaluating their output using classification and segmentation techniques.

Osman et al. [13] used ensemble boosting, making it easier to determine the tumor's stage and whether it was malignant or benign. This will aid doctors in selecting suitable treatment for each tumor stage, allowing cancer patients with this type of tumor to be saved. On several types of UCI breast cancer datasets, the proposed RBFNN approach was tested. The general diagnosis accuracy for the WBC, BCD, BCP, and WBCD UCI datasets was 97.4%, 98.4%, 97.7%, and 97.0%, respectively, based on tenfold cross-validation using the RBFNN approach.

Yari et al. [14] proposed two successful deep transfer learning-based models that improve current state-of-the-art systems in binary and multiclass classification by relying on pre-trained DCNN employing a vast amount of BreakHis dataset. We transfer pre-trained ResNet50 and DesneNet121 weights to ImageNet as initial weights then fine-tune these models with a deep classifier with data augmentation to recognize various cancerous and benign sample tissues in binary and multiclass classification categories.

Cao et al. [15] presented a Multi-Tasking U-shaped Network (MT-UNet) with three fundamental concepts: (1) The proposed U-shaped classification architecture can adapt well to the heterogeneity of breast masses; (2) the proposed truncated normalization method and adaptive histogram equalization method can improve image contrast; (3) training with label smoothing method can alleviate the problem of insufficient training data causing convergence difficulty. The suggested scheme's performance is assessed using the DDSM and INbreast public datasets. The Area Under Curve (AUC) and accuracy (ACC) on the DDSM dataset were 0.9963 and

0.9817, respectively. The AUC and ACC on the INbreast dataset were 0.9767 and 0.9391, respectively.

Das et al. [16] created a CNN framework based on deep Multiple Instance Learning (MIL). Every slide in the framework is represented as a bag of extracted patches. For training purposes, only the bag label is used, hence patchwise labels are not required. The patches take on the label of the bag in which they were found. A WSI image (or bag) is called benign if all of its patches are benign, and malignant if even one patch includes cancer cells. Even with noisy patch labels, learning can be done at the bag level. The accuracy of this technique was tested using the BreakHis, IUPHL, and UCSB breast cancer datasets, with results of 93.06%, 96.63%, and 95.83%, respectively.

Chiu et al. [17] developed a new processing technique based on nine individual variables, including age, BMI, glucose, insulin, and a homeostasis model assessment, to predict breast cancer. To begin, principal component analysis (PCA) was utilized to discover valuable data elements and minimize the data's dimensions. The top five primary components accounted for 99.89% of the total. The Multilayer Perceptron Network (MLP) approach was then used to extract characteristics from the data, and the network's structure was created to allow researchers to see how data changed as the dimensions increased or decreased. The model achieved 86.97% accuracy.

Yu et al. [18] created a precise model for breast abnormalities detection and introduced a novel data augmentation technique called SCDA. They enhanced data augmentation and preprocessing by using the CLAHE contrast enhancement method on patches with the expanded ROIs by scaling before entering the patches obtained from the original mammography images into CNN network. Image contrast measurements reveal a significant improvement in the quality of image patches. They investigated models using cutting-edge connection techniques, such as Inception v3, ResNet, DenseNet, GoogLeNet, and pointwise and depthwise convolution, in order to identify the CNN models that performed the best on the binary classification challenge. According to the experimental findings, ResNet-50 performs the best out of all of those models. After performing a five-fold cross-validation, ResNet-50's performance attained an average accuracy of 95.74%.

Saber et al. [19] proposed characteristics from the Mammographic Image Analysis Society (MIAS) dataset using a pre-trained Convolutional Neural Network (CNN) architecture such as Inception V3, ResNet50, visual geometry group networks (VGG)-19, VGG-16, and Inception-V2 ResNet. Six measures, including F-score, accuracy, sensitivity, specificity, precision, and area under the ROC curve, will be used to evaluate the performance of the proposed model. VGG-16 performs the best out of all of those models. After using a tenfold cross-validation approach, the experimental findings show that the overall accuracy is 98.96%.

Farnoosh et al. [20] used an effective ensemble and transfer learning for the detection of breast cancer. The dataset of CBIS-DDSM has been classified using many models which were pre-trained such as VGG, ResNet, GoogleNet, DenseNet, MobileNet, and EfficientNet. The EfficientNet model was ensemble for better results. After tenfold cross-validation, they obtained significant improvement in accuracy as 96.05%.

Elkorany et al. [21] proposed two optimal approaches for the classification of breast cancer based on a hybridization of the Dragonfly algorithm and Whale optimization algorithm, along with Radial Basis Function Kernel SVMs. These algorithms were tested on the WDBC and WBCD databases and consider various performance matrices such as accuracy, sensitivity, specificity, and the area under the ROC curve. After using tenfold cross-validation process, the accuracy was 97.89% for WDBC and 99.27% for WBCD database.

3 Material and Methods

In this paper, we proposed to do a comparative analysis of the performance of various machine learning models. We have considered various machine learning algorithms—SVM, random forest, Naive Bayes, logistic regression, K-nearest neighbor, ANN, and DNN. We compare the accuracies given by these machine learning models. Figure 1 shows the flowchart for the proposed method.

3.1 Database Description

We have used publicly available Breast Cancer Wisconsin (Diagnostic) datasets from UCI Repository to accurately detect cancer as benign or malignant. There are four types of datasets: Wisconsin Dataset for Diagnostic Breast Cancer (WDBC), Breast Cancer Original (WBC), Breast Cancer Prognostic (BCP), and Breast Cancer Diagnosis (BCD). A digitized image of a fine needle aspirate (FNA) of a breast mass is used to compute features. They describe the properties of the cell nuclei depicted in the image. This database was created using biopsy pictures with 569 rows and 32 columns of data. The dataset's class feature is benign or malignant. In this dataset, 357 instances are benign and 212 are malignant. The 10 integer-valued attributes of dataset are shown in Table 1. There is no sample with a missing attribute value. Table 2 shows all 10 real-valued features that we have used for training our model.

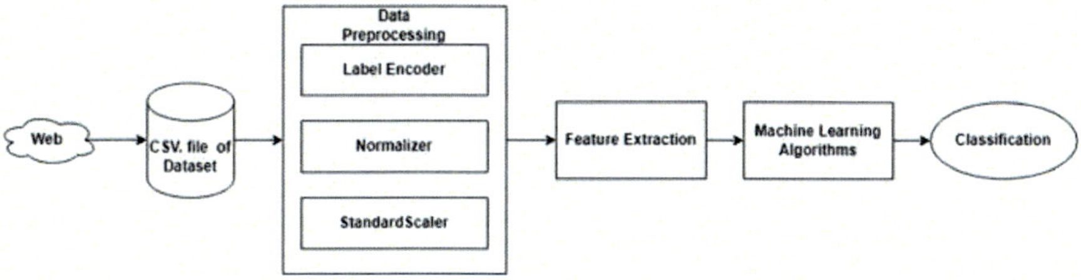

Fig. 1 Breast cancer detection and classification using the proposed method

Table 1 Attributes of wisconsin dataset for diagnostic breast cancer

S.No	Attribute	Range
1	Clump thickness	1–10
2	Uniformity of cell size	1–10
3	Uniformity of cell shape	1–10
4	Marginal adhesion	1–10
5	Single epithelial cell size	1–10
6	Bare nuclei	1–10
7	Bland chromatin	1–10
8	Normal nucleoli	1–10
9	Mitoses	1–10
10	Class	B/M

Table 2 Features of Wisconsin dataset for diagnostic breast cancer

S.No	Feature name	Feature description
1	Radius	Average of distances from center to circumference points
2	Texture	Standard deviation (SD) of gray-scale value
3	Perimeter	Gross distance between the snake points
4	Area	Total number of pixels on the inside of the snake along with one half of the pixels in the circumference
5	Smoothness	Local variance in length of radius, quantified by calculating the length difference
6	Compactness	$\frac{Preimeter^2}{area}$
7	Concavity	Intensity of the contour's concave parts
8	Concave points	The number of contour concavities
9	Symmetry	The difference in length between lines perpendicular to the major axis in both directions to the cell boundary
10	Fractal Dimension	Coastline estimation. A higher value leads to a less normal contour representing a higher risk of malignancy

3.2 Data Preprocessing

This section describes the pre-processing procedures used on breast cancer histopathology images to address issues such as low resolution and noise. Because these difficulties impair model performance during classification, pre-processing approaches are required to remove noise from histopathology images caused by staining operations.

3.2.1 Label Encoder

Label Encoder is efficient tool for encoding the levels of the categorical features into numeric values. Label Encoder encodes labels with values ranging from 0 to -1. All of our category characteristics are encoded. In this paper, we have used 0 and 1 to differentiate between malignant and benign diagnoses. After encoding the dataset, we applied a neural network to it and attained accuracy. However, the achieved accuracy is insufficient for breast cancer diagnosis.

3.2.2 Normalizer

The practice of rescaling one or more attributes to the 0–1 range is known as data normalization. This means that the most significant value for each property is 1 and the least significant value is 0. When you don't know how your data is distributed, normalization is an excellent approach to utilize. After using the Label Encoder technique, your entire dataset is converted to a numeric dataset. Label Encoder converts all string label datasets to numeric datasets. We know that the normalizing method works on numerical datasets. So, after normalizing the data, we applied a neural network to it and reached accuracy. However, until now, precision is insufficient. As a result, we used the following procedure for preprocessing.

3.2.3 Standard Scaler

The process of rescaling one or more attributes so that they have a mean value of 0 and a standard deviation of 1 is known as data standardization. This approach works well when the dataset attribute distribution is Gaussian. After scaling the data, we predicted a model and discovered that its accuracy is superior to that of another method.

3.3 Machine Learning Algorithms for Breast Cancer Prediction

Machine learning is an automatic learning process in which algorithms are built to learn from previous datasets. We input a huge number of data, the machine learning model analysis that data, and we may make predictions about the future based on that train model. Initially, the data will be split into 80% training and validation data before training the various models, with the remaining 20% serving as a holdout set for final testing. For the training data, K-Folds cross-validation is used to ensure a more accurate model and train accordingly. Machine learning algorithms for breast cancer prediction.

3.3.1 Artificial Neural Network

ANN is a sophisticated and powerful machine learning technique that can be applied to a variety of tasks, including classification, regression, voice recognition, and targeted marketing. A perceptron is a single unit of an ANN that was first designed in 1957 by Frank Rosenblat [22], who intended to create something that could learn and modify to match the needs. The way ANNs work is a little complicated but understandable: first, the input nodes are taken, with the number of input nodes equal to the number of independent variables, then these inputs are weighted through synapses and passed on to the next neuron, where activation function is applied, and so on until the weights have reached the maximum weights.

A neural network is made up of three layers: the input layer, the hidden layer, and the output layer. This approach is used to extract patterns that are overly complex. Parallel processing, distributed memory, collective solution, and network design are the foundations of the algorithm. A neural network is employed to represent a nonlinear function, and the K-Fold validation approach is used on ANN algorithms. The correctness of the validation procedure is measured. ANN had a 97.8% accuracy rate.

3.3.2 K-Nearest Neighbor

This algorithm is used to recognize patterns. It is an effective method for predicting breast cancer. Each class has been given equal weight in order to recognize the pattern. KNN finds similar highlighted data in a huge dataset. We classify a large dataset based on feature similarity. KNN was used to predict and diagnose breast cancer risk. We employed the K-Fold cross-validation technique to evaluate the predictive models on WDBC by partitioning the data into training and testing sets. KNN obtained 96.4% accuracy with 0.07 s execution time, which was greater than the other algorithms.

3.3.3 Logistic Regression

It is a supervised learning technique with additional dependent variables. This method generates a binary response. Logistic regression can produce a continuous result for a single dataset. This algorithm is made up of a statistical model and binary variables. The logistics regression approach, which is based on the logistics function, was utilized to calculate the outcome of the independent variable. Logistic regression was 97.6% accurate.

3.3.4 Naive Bayes

This model is used to assume a huge training dataset. The techniques use the Bayesian method to compute probability. It has the maximum accuracy when computing the probability of noisy data as an input. It is an analogy classifier that compares training dataset to training dataset.

We used nested ensemble approaches with classifiers for the automated diagnosis of breast cancer, and K-fold cross-validation, evaluated it using a simple ensemble approach and offered a comparison of the Naive Bayes classifier with Bayes Net. The accuracy of the Bayes Net algorithm was discovered to be 95.7%, which was higher than the accuracy of the Naive Bayes algorithm. We evaluated the NB and Bayes Net with meta-classifiers and found that the third meta-classifier with SV-Naive Bayes performed better than the others.

3.3.5 Support Vector Machine

It is a supervised learning algorithm that can solve classification and regression problems. It is made up of theoretical and numerical functions that are used to solve the regression problem. It has the highest accuracy rate for predicting massive datasets. It is an effective machine learning technique based on 3D and 2D modeling. We employed the SVM, a recursive feature removal technique with a predictive machine learning model, to diagnose breast cancer. The goal is to identify the correct attributes from a collection of normal and cancerous individuals.

The SVM algorithm is being evaluated using the recursive feature removal technique. The performance matrix is intended to evaluate the accuracy rate of the predictive model SVM on various types of kernel. SVM achieved 99% linear kernel accuracy, 98% RBF kernel accuracy, 97% polynomial kernel accuracy, and 84% sigmoid kernel accuracy.

3.3.6 Random Forest

Random forest is a supervised learning technique that can address classification and regression problems. It is a machine learning building element that is used to predict new data based on prior datasets. Nonlinear algorithm, such as random forest, is employed by us for breast cancer prediction. The Bioinformatics and Medical Science categorization technique was utilized by the authors. Random forest was 97.6% accurate.

3.3.7 Deep Neural Network

A DNN is an ANN with multiple hidden layers between the input and output layers. DNN is a computational network stimulated by biological neurons. DNN can create a linear or complex nonlinear mathematical model between input and output. Just like the human brain, DNN can be trained to do some tasks on pragmatic knowledge. It can be trained for classification, recognition, and regression problems. This technique was based on the selection of the best classifier, and a comparison of data mining algorithms was undertaken to choose the best algorithm for prediction. It was 99.4% accurate.

4 Performance Metrics

Several different metrics are used in order to analyze the performance of the model. These include the following.

4.1 Accuracy

One of the most significant metrics in model evaluation is accuracy. It divides the total number of classifications by the number of correct classifications. True positives (TP) and True Negatives (TN) make up correct classifications, while total classifications include both TP and TN, as well as False Positives (FP) and False Negatives (FN). Overall, the following formula is used to calculate classification accuracy:

$$Accuracy = \frac{TP + TN}{TP + TN + FP + FN} \tag{1}$$

4.2 Recall

Recall is a metric for how many positive cases were accurately categorized as malignant diagnoses in this case. Recall is one of the most essential variables in this dataset since a proper diagnosis, in the event that the patient has breast cancer, is critical. The following formula is used to calculate recall:

$$Recall = \frac{TP}{TP + FN} \tag{2}$$

4.3 Precision

Another important parameter in classification issues is precision. It primarily notifies you of the percentage of correctly classified positive cases out of the overall number of positive projected cases, which include both TP and FP. The following formula is used to calculate precision:

$$Precision = \frac{TP}{TP + FP} \tag{3}$$

4.4 F1-Score

F1-Score is a regularly used measure of a test's accuracy in binary classification tasks. The importance of this metric stems from the fact that it calculates the score using both precision and recall values. Equation (4) is used to determine the F1-Score.

$$F1\ Score = 2 * \frac{Precision * Recall}{Precision + Recall} \tag{4}$$

Figure 2 shows the Performance Metrics of applied Machine Learning Algorithms.

4.5 Confusion Matrix

A confusion matrix is generated for the actual and predicted result consisting of TP, FP, TN, and FN for calculating accuracy for each algorithm used below, the meaning of the terms is mentioned and the confusion matrix is shown in Table 3.

TP = True Positive (Accurately Identified).
TN = True Negative (Inaccurately Identified).
FP = False Positive (Accurately Rejected).
FN = False Negative (Inaccurately Rejected).

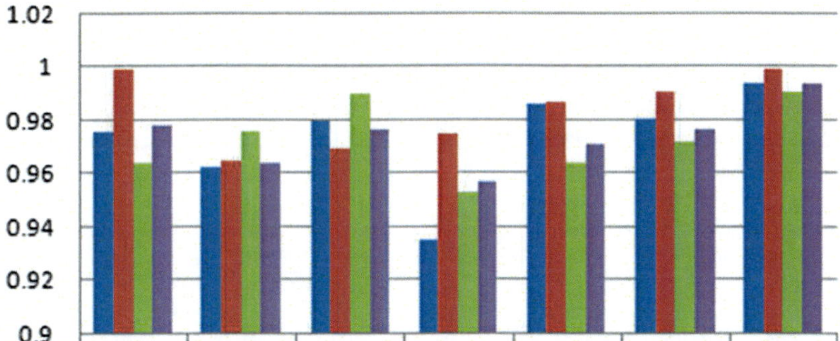

	ANN	KNN	LR	NB	SVM	RF	DNN
F1-Score	0.976	0.963	0.98	0.935	0.986	0.981	0.994
Precision	1	0.965	0.97	0.975	0.987	0.991	1
Recall	0.964	0.976	0.99	0.953	0.964	0.972	0.991
Accuracy	0.9785	0.9642	0.9766	0.9571	0.9714	0.9766	0.9942

Fig. 2 Performance metrics

Table 3 Confusion matrix

	Classified as positive	Classified as negative
Actually positive	TP	FN
Actually negative	FP	TN

Table 4 Comparison between the DNN model and other classifiers in terms of performance metrics

Classifiers	F1-score	Precision	Recall	Accuracy (%)
ANN	0.976	1.000	0.964	97.85
KNN	0.963	0.965	0.976	96.42
LR	0.980	0.970	0.990	97.66
NB	0.935	0.975	0.953	95.71
SVM	0.986	0.987	0.964	97.14
RF	0.981	0.991	0.972	97.66
DNN	0.994	1.000	0.991	99.42

5 Experiment and Results

After creating ML approaches, we evaluated the performance of the algorithms on the dataset in this section. This is performed by running the algorithms on the test dataset that was previously defined. A quarter of the whole dataset was included in the test dataset. Table 4. Shows the result analysis of the used techniques, in which column 1 includes the name of techniques used to predict breast cancer. Columns 2, 3, and 4 show the F1-Score, precision, and recall values of the applied techniques; Column 5 includes the accuracy of the used algorithms. Table 5 shows the comparison between the proposed DNN model and existing models.

According to the statistical analysis provided above in this section, it can be concluded that the suggested DNN model has produced positive results. When compared to existing approaches, it can classify breast cancer patients with more accuracy.

6 Conclusion

This paper covers and illustrates the use of various machine learning algorithms to reliably classify breast cancer as malignant or benign using the Diagnostic Wisconsin Breast Cancer Dataset. Automatic diagnosis and detection systems can do extraordinary help for medical practitioners. Advanced detection and good prediction can help to lower the number of people who die from breast cancer. For the WDBC dataset from the UCI machine learning repository, this study offers a robust DNN model with feature selection strategies that outperforms existing state-of-the-art models (such as

Table 5 Comparison between the proposed DNN model and existing models

Reference no.	Methods	Dataset	Accuracy (%)
Hijab et al. [11]	VGG-16	Ultrasound images	97.00
Osman and Aljahdali [13]	RBFNN	WBC BCD BCP WBCD	97.40 98.40 97.70 97.00
Yari et al. [14]	DCNN	BreakHis	98.00
Cao et al. [15]	MT–Unet	INbreast DDSM	93.91 98.17
Das et al. [16]	CNN-DeepMIL	BreakHis IUPHL UCSB	93.06 96.63 95.83
Chiu et al. [17]	PCA-MLP	BCCD	86.97
Yu et al. [18]	ResNet-50	INbreast	95.74
Saber et al. [19]	VGG-16	MIAS	98.96
Azour and Boukerche [20]	EfficientNet	CBIS-DDSM	96.05
Elkorany et al. [21]	RBFK-SVM	WDBC WBCD	97.89 99.27
	Proposed DNN Model	WDBC	99.42

SVM, KNN, RF, and others). After analyzing all of the algorithms used, the DNNs attained 99.42% accuracy. According to empirical results and statistical analysis, the proposed model for breast cancer diagnosis is very practical and effective. In the future, this accuracy can be enhanced by expanding the data size.

References

1. World Health Organization (2018) Global Health Observatory, "Geneva: World Health Organization," 2018. [Online]. Available: https://www.who.int/data/gho. Accessed 10 July 2019.
2. Gupta P, Kaur Malhi A (2018) "Using deep learning to enhance head andneck cancer diagnosis and classification." In: Proceeding of IEEE international conference on system, computation, automation and networking. New York, NY, USA, pp 1–6
3. Abdel-Zaher AM, Eldeib AM (2016) 'Breast cancer classification using deep belief networks.' Expert Syst Appl 46:139–144
4. Cruz-Roa A, Gilmore H, Basavanhally A, Feldman M, Ganesan S, Shih NNC, Tomaszewski J, González FA, Madabhushi A (2017) "Accurate and reproducible invasive breast cancer detection in whole-slide images: a deep learning approach for quantifying tumor extent." Sci Rep 7(1).

5. Mambou S, Maresova P, Krejcar O, Selamat A, Kuca K (2018) "Breast cancer detection using infrared thermal imaging and a deep learning model." Sensors 18, Aug. 2018
6. Yap MH, Pons G, Marti J, Ganau S, Sentis M, Zwiggelaar R, Davison AK, Marti R (2018) 'Automated breast ultrasound lesions detection using convolutional neural networks.' IEEE J Biomed Health Information 22(4):1218–1226
7. OncoLink Team (2020) All About Breast Cancer. [Online]. Available: https://www.oncolink.org/cancers/breast/all-about-breast-cancer
8. Breast Cancer (2019) Prevention and Control, World Health Org., Geneva, Switzerland
9. Dankwa-Mullan I, Rivo M, Sepulveda M, Park Y, Snowdon J, Rhee K (2019) 'Transforming diabetes care through artificial intelligence: the future is here.' Population Health Manage 22(3):229–242
10. Duraisamy S, Emperumal S (2017) 'Computer-aided mammogram diagnosis system using deep learning convolutional fully complex-valued relaxation neural network classifier.' IET Comput Vis 11(8):656–662
11. Hijab A, Rushdi MA, Gomaa MM, Eldeib A (2019) "Breast cancer classification in ultrasoundimages using transfer learning". In: Fifth international conference on advances in biomedical engineering (ICABME). pp 1–4.
12. Zheng J, Lin D, Gao Z, Wang S, He M, Fan J (2020) 'Deep learning assisted efficient AdaBoost algorithm for breast cancer detection and early diagnosis.' IEEE Access 8:96946–96954
13. Osman AH, Aljahdali HMA (2020) 'An effective of ensemble boosting learning method for breast cancer virtual screening using neural network model.' IEEE Access 8:39165–39174
14. Yari Y, Nguyen TV, Nguyen HT (2020) "Deep learning applied for histological diagnosis ofbreast cancer." IEEE Access 8: 162 432–162 448.
15. Cao H, Pu S, Tan W, Tong J, Zhang D (2020) Multi-tasking U-shaped network for benign and malignant classification of breast masses. IEEE Access 8:223396–223404
16. Das K, Conjeti S, Chatterjee J, Sheet D (2020) 'Detection of breast cancer from whole slide histopathological images using deep multiple instance CNN.' IEEE Access 8:213502–213511
17. Chiu HJ, Li T-HS, Kuo P-H (2020) Breast cancer-detection system using PCA, multilayer perceptron, transfer learning, and support vector machine. IEEE Access 8:204309–204324
18. Yu X, Kang C, Guttery DS, Kadry S, Chen Y, Zhang YD (2021) ResNet-SCDA-50 for breast abnormality classification. IEEE/ACM Trans Comput Biol Bioinform 18(1): 94–102
19. Saber A, Sakr M, Abo-Seida OM, Keshk A, Chen H (2021) A novel deep-learning model for automatic detection and classification of breast cancer using the transfer-learning technique. IEEE Access 9:71194–71209
20. Azour F, Boukerche A (2022) An efficient transfer and ensemble learning based computer aided breast abnormality diagnosis system. IEEE Access.
21. Elkorany AS, Marey M, Almustafa KM, Elsharkawy ZF (2022) Breast cancer diagnosis using support vector machines optimized by whale optimization and dragonfly algorithms. IEEE Access 10:69688–69699
22. Ibanez D (2016) Artificial neural networks–the rosenblatt perceptron.

Applications, Issues, and Deep Learning Analysis of Electronic Health Records

Shahid Ahmad Wani, S. M. K. Quadri, and Gowhar Mohiudin

Abstract Electronic Health Records (EHR) have evolved as a potential tool that has shown significant application in secondary purposes, e.g., research. EHR maintains valuable health information for hospitals in digital form, including demographic details, medical lab reports, procedures, medications, diagnosis codes, and more. Deep learning techniques such as Feed Forward Neural Network (FFNN), Convolution Neural Network (CNN), Recurrent Neural Network (RNN), and more have been utilized to propose models for EHR analysis. In this paper, we present the current state-of-the-art of deep learning application in medical tasks for EHR data. We present the exemplary performance of DL models in healthcare applications such as disease prediction, hospital readmission, future risk prediction, and many more. We also identify various applications of EHR, which include information extraction, representation learning, outcome prediction, computational phenotyping, and medical data de-identification. The study also introduces various issues and challenges related to EHR, and a theoretical comparison of recent deep learning (DL) models for EHR analysis.

Keywords Electronic health record · Deep learning · EHR issues · EHR · Electronic medical record · EMR

1 Introduction

The motivation behind developing EHR at its inception was to automate the billing purpose in hospitals [1]. However, EHRs have evolved as a potential tool that has shown significant application in secondary purposes, e.g., scientific research. EHR maintains substantial health information for hospitals in digital form, including

S. A. Wani (✉) · S. M. K. Quadri
Department of Computer Science, Jamia Millia Islamia, New Delhi 110025, India
e-mail: shahid1910448@st.jmi.ac.in

G. Mohiudin
Lovely Professional University, Punjab 144411, India

A. K. Dubey et al. (eds.), *Advanced IoT Sensors, Networks and Systems*,
Lecture Notes in Electrical Engineering 1027,
https://doi.org/10.1007/978-981-99-1312-1_4

demographics, medical laboratory test results, procedures, medication details, diagnosis codes, and more. Figure 1 shows an example of a typical EHR. EHR has shown great research significance to acquire deep insights of the data to provide better health care. EHR has been used for tasks such as medical concept extraction [2], suicide risk stratification [3], hospital readmission prediction [4], and more. These studies represent the significance of the EHR for secondary purposes [1]. Prediction of medical concepts ahead of their happening can help to provide preventive and personalized medical care for a patient. This could prevent unnecessary admission and expenses which may occur otherwise. EHR has managed to bring significant changes in health care delivery.

A huge amount of medical data has been produced in recent times, including protein sequencing, imaging data like computed topography and Medical Resonance Image (MRI), and EHR. One of the challenges in analyzing such data is the size, structure, heterogeneity, and availability of data. Recent advances in machine learning (ML) and deep learning (DL), which are the branches of artificial intelligence, have shown substantial application in analyzing such high-dimensional data. Deep learning essentially has shown some extraordinary performances on image data

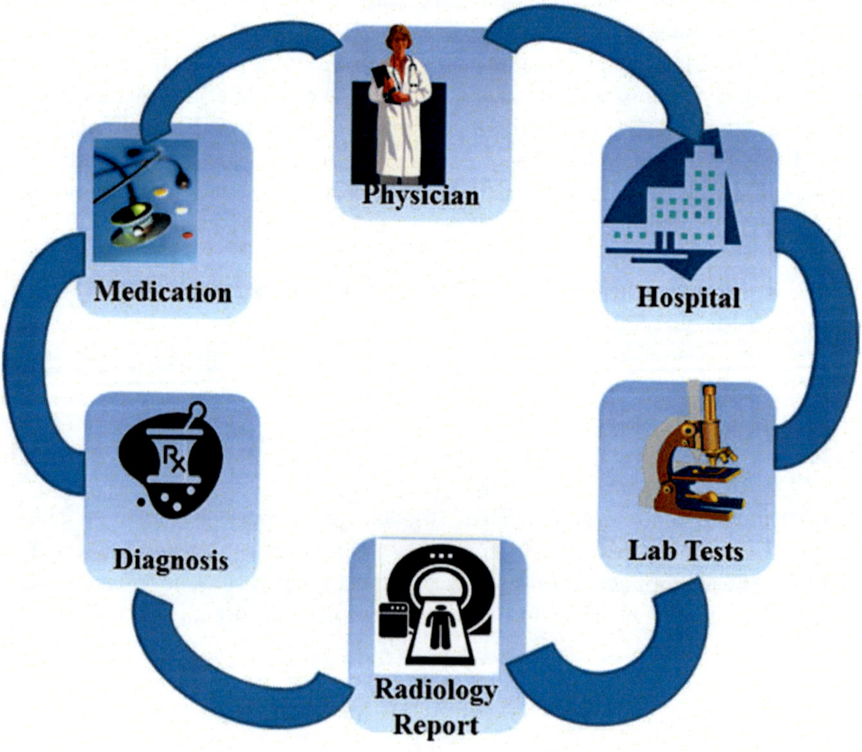

Fig. 1 Electronic health record

by winning the IMAGENET competition [5]. Particularly recurrent neural network has shown success for sequential text data [6].

Several DL models have been proposed for the analysis of EHR data. Deep patient [7], Deepr [4], Med2Vec [8], and DoctorAI [9] are a few examples of the deep learning-based models employed for EHR analysis. Recent advancements in deep EHR learning point to a variety of EHR applications such as information extraction (IE), disease predication (DP), representation learning (RL), phenotyping and de-identification. There are various issues and challenges faced by the EHR analysis such as interpretability, heterogeneity, and interoperability.

Here we present a review of applications and issues related to EHR. We also present a review of the deep learning models available for the analysis of EHR data. We identify the issues and challenges that need to be dealt with for the successful implementation of the EHR.

2 Methodology Used

The focus of this research is to identify the relevant articles regarding the aim of the study. Thus, our research methodology involves identifying articles from digital research data repositories such as scopus, web of science, and more as shown in Fig. 2. We chose scopus, google scholar, and pudmed as the source of finding articles.

Next, we form search strings as S1, S2, S3, ..., Sn. The search strings were formed as the combination of keywords such as electronic health records, EHR, electronic medical records, EMR, deep healthcare, and more. We connected these keywords with AND and OR operators to form more complex search strings, for example, "electronic health records AND electronic medical records OR EHR", "EHR OR EMR OR deep health care", and so on. A total of 206 relevant articles were identified from all the databases. In screening step, first 97 duplicates were removed and then 63 articles were found irrelevant after abstract reading, thus removed. Next, we apply the inclusion and exclusion criteria to shortlist the papers for final selection of the article. The inclusion criteria, for example, is that, study must be published in English language, must support the aim of the study, should be experimental work, whereas the exclusion criteria applied is studies published in other than English language, not fulfilling the aim of study, study only presents a theoretical perspective. In final step, we perform quality assessment to evaluate the credibility and relevance of the articles. Finally, 12 relevant articles were selected to be considered for the study.

3 Applications of DL for EHR Analysis

Applications of DL for the analysis of EHR include information extraction, concept learning, phenotyping, and data de-identification. In hospitals usually the clinical notes and medical codes serve as the primary source of data for EHR. This data is

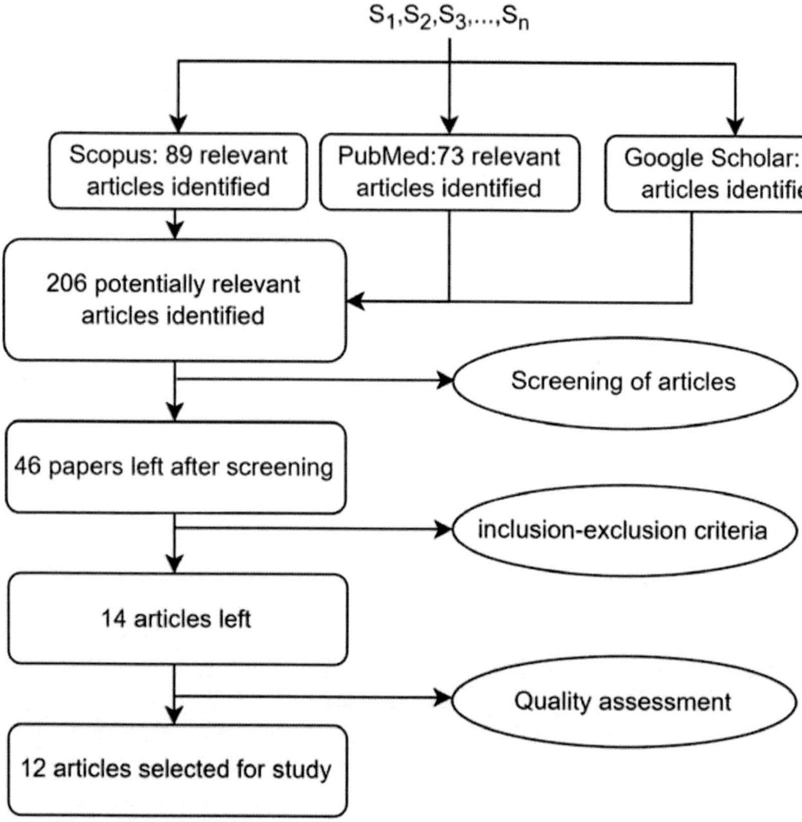

Fig. 2 Methodology used

processed and exploited by the DL models for a particular application such as disease prediction. Several DL and ML models have been proposed for tasks such as heart failure prediction, suicide stratification, and more. The process of DL analysis for EHR is shown in Fig. 3.

3.1 Information Extraction

Information Extraction is a technique for extracting and encoding the clinical information from textual data [10]. Typically, in text mining and natural language processing (NLP), it is referred to as the automatic extraction of entities, concepts, events, and also the relations and their linked attributes from the free text. The extraction of medical concepts such as disease, therapies, and procedures from clinical free texts is one of the most basic responsibilities [11]. Clinical notes are the primary source of IE for EHR. It consists of information regarding admission, medication,

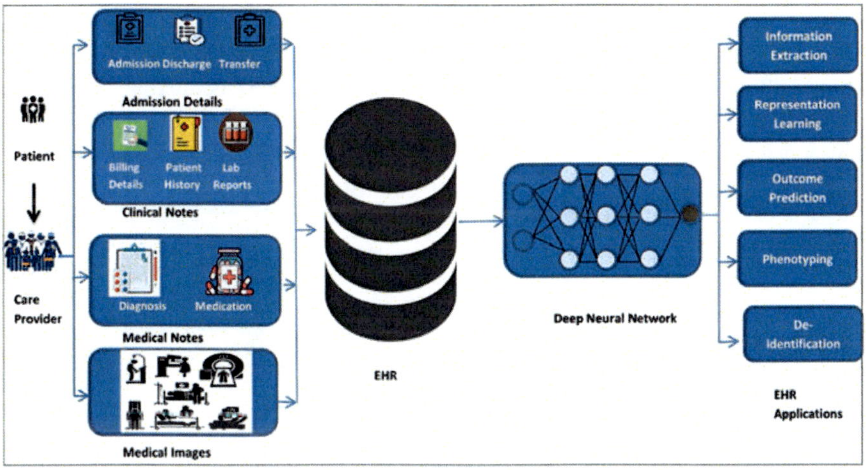

Fig. 3 Deep learning analysis for EHR

drug dosage, discharge, and transfer orders. IE typically involves sub-tasks such as concept extraction that pinpoints concept mentions (such as disease and medication), time-stamped-based information, relation extraction that identifies information between concepts, entities, and attributes. Thus, IE is classified as the following sub-tasks single concept extraction [11], temporal event extraction [12], relation extraction [13], and abbreviation expansion [14].

Single Concept Extraction. Single concept extraction involves the fundamental activity of labeling the standardized individual concepts such as disease, treatment, medication, and procedures. For example, a few major studies conducted are heart failure, cancer disease, and MedEx developed for providing the medication details [11, 15]. XU [16] proposed a model to extract information from clinical notes to describe the details of daily medication.

Temporal Extraction. Temporal information describes the details of time of an event happening, the time for which the event lasted, and the frequency of event [12]. The clinical encounters between the patient and the care provider, i.e., a physician, happen at different times. Temporal Extraction connects medical events to the time period or date in which they occurred. Thus, the clinical notes have strong temporal relations and have huge impact on decision-making.

Relation Extraction. A relation describes the effects of treatment of one entity, e.g., a medical condition over other entities; such as if the application of drug improves, worsens, or causes some other condition. The extraction of relation information is important in future decision-making. For example, if the extracted information suggests negative relation between two entities such as drug dosage and medical condition, the physician would avoid such a treatment in future. Another example of relation extraction would be a medical test revealing the severity of a medical problem. Roberts [17] proposed a Support vector machine (SVM) method to capture the relation information from clinical texts.

Abbreviation Expansion. Physicians use abbreviations on clinical texts to save time. The correct understanding of such abbreviations requires expansion before mapping to a medical concept. There are more than 197,000 medical abbreviations and thus the expansion is necessary for correct mapping which is a challenging task [18]. Chang [18] proposed a supervised ML framework to identify abbreviations related to their expansions from MEDLINE abstracts. Liu [19] proposed a statistical expert system model for extracting abbreviations with their corresponding expansions from the biomedical text.

3.2 EHR Representation Learning

The patient-doctor meeting is represented through distinct medical codes by an EHR. The manual mapping between the medical concepts and medical codes does not quantify the similarities between them. Recently, deep learning approaches have been used to get more detailed representation between the concept and medical codes by projecting discrete medical codes into a vector space.

Concept Representation. A medical concept is a technical description of the medical problem, treatment, or description of a medical document. An example of medical concept is a diagnosed "congestive heart failure". Concept representation is a process of identification and extraction of the medical concepts from the clinical notes. There is significant unknown relation between each concept. It is a critical task to represent such relationships with word-level representation such as one-hot coding or natural language token representation [13]. Deep learning techniques have been employed to capture the similarities between medical concepts for deriving the EHR concept vectors. Distributed embedding and latent encoding are the two techniques which have been commonly used for the representation of EHR concepts [13].

Patient Representation. In literature, studies have been proposed that utilize DL techniques to derive the vector representation of patients. These DL methods include dimensionality reduction techniques such as AE and NLP techniques. Choi [15], Deepr [4], DeepCare [20] models use RNN, CNN, and vector joining for the patient representation in EHR.

3.3 Outcome Prediction

Clinical outcome prediction is a task to predict patient's development from EHR. It is a difficult task because of the temporal data captured in different time settings. It is usually classified into two categories which are static or one-time and temporal outcome prediction. For example, ICU admission within next 6 months and heart failure prediction within a given period of time are the examples of single and temporal outcome prediction, respectively. Liang [21] and Esteban [22] presented deep learning models for disease classification.

3.4 Computational Phenotyping (CP)

CP is the process of mining significant and meaningful descriptors or phenotypes from EHR. It has the capability to discover latent phenotypes from the available raw data which are not possible to discover manually. Miotto [7] implemented auto-encoder for phenotype-based representation of 704,587 patients. Lipton [23] uses time series with 13 variables and defines the multi-class problem for phenotype prediction. It is broadly classified into two categories as new phenotype discovery and improving existing definition.

3.5 Clinical Data De-identification

It is important to keep safe the privacy of the patient or any person in general. Clinical or medical notes carry significant information of a patient such as name, gender, residence, Identification number, and illness type. It is critical to not let this information fall into the hands of the people who could use it for their personal benefit. Thus, it is imperative to handle such sensitive data carefully and protect the data from being misused. In this regard, United Stated of America enacted the Health Insurance Portability and Accountability Act (HIPAA). According to HIPAA, all the clinical notes released by hospitals should be free from critical Information. This makes the distribution of health data sets a little problematic and hence a hindrance in research. Thus, the de-identification of the personal data on the clinical notes becomes necessary. Dernoncourt [24] created a model for automatic de-identification of clinical using long short term memory (LSTM) networks. Shweta [25] performed an analysis of recurrent neural network (RNN) techniques for identifying named entities in medical scripts.

4 Issues and Challenges for Analysis of EHR

There are several issues and challenges associated with the analysis of EHR. Some critical issues and challenges are described in this section.

4.1 Data Privacy and Integrity

Care givers have found EHR very significant in providing better care however the patients are in constant fear of the privacy and integrity of the personal health information. Data privacy is a challenge in any electronic health record to maintain a trust between care giver and the patient. If the privacy of patient is not addressed properly, then he may be reluctant in sharing the medical history or personal information critical in diagnosing an illness. It is very important to address this challenge so that enough data can be obtained during the treatment of the illness. There are many ways which leads to compromise of the privacy of data which includes data misuse, accidental damage of data, data loss of hospital reports, etc. With time most of the countries are implementing the EHR, the chances of data breaches will supposedly raise if proper care is not taken. The list of health data privacy acts implemented in various countries is mentioned in Table 1.

Table 1 Health data privacy acts of countries

Country	Law	Description
Australia	Privacy act 1988	It specifies how health organizations handle personal health information
European union	General data protection regulation(GDPR)	It considers privacy as fundamental human right and prohibits unauthorized and illegal access of personal information
United Kingdom	GDPR extension of Data Protection act (DPA)	To collect, store, and share health data in a secure and legal way
United states of America (USA)	Health insurance portability and accountability act of 1996 (HIPPA)	It ensures patient access, consent, and privacy before the information is collected or shared. It provides a comprehensive data protection
India	Information technology act 2000 personal data protection bill 2019	It provides safety of personal information and to setup a data protection authority
Russia	Russian federal law on personal data	It provides rights and liberties for privacy and secrecy of personal information
South Korea	Personal information protection act (PIPA)	It specifies the use of data such that the privacy, rights, and interests of citizens are protected
China	Personal information protection law (PIPL)	Restrictions on sharing data and regulates imposition of fine on violation of PIPL

4.2 *Interoperability*

Interoperability is one of the crucial issues in implementation of EHR. It allows the sharing and interpretation of the data among the various stake holders [26]. Interoperability issues are classified into three categories syntactic, semantic, and organizational. Syntactic interoperability implies that a system using different data and information formats such as XML and SQL should be able to share information easily. Semantic interoperability means that data shared should be interpreted by the systems to utilize data in an efficient manner. Organizational interoperability refers to the collaboration of socio-political, and legal bodies for the greater good or the sharing of information.

4.3 *Adversarial Attack*

ML and DL models have shown a great potential to analyze health data for several functionalities such as disease prediction, early diagnosis, hospital readmission, and more. However, these models face a great challenge in their deployment because of adversarial examples, which is an input that manipulates DL and ML systems to cause misclassification of the problem [27]. For example, an adversary can cause a ML or DL model wrongly classify a benign disease as malignant. This can lead to the wrong treatment procedures thus deteriorating a patient's health. There is a pressing demand for the development of ML and DL models which could address such a critical issue.

4.4 *Usability*

Usability is defined as the measure of effectiveness of a product. Poor usability of EHR has direct effects on the performance of the care provider. The time spent to enter data and other stuff in EHR could be utilized with patient that can help in providing better treatment. So solving usability issue of EHR is imperative for its high acceptance among the users of the system. Studies report that poorly managed EHR can pose a potential threat to the safety of the patient. Bloom [28] performed a survey to identify the usability score of the various EHRs in UK. They found that not a single EHR system could achieve the median usability score and meet the standards of acceptable usability.

4.5 *Heterogeneity*

The EHR is promising to a great extent in improving healthcare delivery and advancing medicine. However, the heterogeneity of the EHR makes the clinical research questionable because of poor research reproducibility [11]. EHR comprises different data formats such as clinical notes, radiological reports, medical lab tests, disease codes, medication dosages, and demographic information. The heterogeneity of EHR data makes it unique, although difficult to use and complex for clinical research. Discrete medical codes related to each patient are examined for finding a tractable way to handle EHR events comprising of unstructured data types. The models proposed for unplanned readmission prediction [4], risk prediction for future [9], stratifying suicide risk [3], and more exploit the code representation for handling heterogeneous data.

5 DL Models for Electronic Health Records

Electronic health records are used in most of the healthcare delivery setting or hospitals nowadays. The amount of data kept in EHR has increased dramatically in recent years. EHR was initially used for primary use such as billing, invoices, medical lab records, radiology images, demographic information, and more. Now the DL have shown promising results for analysis of EHR for secondary purpose. Here we present a review of a few of the state-of-the-art models for EHR analysis. An EHR data has been used for various research purposes such as extraction of medical concept, heart failure prediction, multi-outcome prediction, concept representation, and more described briefly in Table 2.

Tran [3] reported a model with restricted Boltzmann Machine (RBM) to represent medical concepts by embedding them in compressed space for the downstream analysis such as visualization and clustering. The model was implemented on a data set of patients for the assessment of having suicidal tendencies.

Miotto [7] presented deep patient a feature learning method for patient representation to help physicians make better clinical decisions. A denoising auto-encoder was applied to capture hierarchical regularities from EHR of 700,000 patients. The model was used as multi-classifier to predict development of future health conditions of about 78 diseases in a patient.

Choi [9] developed an RNN model called as DoctorAI, for prediction of physician diagnoses and medication orders. The dataset for heart failure was extracted from primary care patients in a large healthcare group with 263,706 patients. The model outperformed other baseline models by a huge margin and achieved a recall of 79.58%.

Choi [8] proposed Med2Vec framework for the representation of the EHR medical records such as medication, diagnosis, procedure codes, and more. Med2Vec learns the representation of medical codes and visits from a huge dataset of more than

Table 2 Deep learning models for electronic health records

Year and author	Model/architecture	Dataset with size	Objective task	Evaluation metrics	Significance	Predictors considered
Tran [3]	eNRBM/RBM	Barwon Mental Health Cohort Australia 7,578 patients	Suicide risk stratification	F-score recall precision	Object grouping and risk stratification	Diagnosis procedures
Miotto [7]	Deep patient/SDA	Not defined 704,587 patients	Future disease prediction	AUROC, F-score accuracy	Personalized medicine	Demographic variables diagnosis medication
Choi [9]	DoctorAI/RNN	Primary care patients from Sutter Health Medical Foundation 263,706 patients	Predict diagnosis and medication orders	Recall	Anticipating patient status and providing useful data to physician	Diagnosis procedures medication
Choi [8]	Med2vec/FFNN	Children Health of Atlanta (CHOA) 3,359,240 visits 550,339 patients and Center for Medical and Medicare Services (CMS) 5,464,950 visits 831,210 patients	Predict medical codes	AUROC recall	Representation learning of medication, diagnosis, and more	Diagnosis procedures medication
Choi [29]	RETAIN/RNN	Sutter Health EHR 32787 patients	Heart failure prediction	AUROC	Improved predictive accuracy and interpretability	Diagnosis procedures medication

(continued)

Table 2 (continued)

Year and author	Model/architecture	Dataset with size	Objective task	Evaluation metrics	Significance	Predictors considered
Nyugen [4]	Deepr/CNN	Data from Private Hospital Australia 590,546 records of 300,000 patients	Hospital readmission prediction	AUROC	Detects motifs and underlying disease structure	Diagnosis procedure
Pham [20]	DeepCare/RNN	Regional Hospital Australia 7,191 diabetes and 6,109 Mental Health Patients	Diabetes and mental health prediction	Precision F-score recall	Personalized medicine and care	Diagnosis procedures medication
Landi [30]	ConvAE/CNN & AE	Mount Sinai Health System 1,608,705 patients	Patient stratification	Entropy purity average score	Personalized medicine	Diagnosis
Li [31]	BEHRT/transformers	Clinical Practice Research Datalink (CPRD) primary care 1,609,024 patients	Predict medical condition	Precision recall AUROC	Personalized and preventive medicine	Diagnosis procedures medication
Li [32]	Hi-BEHRT/transformers	CPRD primary care data 2,438,352 patients for training	Clinical event prediction	AUROC AUPRC	Prediction with less training data available	Diagnosis medication procedures
Ramsy [33]	Med-BERT/transformers	Cerner Health EHR 28,490,650 patients	Disease prediction	AUC	Disease prediction transfer learning	Diagnosis procedures medication

one million visits. A multi-layer perceptron (MLP) generates a visit representation corresponding to a specific visit.

Choi [29] proposed a Reverse Time AttentIoN (RETAIN) model for the analysis of EHR. It addresses data accuracy and interpretability challenges related to health data for prediction. The model is developed as a two-layer neural attention model that detects past visit by patients and substantial variables such as diagnosis in those visits. The model was trained with data set of 14 million visits records from 32787 patients.

Nguyen [4] proposed Deepr (or deep record) a feature extraction method based on CNN to predict future risk of getting unplanned readmission after being discharged from a hospital. Medical Records containing several coded objects like medication, diagnoses, etc., are sequenced as words separated by coded time gaps. Words are embedded into continuous vectors to detect convoluted local motifs using relu function. The motifs are max-pooled to derive a feature vector and a classifier at top for prediction. The model predicts readmissions successfully and thus could be promising for predictive medicine.

Pham [20] introduced DeepCare a neural network (NN) model that reads medical records to store earlier medical history and infer current medical state of patient to predict future medical outcome. It is a three-layer deep NN using LSTM that takes as input the records of admissions as feature vectors at bottom layer. Middle layer of LSTM computes corresponding illness states from admission sequences by means of pooling. The top layer is a NN which estimates the outcome probability. The development of disease and its advancement, treatment suggestion, and long-term risk forecasting are reported from two different cohorts of mental health and diabetes.

Landi [30] proposed an unsupervised deep learning model ConvAE, to facilitate patient stratification through representation. The model based on word embeddings, CNN, and AE is a representation learning model to transform trajectories of patients into latent vectors of low dimension, and considers a data set of 1,608,705 patients. It was reported that ConvAE performed better in clustering task as compared to other models. The significance of model is highly recommended in personalized medicine.

Li [31] proposed BEHRT a deep neural sequence model for EHR analysis. The model predicts the possibility of 301 medical condition in future visit of a patients. This could help to provide personalized and preventive medical care for patients. The model was trained and evaluated over a dataset of 1.6 million records showing very good results under the average precision evaluation metrics. The model includes the heterogeneous concepts such as medication, diagnosis, lab measurements, and more to improve the prediction.

Li [32] presented Hi-BEHRT to overcome the limitations of DL models in processing long sequences of unstructured multimodal longitudinal data of EHR. It predicts clinical events such as heart failure, diabetes, chronic kidney disease, risk of a stroke, and more. The model performed performance shows improved results for AUROC and area under the precision recall curve (AUPRC).

Ramsy [33] recently presented a predictive model based on bidirectional encoder representations from transformers (BERT) which they called as Med-BERT. The pretraining of Med-BERT on a very large dataset of 28,490,650 patients generates contextualized embedding. The Med-BERT with fine-tuning significantly improves prediction accuracy and AUROC in disease prediction.

6 Conclusion

EHR has shown great success for secondary usage to improve healthcare by providing help to care providers such as doctors in decision-making. In this review, we present applications of EHR where the new entrants in the research can advance the knowledge. We also identify the issues and challenges that need to be overcome for the proper and efficient implementation of EHR. Interoperability, privacy, de-identification, and heterogeneity are a few critical challenges which need to be addressed. EHR contains a significant amount of information that has helped in disease prediction, detection, future risk prediction, next visit medication, personalized care, precision treatment, and more. Deep learning models help to gain more insights in data to achieve these objectives. This review presents a brief discussion of DL methods for EHR analysis. Although EHR implementation has increased due to lack of common and standard framework, organizations are still reluctant for its use. Thus, we suggest a standard format is imperative.

References

1. Yadav P, Steinbach M, Kumar V, Simon G (2018) Mining electronic health records (EHRs) a survey. ACM Comput Surv (CSUR) 50:1–40
2. Meystre S, Savova G, Kipper-Schuler K, Hurdle J (2008) Extracting information from textual documents in the electronic health record: a review of recent research. Yearb Med Inform 17:128–144
3. Tran T, Nguyen T, Phung D, Venkatesh S (2015) Learning vector representation of medical objects via EMR-driven nonnegative restricted Boltzmann machines (eNRBM). J Biomed Inform 54:96–105
4. Nguyen P, Tran T, Wickramasinghe N, Venkatesh S (2016) Deepr: a convolutional net for medical records. IEEE J Biomed Health Inform 21:22–30
5. Creswell A, White T, Dumoulin V, Arulkumaran K, Sengupta B, Bharath A (2018) Generative adversarial networks: an overview. IEEE Signal Process Mag 35:53–65
6. Jagannatha A, Yu H (2006) Bidirectional RNN for medical event detection in electronic health records. In: Proceedings of the Conference. Association for Computational Linguistics, North American Chapter, Meeting, pp 473.
7. Miotto R, Li L, Kidd B, Dudley J (2016) Deep patient: an unsupervised representation to predict the future of patients from the electronic health records. Sci Rep 6:1–10
8. Choi E, Bahadori M, Searles E, Coffey C, Thompson M, Bost J, Tejedor-Sojo J, Sun J (2016) Multi-layer representation learning for medical concepts. In: Proceedings of the 22nd ACM SIGKDD. International Conference on Knowledge Discovery and Data Mining, Washington D.C. USA, pp 1495-1504.
9. Choi E, Bahadori M, Schuetz A, Stewart W, Sun J (2016) Doctor ai: predicting clinical events via recurrent neural networks. In: Machine Learning For Healthcare Conference, PMLR. pp. 301-318

10. Wang Y, Wang L, Rastegar-Mojarad M, Moon S, Shen F, Afzal N, Liu S, Zeng Y, Mehrabi S, Sohn S (2018) Others: clinical information extraction applications: a literature review. J Biomed Inform (77): 34-49.
11. Shickel B, Tighe P, Bihorac A, Rashidi P (2017) Deep EHR: a survey of recent advances in deep learning techniques for electronic health record (EHR) analysis. IEEE J Biomed Health Inform 22:1589–1604
12. Chikka V (2016) Extraction of temporal information from clinical documents using machine learning techniques. In: Proceedings of the 10th International Workshop On Semantic Evaluation. pp. 1237-1240.
13. Lv X, Guan Y, Yang J, Wu J (2016) Clinical relation extraction with deep learning. Int J Hybrid Inf Technol 9:237–248
14. Liu Y, Ge T, Mathews K, Ji H, McGuinness D (2018) Exploiting task-oriented resources to learn word embeddings for clinical abbreviation expansion. ArXiv Preprint ArXiv:1804.04225
15. Choi E, Schuetz A, Stewart W, Sun J (2016) Medical concept representation learning from electronic health records and its application on heart failure prediction. ArXiv Preprint ArXiv:1602.03686.
16. Xu H, Stenner S, Doan S, Johnson K, Waitman L, Denny J (2010) MedEx: a medication information extraction system for clinical narratives. J Am Med Inform Assoc 17:19–24
17. Roberts A, Gaizauskas R, Hepple M, Guo Y (2008) Mining clinical relationships from patient narratives. MC Bioinform 9:1–17
18. Chang J, Schütze H, Altman R (2002) Creating an online dictionary of abbreviations from Medline. J Am Med Inform Assoc 9:612–620
19. Liu H, Friedman C (2002) Mining terminological knowledge in large biomedical corpora. In: Biocomputing 2003. World scientific, pp 415-426.
20. Pham T, Tran T, Phung D, Venkatesh S (2016) Deepcare: a deep dynamic memory model for predictive medicine. In: Pacific-Asia Conference on Knowledge Discovery and Data Mining, LNCS, Springer, Vol 9652, pp 30-41.
21. Liang Z, Zhang G, Huang J, Hu Q (2014) Deep learning for healthcare decision making with EMRs. In: 2014 IEEE International Conference on Bioinformatics and Biomedicine (BIBM). IEEE, Belfast UK, pp 556-559.
22. Esteban C, Staeck O, Baier S, Yang Y, Tresp V (2016) Predicting clinical events by combining static and dynamic information using recurrent neural networks. In: 2016 IEEE International Conference on Healthcare Informatics (ICHI). IEEE Chicago, IL, USA, pp. 93-101.
23. Lipton Z, Kale D, Elkan C, Wetzel R (2015) Learning to diagnose with LSTM recurrent neural networks. Arxiv preprint arXiv:1511.03677
24. Dernoncourt F, Lee J, Uzuner O, Szolovits P (2017) De-identification of patient notes with recurrent neural networks. J Am Med Inform Assoc 24:596–606
25. Yadav S, Ekbal A, Saha S, Bhattacharyya P (2016) Deep learning architecture for patient data de-identification in clinical records. In: Proceedings of the Clinical Natural Language Processing Workshop (ClinicalNLP). The COLING 2016 Organizing Committee, Osaka, Japan, pp 32-41.
26. Sreenivasan M, Chacko A (2021) Interoperability issues in EHR systems: research directions. Data Anal Biomed Eng Healthc. 13-28.
27. Finlayson S, Chung H, Kohane I, Beam A (2018) Adversarial attacks against medical deep learning systems. ArXiv Preprint ArXiv:1804.05296.
28. Bloom B, Pott J, Thomas S, Gaunt D, Hughes T (2021) Usability of electronic health record systems in UK EDs. Emerg Med J 38:410–415
29. Choi E, Bahadori M, Kulas J, Schuetz A, Stewart W, Sun J (2016) Retain: an interpretable predictive model for healthcare using reverse time attention mechanism. ArXiv Preprint ArXiv:1608.05745.
30. Landi I, Glicksberg B, Lee H, Cherng S, Landi G, Danieletto M, Dudley J, Furlanello C, Miotto R (2020) Deep representation learning of electronic health records to unlock patient stratification at scale. NPJ Digit Med 3:1–11

31. Li Y, Rao S, Solares J, Hassaine A, Ramakrishnan R, Canoy D, Zhu Y, Rahimi K, Salimi-Khorshidi G (2020) BEHRT: transformer for electronic health records. Sci Rep 10:1–12
32. Li Y, Mamouei M, Salimi-Khorshidi G, Rao S, Hassaine A, Canoy D, Lukasiewicz T, Rahimi K (2021) Hi-BEHRT: hierarchical transformer-based model for accurate prediction of clinical events using multimodal longitudinal electronic health records. ArXiv Preprint ArXiv:2106.11360
33. Rasmy L, Xiang Y, Xie Z, Tao C, Zhi D (2021) Med-BERT: pretrained contextualized embeddings on large-scale structured electronic health records for disease prediction. NPJ Digit Med 4:1–13

Design and Implementation of Multiple-User Variable Rate Channel Assignment Method in Cognitive Radio System

Gowda C. P. Mallikarjuna, Kuber Vijaykumar Bellad, P. Prajwal, V. Hima, and Niharika Suresh

Abstract With the ever-growing demand for technological advancement in the field of wireless communication to accommodate the huge increase in users, research on cognitive radio has been flourishing. Apart from developing ways to utilize the underused portions of the spectrum, it is equally important to explore the unused portion to expand its utility. In this paper, a technique for channel assignment is proposed which takes the available data rates of the vacant channels as well as the data rate required by the secondary users. For the purpose of comparing the effectiveness of the proposed method, simulation of this process is carried out in comparison with the sequential method. The number of secondary users served for a range of busy probabilities is found out and the obtained results are used to plot a graph which can be used to determine how the variable data rate assignment method utilizes the available channels better than the existing sequential method.

Keywords Cognitive radio · Busy probability · Sequential channel assignment · Multiple-user Variable rate channel assignment

G. C. P. Mallikarjuna · K. V. Bellad (✉) · P. Prajwal · V. Hima · N. Suresh
Department of Electronics and Telecommunication Engineering, BMS Institute of Technology and Management, Bengaluru, Affiliated to VTU, Belagavi, Bangalore, India
e-mail: kuber.bellad4@gmail.com

G. C. P. Mallikarjuna
e-mail: cpmallikarjunagowda@bmsit.in

V. Hima
e-mail: hima.himav@gmail.com

N. Suresh
e-mail: niharikasuresh20@gmail.com

A. K. Dubey et al. (eds.), *Advanced IoT Sensors, Networks and Systems*,
Lecture Notes in Electrical Engineering 1027,
https://doi.org/10.1007/978-981-99-1312-1_5

1 Introduction

Cognitive Radio (CR) is a smart technology that allows devices/users to make use of a particular part of the spectrum for a given period of time. It helps the Secondary Users (SUs) to make use of an empty portion of the spectrum when it is not used by the Primary Users (PUs). At present, efficient use of the spectrum is one of the most important aspects to overcome the spectrum congestion which is increasing day by day as the regulatory bodies that are responsible for assigning spectrums to different vendors are facing spectrum scarcity.

As per reports from the Federal Communications Commission (FCC), only 6% of the allocated spectrums are being used efficiently by the PUs, whereas the remaining 94% is either vacant or underutilized [2]. Hence, it becomes essential to identify the vacant portions of the spectrum that can be exploited to maximize its usage. These vacancies can also be referred to as spectrum holes in a licensed spectrum that can be made used by secondary users.

A lot of research is being done in the area of the design of different algorithms that could be put into use for the effective utilization of the spectrum. Such an approach is needed to serve multiple SUs trying to access the spectrum as per the data rate demands. We aim to design and implement two algorithms with different approaches to channel assignment and also analyze which is more efficient. Many algorithms assume that the spectrums have a fixed data rate while developing algorithms, but it is not so. Practically, each user has a specified data rate that has to be met in order to transfer the information effectively which often keeps varying. Along with this, every channel has a particular data rate to which SUs can be assigned when PUs are inactive.

The CR can be seen as an intelligent wireless radio communication device, which initially senses the spectrums that are left unused, after which it assigns SUs to these idle bands to achieve effective and optimized wireless communication [1]. Hence, it can be referred to as a predator–prey relationship in which the predator refers to PUs and SUs which are fed by prey which are spectrum resources. As seen in Fig. 1, the PUs and SUs can be referred to as predators; and the spectrum resource can be called prey. These two entities try to hunt the Spectrum Resource whenever necessary, as per requirement in the allocated spectrum. When it is not actively used by the PUs, it can be utilized by the SUs. Apart from the actively unused portions of the spectrum, the SU can also try to access the unused channels which are not occupied by the licensed users [3].

Cognitive radio networks make use of the licensed idle spectrums. As a result, the performance will be now dependent on the activity of the PUs. PU activity modeling is very crucial with respect to the performance of Cognitive Radio Networks (CRNs). If all PU activity models are identified by SUs with the usage of the right models, then the SUs will be able to enhance performance of their network [4].

Spectrum sensing is the most basic step through which the SU can identify the channel characteristics of the radio, i.e., to check whether the spectrum is used by a PU or not, what the interference level in the channel is, and at last identification of the

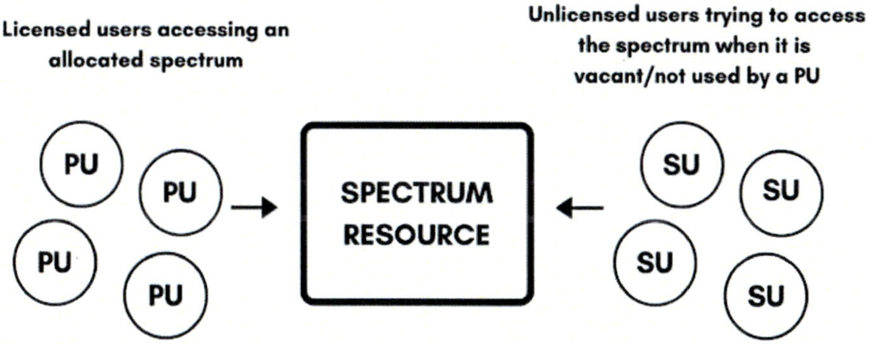

Fig. 1 Predator–prey model

spectrum holes whether white or gray spaces for its transmission. After the RF Signal is received, spectrum sniffing is performed. Spectrum sensing can be done with the help of these two methods: Energy and Cyclostationary Feature Detection-based Spectrum Sensing.

2 Literature Survey

The authors of [1] discuss energy detection technique. In order to allow opportunistic access, this technique detects signals in a given band. Through simulations, it was learnt that depending on the threshold of a single-user energy detector, performance improves over a non-fading channel (AWGN). It was also observed that as SNR increases, detection probability also increases for a single-user detector node in a channel without fading. Fewer samples produce better performance for non-fading channels in the case of a single secondary user. To improve system performance, MIMO-STBC coding can be implemented to provide spatial diversity, providing efficiency and reliability to the system.

The authors of [2] speak about spectrum holes in the licensed bands, the concepts of cognitive radio, its features, and the transceiver details. The functionality of the cognitive radio cycle is explained along with the description of methodologies for spectrum sensing, spectrum decision, spectrum sharing, and spectrum mobility. Rapidly growing users of wireless communication has led to a hike in spectrum demands. The problem of spectral scarcity and resource congestion can be resolved using programmable cognitive radios that dynamically operate in underutilized spectral bands used to sustain multiple communications concurrently. In this paper, an account of what is Cognitive Radio and how it adaptively accesses PU resources is described. Various techniques for the operation of cognitive radio are discussed.

The authors of [3] observed that the system behavior is very similar to interaction among different species coexisting in an ecosystem. Therefore, the well-known predator–prey model is made use of, in which two species of predator species model, representing heterogeneous users, are fed by the preys, representing the communication resources. By analyzing the population dynamics in the spectrum-sharing ecosystem, the transient behavior of the CRs is analyzed, and efficient ways to assess the stability of the CR systems are developed.

The performance of CRNs is strongly dependent on PU activity modeling. The spectrum range permitted for the operation of PR users' networks varies from kHz to GHz. Activity models used in CRNs are classified based on different models and processes such as queuing theory, Markov process, ON/OFF model, time series, and miscellaneous models. A single power threshold for spectrum occupancy detection is used instead of the actual energy detector (ED) corresponding to a channel in CRNs in Spectrum occupancy measurement techniques. If real traces are used in the modeling then the accuracy of PU activity modeling can be certainly increased [4].

There are multiple approaches to increase the efficiency of the channel assignment, to serve the SUs. A fairness problem is addressed, wherein if some of the SUs utilize most of the idle channels then the other SUs will be left with no idle channels, which will ultimately increase the blocking probability. A Fair Multiple Channel Access (FMCA) is proposed. For the optimization of the performance of the FMCA scheme & MAC framework, a fairness criterion is developed known as Jain's Fairness criterion that would evaluate how fair the resource allocation is among the traffic and it was observed that there is a good trade-off between the throughput and fairness compared to that of the existing spectrum access schemes [5].

For cognitive radio ad hoc networks with spectrum aggregation, cross-layer optimization is necessary, which should be able to increase network performance under network resource constraints. To meet this demand, researchers are looking at joint optimization for channel allocation, power control, and routing under SINR. The joint cross-layer optimization problem is divided into two parts: physical layer resource allocation and network layer throughput optimization. When six channels are aggregated, a network throughput of 40% is attained, compared to only two channels when two channels are combined [6].

Spectrum shortage does not occur due to the increase in the number of users but rather due to the inefficient use of spectral resources. Hence, "White Space" can be defined and exploited in every possible perspective to improve the sensing of the idle bands [7].

In paper [8], the authors have proposed a channel assignment algorithm which is based on a per block basis. It takes into consideration the time-varying channel conditions for each and every CR-IoT transmission over each idle band channel by assigning a minimal number of vacant channels to each user depending on the data rate demands and interference constraints thereby improving the spectrum efficiency.

Improving the network throughput and communication reliability is important to enhance the system performance while MIMO operating modes which are the best. For this, a joint channel assignment and adaptive Multi-Input–Multi-Output (MIMO) mode selection protocol can be utilized [9].

The efficiency of the spectrum can be increased by using opportunistic Medium Access Control (MAC). CRNs will benefit from a probabilistic channel quality and availability-aware MAC protocol (MAX-PSMAC). Through a channel assignment system, MAX-PS-MAC is responsible for increasing CRN throughput. Channel quality and channel availability information are utilized, according to the proposal. This was done to boost system throughput by increasing the packet success probability of a given CR transmission. Various simulations have backed up this claim [10].

In paper [11], a nature-based technique called ADALINE is used which consists of a feedback mechanism that periodically examines the achievable and achieved transmission parameters and adapts itself to minimize this difference making it as low as possible. This ensures that the available spaces in the spectrum are being effectively used which are otherwise left either unused or underutilized by the licensed users.

As mentioned in the above references, several techniques are present for the channel assignment of secondary users in the presence of primary users. These adopt a sequential channel assignment method which leads to wastage of the available spectrum as the channels that do not satisfy the data rate requirements of the secondary user are left unused. This results in inefficient usage of the spectrum and fails to facilitate the cognitive radio users. One way to enhance this method is by assigning multiple users where the SUs try to find the ideal vacant channel. For practical use, it is necessary to consider the spectrum specifications like data rate which is different for different channels as well as the users. Hence, the technique proposed takes allocation of multiple users into consideration where the data rate requirement is different for each user. Multiple users are assigned to utilize the vacant channels to the maximum extent possible subject to the data rate requirement.

3 Proposed Methodology

An algorithm for the simulation to be carried out is written to understand the working of the two methods of channel assignment and compare their effectiveness. The process is initiated by taking the necessary inputs such as the number of channels, data rates, etc. In the case of sequential assignment, based on the data rate of the sequentially available channel, the secondary user is allocated if the required rate is met by it. For variable data rate channel assignment, the data rates required by the secondary users are all taken into consideration to allocate the best suitable channel such that the channel can be utilized to the best extent. An important measure for the evaluation of channel assignment is Busy Probability (P_{busy}) which gives the information on the number of channels available to the ones occupied. This parameter is used to plot the graph to compare the number of secondary users served using the sequential and variable data rate methods.

4 Implementation

For the implementation of the above-proposed algorithms, we make use of Python Programming Language. We have used the PyCharm Community Edition 2021.1.2 IDE to write and execute the code. Initially, we install a library known as Matplotlib which is mainly used for the purpose of plotting different types of graphs.

After all the basic inputs are provided, we compute a parameter known as Busy Probability (Pbusy). The Pbusy is given by the equation

$$P_{busy} = \frac{Number\ of\ channels\ that\ are\ occupied\ in\ a\ Spectrum}{Total\ Number\ of\ channels\ in\ a\ Spectrum} \tag{1}$$

We initiate the channel assignment process by taking inputs from the user which includes the number of channels and their respective data rates followed by the number of users and their respective data rates. Note that the data rates are provided in Mbps. Now, the sequential channel assignment begins with a "for" loop which is initialized to 9. This is done to iterate the channel assignment process for nine different busy probabilities (starting from 0.1 to 0.9). In sequential assignment, firstly, we ask the user to input the number of vacant channels using which the busy probability is calculated. This busy probability will then be stored in a separate list which will be used for plotting the comparison graph. Now, for the given user, the corresponding vacant channel's data rate will be checked and if the vacant channel data rate is greater than the user data rate, that particular channel will be assigned to the user. Otherwise, an error message will be displayed. Note that the number of users served for the given busy probability will be stored in a separate list which will be used with busy probability to plot the comparison graph. This ends the sequential channel assignment.

The next channel assignment is the multiple-user variable rate channel assignment for the same set of users and channels. Similar to sequential channel assignment the multiple-user variable rate channel assignment will also begin with a "for" loop initialized to 9 for the same purpose as mentioned above. We will once again prompt the user to input the number of vacant channels and calculate the busy probability and store it separately. Now, the given user's data rate will be compared with the data rates of all the vacant channels as opposed to sequential assignment and the best channel with suitable data rate will be assigned to the user. If no suitable channel is found an error message will be displayed. Similar to sequential assignment, the number of served users for the given busy probability will be stored separately. The Flowchart for both the channel assignments is present in Fig. 2.

Finally, a graph will be generated to show the number of users served for a given busy probability for both sequential channel assignment and multiple-user variable rate channel assignment. This will enable us to analyze the performance and efficiency of both the channel assignment methods.

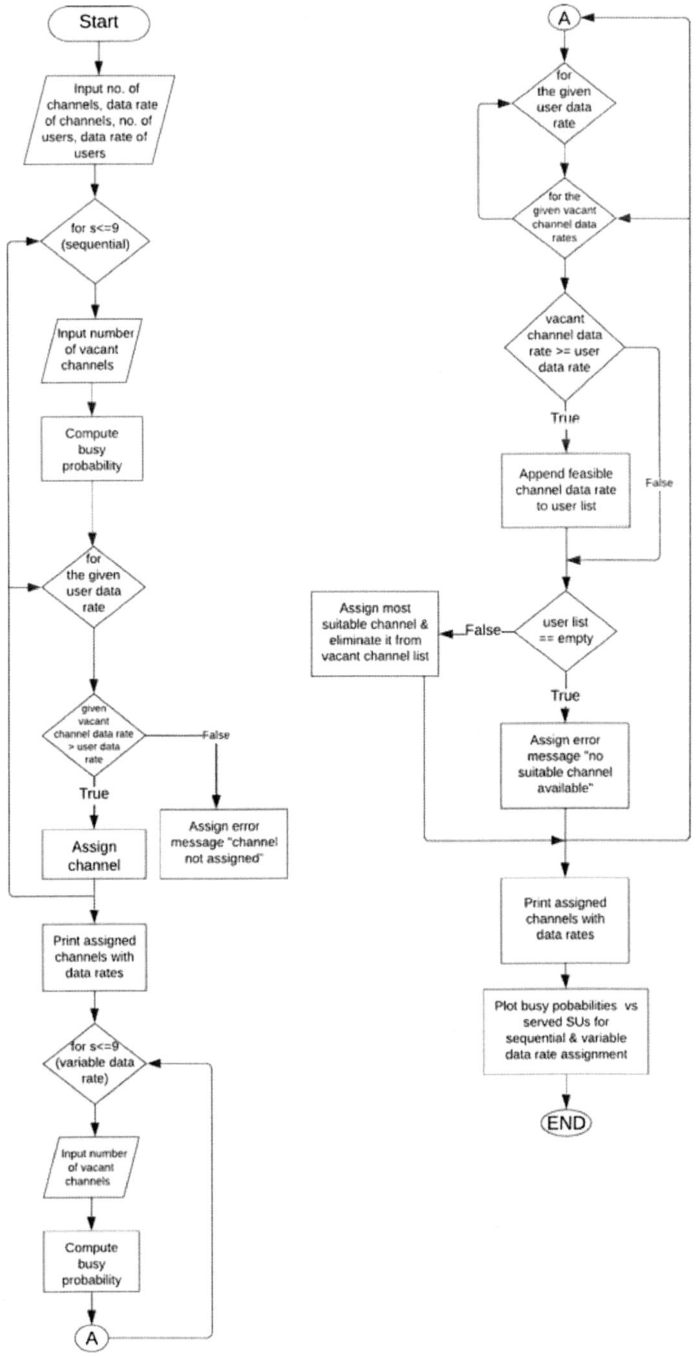

Fig. 2 Flowchart of sequential and multiple-user variable rate (MUVR) channel assignment methods

5 Results

A communication system which consists of primary users, as well as secondary users seeking to avail channels, is considered. The algorithm is implemented for a spectrum of 20 channels that can be used by primary users as well as secondary ones. The channels can be used by the secondary users on an opportunistic basis according to the availability. While the primary users can occupy the channels as licensed users, the secondary users are the unlicensed users that use the spaces that are left unutilized by the primary users. Based on the data rate required by the user and the rate of the channels left unutilized by primary users, allocation is done according to the procedure of the two assignment techniques.

Figure 3 shows the comparison between the two channel assignment techniques based on the number of users served for a given busy probability. Here, the number of secondary users trying to access the spectrum is 10. For each busy probability, the number of users served (i.e., the number of users who were successfully assigned a suitable channel) using the MUVR technique is considerably higher compared to that of the sequential channel assignment technique. Notably, when the busy probability was in the range of $0.1 - 0.4$, there was approximately a 110% increase in the number of users served using the MUVR technique when compared to the sequential assignment technique. As the busy probability increases (i.e., the channel gets more congested), it is obvious that the number of users that can be served reduces substantially. In fact, when the busy probability is maximum (i.e., 0.9), the sequential assignment technique is unable to serve even a single user, whereas the proposed MUVR technique was able to assign channels to two secondary users successfully.

Similarly, the algorithm was simulated for 15 secondary users trying to access the spectrum as shown in Fig. 4. It was seen that even with the increase in the number of secondary users trying to access the channel, the MUVR technique was able to serve more users compared to the sequential assignment technique. In particular, when the

Fig. 3 Busy probability (P_{busy}) versus served cognitive radio (CR) users for 10 secondary users

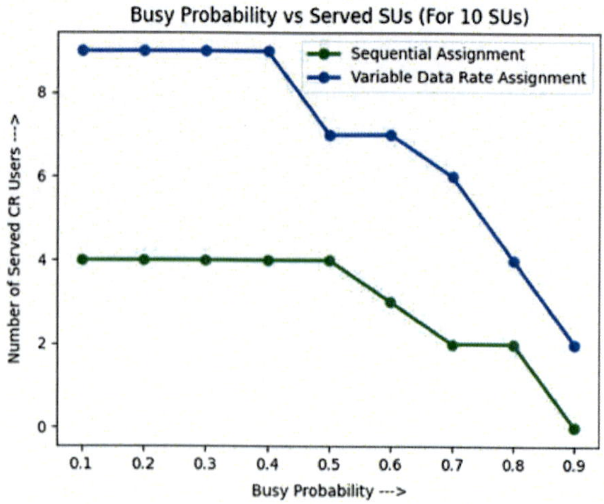

Fig. 4 Busy probability (P_{busy}) versus served cognitive radio (CR) users for 15 secondary users

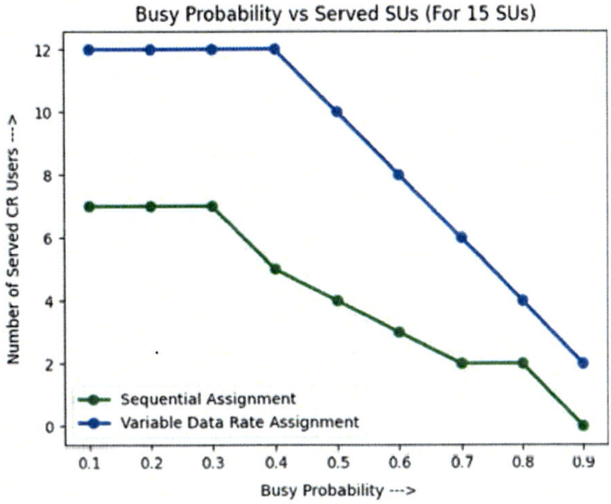

probability was in the range of $0.1 - 0.3$ we could see that there was almost a 70% increase in the number of users served using the MUVR technique when compared to that of the sequential method. Also, when the busy probability was 0.9, the MUVR technique was able to serve two users while the sequential assignment technique failed to serve even a single user.

To verify that the proposed method is versatile with respect to the number of secondary users, a scenario was simulated where the number of channels available in the spectrum was equal to the number of secondary users trying to access the spectrum as seen in Fig. 5. Even though the number of users per spectrum increased substantially, the performance of the MUVR method was better than that of the sequential method. It is obvious that as the number of secondary users increased, the performance of both the techniques degraded but the degradation was comparatively less in MUVR technique. Especially, when the probability was in the range of $0.1 - 0.2$ there was approximately a 87% increase in the number of users served using the MUVR method compared to the sequential assignment. Similar to previous cases, when the busy probability was maximum, the MUVR method served two users, while the sequential assignment method was not able to assign even a single user with a channel successfully.

From the above discussions, it is clear that the proposed multiple-user variable rate technique is way more efficient in all scenarios considering a varying number of secondary users as well as varying busy probability when compared to the sequential method.

The same concept can be used in channel assignment for a greater number of users considering more than 20 channels and the performance remains the same.

Table 1 summarizes the number of SUs served for different busy probabilities using both the algorithms. Note: SCA corresponds to Sequential Channel Assignment.

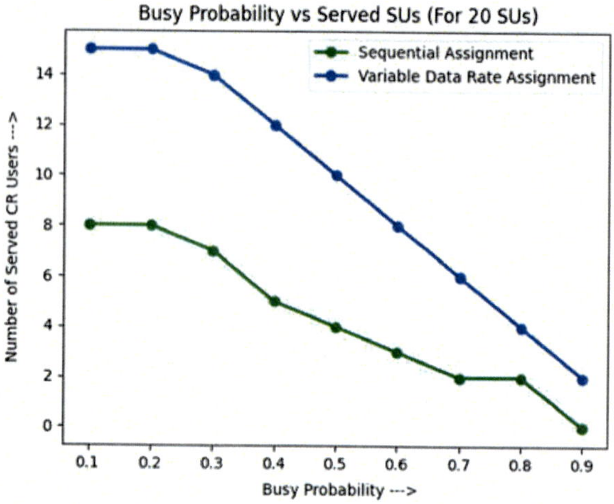

Fig. 5 Busy probability (P_{busy}) versus served cognitive radio (CR) users for 20 secondary users

Table 1 Number of served cognitive radio users

P_{busy}	Number of served CR users					
	10 SUs		15 SUs		20 SUs	
	SCA	MUVR	SCA	MUVR	SCA	MUVR
0.1	4	9	7	12	8	15
0.2	4	9	7	12	8	15
0.3	4	9	7	12	7	14
0.4	4	9	5	12	5	12
0.5	4	7	4	10	4	10
0.5	3	7	3	8	3	8
0.7	2	6	2	6	2	6
0.8	2	4	2	4	2	4
0.9	0	2	0	2	0	2

6 Conclusion

The paper presents the design, implementation, and comparison of existing sequential channel assignment method and proposed multiple-user variable rate channel assignment method algorithms which was carried out successfully via the PyCharm IDE platform using Python Programming Language. The proposed algorithm is tested in an environment where 20 channels are present wherein the inputs are given by the user. After the algorithm was simulated, we would infer that the proposed method multiple-user variable rate algorithm showed better results than the existing sequential channel assignment. At times, there was a 100% increase in the SUs being assigned to the vacant channels when multiple-user variable rate algorithm was simulated.

This system restricts itself to an environment which is free from guard bands. Although the usage of guard band reduces the efficiency of the spectrum, it still becomes important to consider this constraint so that adjacent channel interference can be mitigated. In the future, this constraint can be added to the primary users before any kind of channel assignment takes place.

References

1. Pandya P, Durvesh A, Parekh N (2015) Energy detection based spectrum sensing for cognitive radio network. In: 2015 Fifth International Conference on Communication Systems and Network Technologies. pp 201–206. https://doi.org/10.1109/CSNT.2015.264
2. Ridhima, Buttar AS (2019) Fundamental operations of cognitive radio: a survey. In: 2019 IEEE International Conference on Electrical, Computer and Communication Technologies (ICECCT). pp. 1–5. https://doi.org/10.1109/ICECCT.2019.8869190.
3. Liau D, Chen K, Cheng S (2013) A predator-prey model for dynamics of cognitive Rradios. IEEE Commun Lett 17(3):467–470. https://doi.org/10.1109/LCOMM.2013.020513.122120
4. Saleem Y, Rehmani MH (2014) Primary radio user activity models for cognitive radio networks: A survey. J Netw Comput Appl 43: 1–16. ISSN 1084–8045.
5. Wei Z, Hu B (2018) A fair multi-channel assignment algorithm with practical implementation in distributed cognitive radio networks. IEEE Access 6:14255–14267. https://doi.org/10.1109/ACCESS.2018.2808479
6. Chen J, Ping S, Jia J, Deng Y, Dohler M, Aghvami H (2016) Cross-layer optimization for spectrum aggregation-based cognitive radio ad-hoc networks. In: 2016 IEEE Global Communications Conference (GLOBECOM). pp 1–6. https://doi.org/10.1109/GLOCOM.2016.7841708
7. Akhtar F, Rehmani MH, Reisslein M (2016) White space: definitional perspectives and their role in exploiting spectrum opportunities. Telecommun Policy 40(4): 319–331. ISSN 0308–5961.
8. Salameh HAB, Al-Masri S, Benkhelifa E, Lloret J (2019) Spectrum assignment in hardware-constrained cognitive radio IoT networks under varying channel-quality conditions. IEEE Access 7:42816–42825. https://doi.org/10.1109/ACCESS.2019.2901902
9. Bany Salameh H, Shraideh S, Alshamali A (2020) Joint channel assignment and adaptive mode selection in MIMO-based cognitive radio networks. Arab J Sci Eng 45: 10233–10244.
10. Salameh HB, Badarneh OS (2013) Opportunistic medium access control for maximizing packet delivery rate in dynamic access networks. J Netw Comput Appl 36(1): 523–532. ISSN 1084–8045.
11. Ridhima, Buttar AS (2019) Proficient pesource allocation to secondary user using Nnature inspired technique in cognitive radio Nnetwork. In: 2019 International Conference on Communication and Electronics Systems (ICCES). pp. 1755–1761. https://doi.org/10.1109/ICCES45898.2019.9002174.

Wireless Communication, WSN and Adhoc Networks

Design of Frequency Beam Scanning Antenna Using SIW for mm-wave Applications

Arman Singh, Arun Kumar, Arun Singh, Ayush Jaiswal, and Ankit Sharma

Abstract In this work, a novel substrate integrated waveguide (SIW) leaky wave antenna (LWA) is proposed. The proposed antenna offers wide impedance bandwidth and continuous backward to forward beam scanning. The unit cell of the proposed CRLH-based LWA consists of a pair of roller-shaped slot on the top metallic layer of SIW and has been developed with higher modes. Contrasted among different published papers of transversely slotted SIW LWA created with the fundamental mode (TE10 approach), this proposed structure displays consistent gain values because of using higher modes. Besides this, the proposed antenna bears a wide scan range within a fixed bandwidth. The beam scanning of the proposed leaky wave antenna is precisely adjusted by alternating the gap connecting two joining units, keeping frequency band unchanged. To tone down the (OSB) open stopband and to enhance the impedance matching, the slot length and gaping between the slots are optimized. The main beam of the proposed antenna scans continuously from $-35°$ to $+35°$ in the range from 26.5 to 28.5 GHz. Besides, the SIW LWA provides a gain of 12.308 dBi and radiation efficiency is 92.41 at 27.5 GHz and also side-lobe level is smaller than -10 dB at every measured frequency. The proposed antenna can be a good candidate for mm-wave beam scanning, satellite Communication, automobile radar, and military applications.

Keywords BR—Broadside Radiation · TE—Transverse Electric Mode · SIW—Substrate Integrated Waveguide · LWA—Leaky Wave Antenna · CP—Cross Polarization · CRLH—Composite Right Left Handed

A. Singh (✉) · A. Kumar · A. Singh · A. Jaiswal · A. Sharma
Department of Electronics and Communication Engineering, Galgotia College of Engineering and Technology, Greater Noida, India
e-mail: nv.armansingh@gmail.com

A. K. Dubey et al. (eds.), *Advanced IoT Sensors, Networks and Systems*,
Lecture Notes in Electrical Engineering 1027,
https://doi.org/10.1007/978-981-99-1312-1_6

1 Introduction

Substrate integrated waveguide-based LWA certainly stand out, because of the advantages like—low profile, minimal expense, and simple fabrication [1].

They offer a beam scanning in the range of frequency in which the beam width is restricted along the examining region because of the long opening in the longitude bearing.

The transverse-shaped slotted SIW structure of LWA is considered an interesting semi-uniform LWA that performs scan task from near broadside to end-fire because of identical attractive current sources. The SIW upholds TEm0 modes along with practically similar propagation attributes as the TEm0 modes in the equivalent rectangular waveguide. The vast majority of researched SIW leaky wave antennas depend on the fundamental mode (TE10 mode) [2].

In this paper [3], another sort of broadband circularly polarized leaky wave antenna (LWA) with high gain is proposed and researched utilizing a layered substrate integrated waveguide (SIW) development. The proposed LWA gives an exceptional scanning capability all through the Ka-band.

In paper [4] explains that the parallel size that is normally not exactly a large portion of the free-space wavelength, a SIW LWA in view of the TE10 mode normally creates fan-shaped beam. Several methods have been described for increasing the gain of LWA.

In addition, a linear array of leaky wave antennas, exhibit scanning in 2 planes by changing the frequency and the scanning capability in the azimuth plane is accomplished through changing the phase difference among the consecutive excitation [5]. An extra complex feeding mechanism is required by leaky wave antenna, which consumes additional space and may result in unnecessary radiation. Leaky-wave antenna belongs to the category of traveling wave antenna that allows to release electromagnetic wave along the propagation direction [6].

The main benefit of this antenna is that it bears scanning capacity and also does not requires necessity of presence of complicated feeding method. These antennas are utilized for beam examining, direction of arrival approximation, and indoor interchanges, etc. [7].

Leaky-wave antennas can be acknowledged in substrate integrated waveguide (SIW) structure. SIW-based LWA gives several benefits like—simple to manufacture, low loss, high power handling capacities, and small beam width. Due to these benefits, Substrate integrated waveguide-based LWAs have gained extensive consideration among the researchers [8].

In [9], a phase-adjusting metallic grinding cover is utilized for gain improvement of a SIW-based leaky wave antenna, where the grinding cover is set at a specific height over the antenna. Regardless, the work of a grinding cover brings about an expanded profile and a complex structure. A particular leaky wave structure with a wide transverse radiation gap can likewise accomplish an improved gain.

In [10], a LWA based on spoof surface plasmon polarization can accomplish an improved gain, with extended grooves. Frequency-scanning LWAs belongs to a category of antennas which can be used for radar applications.

Examining scanning is a figure of legitimacy for frequency scanning antennas. Augmenting scanning range is valuable for radar and military applications. Some mostly used types of frequency scanning antennas consist of one-dimensional planar metamaterials. Therefore, these antennas can intend to scan beam from broadside along with steady gain but are burdened with their little frequency scanning sensitivity [7].

Picking electrical deferral among transmitting components with the waveguide takes command on sensitivity of scanning into consideration. At the point when these antennas are executed in planar structure, they can be restricted by losses [11].

In this paper [12], based on a CRLH-SIW, a little single-radiator CP LWA is given. for CRLH executions and CP radiation, T-molded spaces are utilized.

A little circularly constructed CRLH LWA with a constant sweep from $-19°$ to $+84°$ all through the recurrence scope of 7.35–10.15 GHz has been made utilizing the CP radiation approach. The arranged CP LWA has a basic development since it doesn't need a subsequent radiator or a coupler [13].

The SIW is basically used for the millimeter wave application and designed to generate the topic and resultant of the product. It is easy to use in the fabrication of the design antenna works and define different type of task [14].

This antenna is suitable for LWA structure which is based on SIW. For the first time when SIW was introduced. The confinement of the fields was achieved by placing posts in a parallel plate waveguide [15].

LWAs have a number of drawbacks, including the difficulty of adjusting and independently, a LWA's intrinsic frequency beam-squint, and lesser gain when compared to phased array antennas. Frequency, mechanical, or electronic beam scanning are all options for beam scanning in the LWA. Numerous investigations on LWAs have been conducted since the late 1950s [16].

The guided structures for creating a LWA are often waveguide and microstrip structures. A microstrip structure has a high dielectric loss by definition. Waveguides, on the other hand, are large and unsuitable for integration and downsizing. Because of its integration capabilities, SIW has been routinely used to build LWAs [17].

For CRLH-SIW executions and CP radiation, T-molded spaces are utilized. A little circularly captivated CRLH LWA with a constant sweep from $-19°$ to $+ 84°$ all through the recurrence scope of 7.35–10.15 GHz has been made utilizing the CP radiation approach [18].

The creation of a layered SIW is presented and used to investigate a different kind of broadband circularly polarized leaky wave antenna (LWA) with high gain. The proposed LWA has outstanding scanning capabilities over the whole Ka-band [19].

A metallic cover with phase adjustment is used to increase the gain of a leaky wave antenna based on SIW, and it is positioned at a certain height above the antenna [20].

2 Designed Antenna Unit Cell and Array Configurations

2.1 Unit Cell Structure and Dispersion Diagram Analysis on HFSS

The word unit cell and unit cell structure both are the same things both signifies the unit cell which consists of a roller shape slot in the pair, the whole pair is considered as a unit cell as shown in Fig. 1. The Creation of every Unit cell has been performed with simulator named HFSS. Here, in this unit cell design H is the width of feed, F is the length of the unit cell and G is the width of the unit cell, respectively. The values of F, G, H, I, J, and K are 14 mm, 16 mm, 2 mm, 10 mm, 10 mm, and 4.2 mm, respectively, as shown in Fig. 1. Unit Cell structure is formed by combining rectangular slot in which one slot has thin width and thin length and another rectangular slot which has thick width and thick length. After combining these two slots make a roller shape type structure and further combining these two-roller shape parallel structure make a unit cell. The roller shape pair consists of inductive posts and 1 longitude slot kept opposite by the middle of SIW. The SIW-based unit cell consists of an inductive post and 1 longitude slot kept opposite by the middle of SIW as shown in Fig. 1. Longitude slot and Longitude posts show capacitive and inductive effect individually. Thus by adding these 2, impedance matching has been attained. The slot and post unit cell radiates from backward direction to forward direction by broadside. The unit cell dispersion diagram predicts the frequency with which the antenna emits radiation at broadside direction.

The shunt elements and series elements possessed by the network are computed through extraction of S-parameters. Thus, substrate integrated waveguide with longitude slot shows capacitive loading issues. The real components and imaginary components of the explained network are zero. The only exception is the imaginary component of shunt element. Thus, this shows that the SIW loaded with post-possess inductive loading. The matching of impedance can be achieved by making admittance approximately equal to zero. The slots dimensions and positions should be increased for attaining shunt admittance equal to zero. This results in an open circuit and

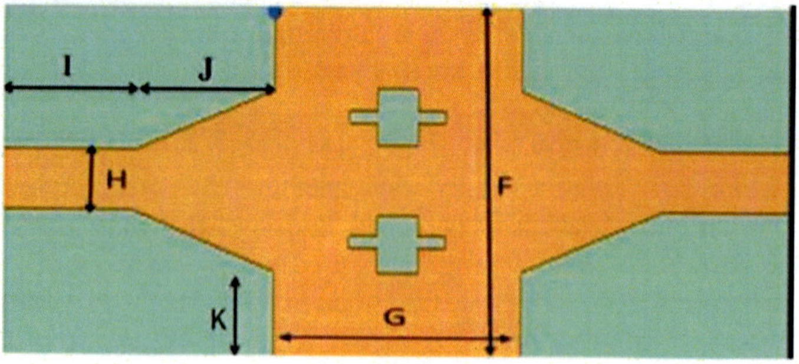

Fig. 1 Top view of unit cell

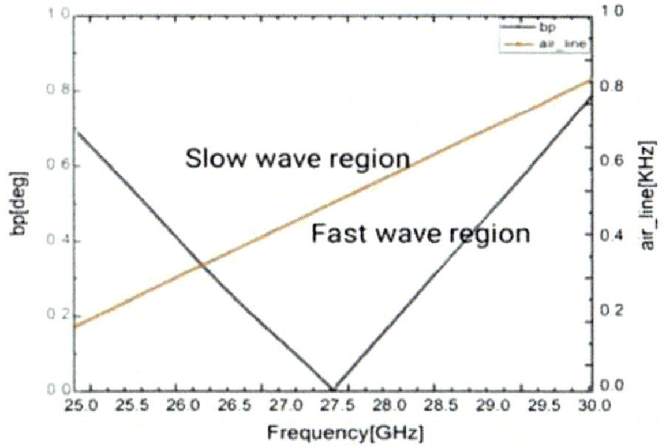

Fig. 2 Phase velocity (bp) versus frequency graph

parallel matching of impedance in the unit cell. The open stop-band elimination is demonstrated by the support of the dispersion result of the simulated unit-cell.

Figure 2 depicts dispersion graph of a unit cell structure which has pair of roller-shaped slots. The behavior of the complex propagation constant within the operational frequency band is known as dispersion response. The open stopband can be suppressed by a unit cell having inductive and longitude post slots. This suppresses the open stopband. Therefore, the proposed unit cell can perform scanning smoothly and constantly from backward to forward region through the broadside. Dispersion diagram of unit cell shows that broadside radiation occurs at 27 GHz. The simulated results of unit cell show that it has maximum return loss of −60 dB at frequency of 27.6 GHz. The slot radiates from backward direction to the forward direction by the broadside.

2.2 *Antenna Array Considerations on CST*

In the proposed antenna structure as shown in Fig. 3, the waveguide is formed by substrate which is merged along with conductor at the upper and lower sides along with 2 rows of vias of metal at the corners. The resulted structure is repeated and an array of 6 unit cell design is formed. In the proposed antenna, the diameter of vias is marked as d and its value is 0.8 mm. A three-dimensional viewpoint including upper perspective on the proposed structure is displayed in Fig. 3. The name of the substrate selected for the proposed design is Rogers RO4350 ($\varepsilon r = 3.66$ and $\tan \delta = 0.004$) and its height is 0.762 mm. The distance between two consecutive slots is 8 mm. The length and width of the substrate is denoted by Lsub and Wsub, respectively, as shown in Fig. 4 and their values are mentioned in Table 1. The slots on the patch are cut in a way that they resemble shape similar to roller (Table 2).

In beginning, the unit cells have been positioned parallel to each side creating a leaky-wave periodic design. The array bears an issue of low radiation due to lesser number of longitude slots as shown in Fig. 4. Therefore, it likewise experiences high

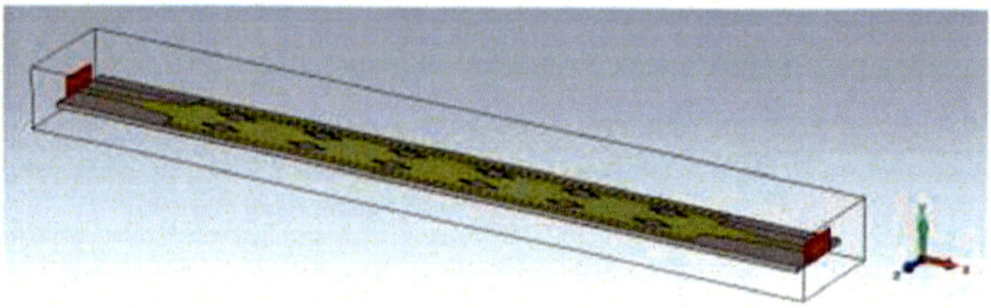

Fig. 3. 3-Dimensional view of the proposed antenna

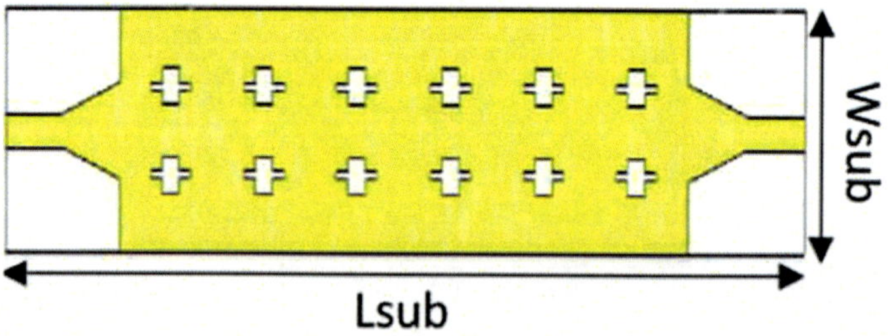

Fig. 4 Top view of the proposed antenna

Table 1 Configurations of the unit cell and slot on the patch

Parameter	Value (mm)	Parameter	Value (mm)
W	10.5	Lu	3
Ws	0.45	Ls	7
Wu	2.15	s	1.6
P	16	d	0.8
H	2	Wsub	4
G	16	Lsub	6

Table 2 A comparison between some latest papers and the proposed work

Reference	Scanning range	Max. return loss	Gain	Scanning rate
17	−10° to 65°	−24 dB at 28.7 GHz	9 dBi	25
18	−34° to 33°	−18 dB at 24.25 GHz	9.27 dBi	13.4
19	−23° to 19°	−38 dB at 23 GHz	15.27 dBi	5.25
20	−69° to −10°	−23 dB at 26.8 GHz	16.2 dBi	8.194
Proposed work	−35° to +35°	−58 dB at 28 GHz frequency	12.308 dBi	35

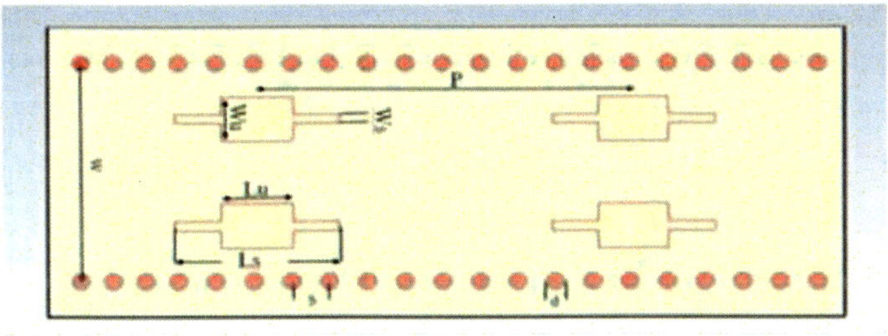

Fig. 5 Top view of 2 unit cell

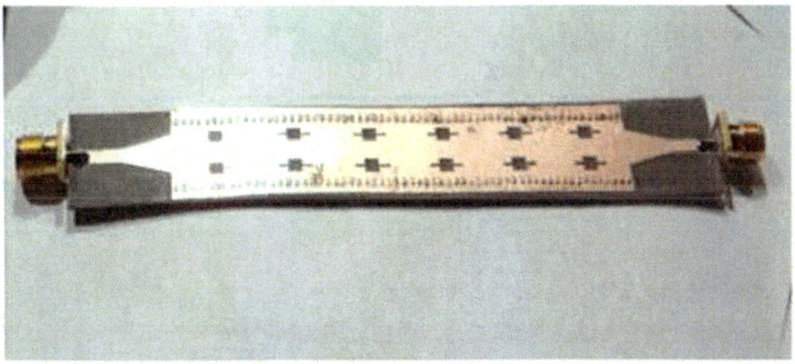

Fig. 6 Fabricated antenna top view

CP (cross-polarization) level because of the connecting impact. This issue was solved by altering the array design. The altered slots have been positioned randomly so the posts and slots in adjoining cells have been placed in opposite manner. Because of this course of arrangement, the unit cell phase delay decreases to π. Furthermore, to get in-stage radiation the unit cell period is decreased considerably by quarter. Every measurement of array has been acclimated to obtain ideal outcomes utilizing CST software. The overall length of the proposed and fabricated antenna is 138 mm as shown in Figs. 5, 6, and 7, respectively. The value of dimensions of slot is explained in Table 1.

3 Simulated and Measured Results

3.1 Reflection Coefficient

The return loss of an antenna is the percentage of electromagnetic radiation arriving at the antennas input that are rejected in comparison to those that are accepted. It's measured in decibels (dB) in comparison to a short circuit (100 percent rejection). The simulated antenna shows a maximum return loss of −58 dB at 28 GHz frequency as

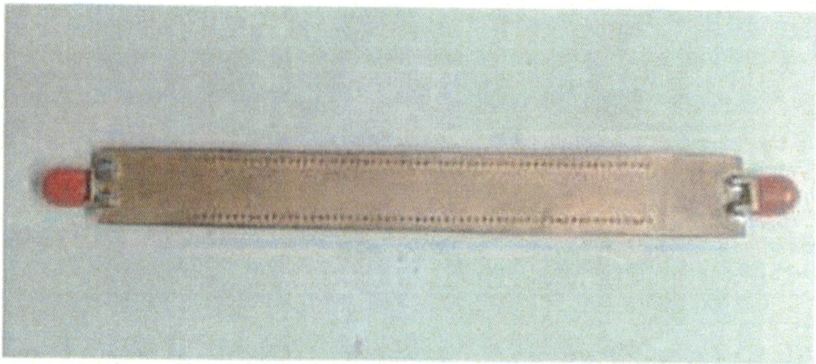

Fig. 7 Fabricated antenna back view

Fig. 8 Return loss characteristics of simulated and measured design (S11, S21)

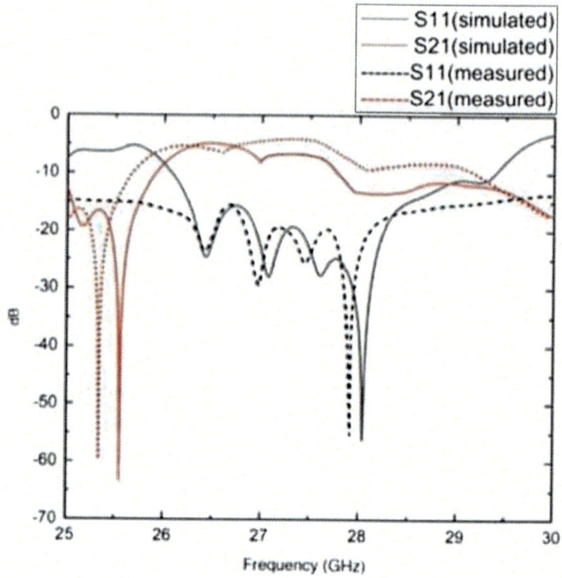

shown in Fig. 8, allowing it to produce strong radiation power. Also the simulated S11 of the SIW LWA is below −10 dB from 26.08 to 29.38 GHz which is quite feasible for mm-wave applications. The measured maximum return loss of the fabricated antenna is −60.01 dB at 27.98 GHz as shown in Fig. 8 which is approximately the same as the simulated result.

3.2 *Gain*

The gain of an antenna in a given direction is defined as the ratio of that direction's intensity to the radiation intensity that would be reached if the antenna's power were

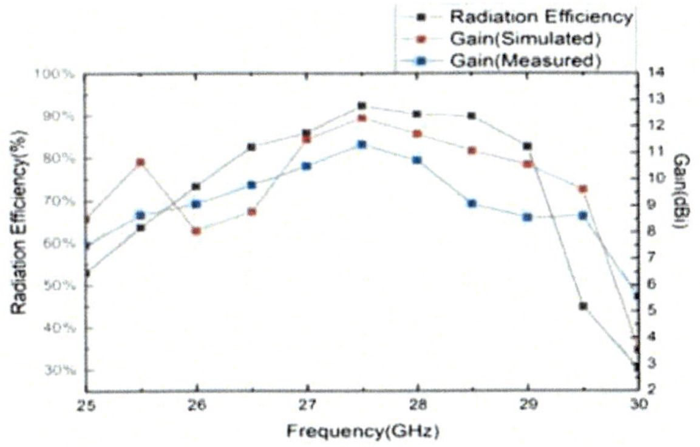

Fig. 9 Plot for gain versus frequency of simulated and measured design

emitted isotopically. The gain is believed to refer to the highest value of the gain, the gain in the direction of the antenna's primary lobe, when no direction is provided. The term antenna gain refers to the amount of power transferred from an isotropic source in the direction of peak radiation. Because it accounts for actual losses, antenna gain is more typically reported in an antenna's specification sheet than directivity. The gain of our simulated antenna simulated is 12.308 dBi at 27.5 GHz frequency as shown in Fig. 9. The radiation efficiency of simulated antenna as shown in Fig. 9 is 92.41% at 27.5 GHz. The measured gain of the fabricated antenna is 11.04 dB at 27.5 GHz.

3.3 *Radiation Pattern*

The term "radiation" is used to describe the emission or receipt of a wavefront at an antenna, as well as its intensity. The radiation pattern of an antenna is the sketch created to illustrate its radiation in any image. An antenna's function and directivity may be easily deduced by looking at its emission pattern. The energy emitted by an antenna is represented by its radiation pattern. The radiation patterns are graphic representations of the distribution of radiated energy in space as a function of direction. The simulated antenna is made up of an array of unit cells that are impedance matched in propagation directions and have a scan angle from −35° to +35° in the frequency range from 26.5 to 28.5 GHz as shown in Fig. 10. The proposed antenna shows the capability to scan from backward direction to forward direction. The measured scan angle of the fabricated antenna is from −37° to +32° in the

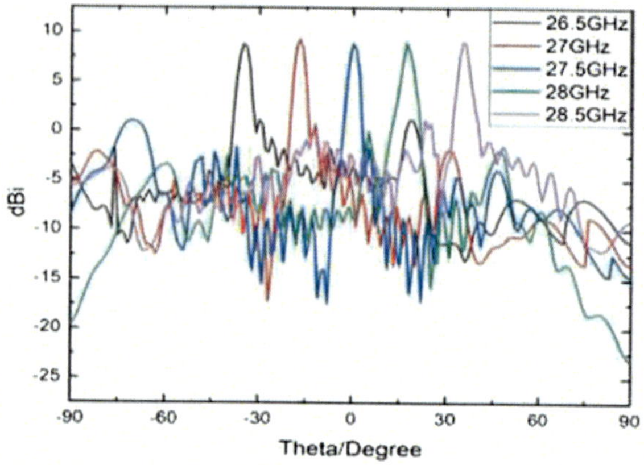

Fig. 10 Simulated antenna radiation pattern plot at 26.5–28.5 GHz

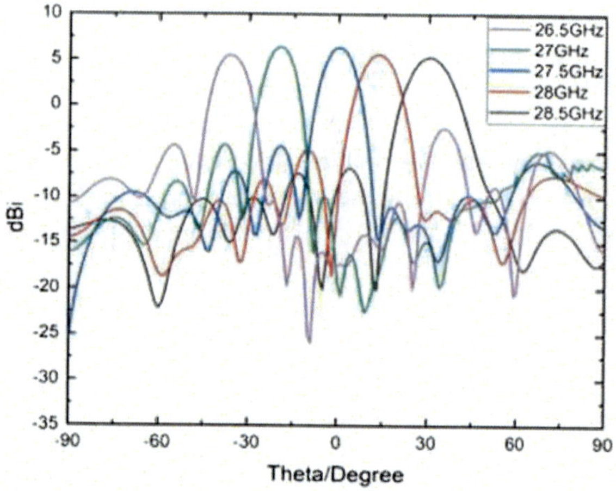

Fig. 11 Radiation pattern plot at 26.5–28.5 GHz (measured)

frequency range from 26.5 to 28.5 GHz as shown in Fig. 11 which is approximately similar to the simulated design scan angle result. The frequency shift in measured and simulated result are due to the fabrication losses that occur in the antenna it may be due to the fma connector losses that's why frequency shift occur.

4 Conclusion

The proposed LWA radiates close to broadside with less frequency bandwidth from 26.08 to 29.38 GHz, which is valuable for point-to-point communication or radar-based framework. The primary commitment of the proposed plan is the very low return loss, high gain, and continuous beam scanning among the other published papers with similar antenna length and operating frequency without expanding the size of the antenna or adding additional segments to the design. We have proposed a novel SIW-based LWA along with a unit cell structure that comprises roller-shaped slots giving a low return loss, high radiation efficiency, and constant beam scanning. The predominant contribution of the proposed structure is that it gives good results without elevating the dimensions of the antenna and without including greater dimensions to the shape. The simulated antenna involves an array design of unit cells with impedance matched in directions of propagation which gives a consistent scanning range from −35° to 35°. The antenna provides the 12.308 dBi gain at 27.5 GHz. The antenna scans from backward direction to forward direction. The Maximum return loss of −58 dB at 28 GHz frequency is shown by simulated antenna and also the S11 of the antenna is below −10 dB from 25 to 30 GHz. The measured maximum return loss of the fabricated antenna is −60.01 dB at 27.98 GHz which is approximately the same as the simulated result. The antenna is suitable for different millimeter-wave applications due to the above-mentioned benefits which are mostly needed in a feasible millimeter-wave antenna.

References

1. Sabahi MM, Heidari AA, Movahhedi M (2018) A compact CRLH circularly polarized leaky-wave antenna based on substrate-integrated waveguide. IEEE Trans Antennas Propag 66(9):4407–4414
2. Xu J, Hong W, Tang H, Kuai Z, Wu K (2008) Half-mode substrate integrated waveguide (HMSIW) leaky-wave antenna for millimeter-wave applications. IEEE Antennas Wirel Propag Lett 7:85–88
3. Bai X, Qu SW, Ng KB, Chan CH (2015) Sinusoidally modulated leaky-wave antenna for millimeter-wave application. IEEE Trans Antennas Propag 64(3):849–855
4. Liu J, Liang J (2021) Gain enhancement of transversely slotted substrate integrated wave- guide leaky-wave antennas based on higher modes. IEEE Trans Antennas Propag 69(8):4423–4438
5. Luo P, He W, Zhang Y, Liu H, Forsberg E, He S (2020) Leaky-wave antenna with wide scanning range based on double-layer substrate integrated waveguide. IEEE Access 8:199899–199908
6. Zhong T, Zhang H, Chen Q, Niu Z (2019) High-efficiency leaky-wave antenna with continuous wide-angle scanning using hybrid composite right/left handed structure. Int J RF Microwave Comput-Aided Eng 29(11):e21931
7. Jiang H, Xu K, Zhang Q, Yang Y, Karmokar DK, Chen S, Peng L (2020) Backward-to-forward wide-angle fast beam-scanning leaky-wave antenna with consistent gain. IEEE Trans Antennas Propag 69(5):2987–2992
8. Chaudhary D, Agrawal R, Gupta SC (2020) Scanning rate enhancement in substrate integrated waveguide periodic leaky wave antenna with continuous beam scanning using delay lines. Frequenz 74(11–12):377–382
9. Zheng D, Lyu YL, Wu K (2019) A quasi-uniform transversely slotted SIW leaky-wave structure with enhanced beam-scanning rate for millimeter-wave applications. In 2019 IEEE MTT-S international microwave symposium (IMS). IEEE, pp 885–888
10. Zheng D, Lyu YL, Wu K (2020) Transversely slotted SIW leaky-wave antenna featuring rapid beam-scanning for millimeter-wave applications. IEEE Trans Antennas Propag 68(6):4172–4185
11. Attar A, Sebak AR (2020) High gain periodic 2-D leaky-wave antenna with backward radiation for millimeter-wave band. IEEE Open J Antennas Propag 2:49–61

12. Zhang Y, Liu H, Meng C, Lin Y, Zhang Y, Forsberg E, He S (2020) A broad-band high-gain circularly polarized wide beam scanning leaky-wave antenna. IEEE Access 8:171091–171099
13. Ali SMA, Ahmed Z, Ihsan MB (2015) Beam steering in HMSIW LWA at fixed millimeter wave frequency. In 2015 12th international conference on high- capacity optical networks and enabling/emerging technologies (HONET). IEEE, pp 1–2
14. Mondal P, Wu K (2016) A leaky-wave antenna using periodic dielectric perforation for millimeter-wave applications. IEEE Trans Antennas Propag 64(12):5492–5495
15. Lopez DG, Filipovic DS (2016) On the design of milimeter-wave antennas for amplitude-only direction finding. In 2016 IEEE international symposium on phased array systems and technology (PAST). IEEE, pp 1–8
16. Javanbakht N, Amaya RE, Shaker J, Syrett B (2021) Reconfigurable SIW-based leaky-wave antenna composed of longitudinal cells. IEEE Access 9:102304–102311
17. Li H, Li Y, Chang L, Sun W, Qin X, Wang H (2020) A wideband dual-polarized endfire antenna array with overlapped apertures and small clearance for 5G millimeter-wave applications. IEEE Trans Antennas Propag 69(2):815–824
18. Zhong J, Rashid AK, Zhang Q (2020) 45° linearly polarized and circularly polarized high-scanning-rate leaky-wave antennas based on slotted substrate integrated waveguide. IEEE Access 8:82162–82172
19. Chiwande SS, Meshram P, Charde A, Bhave S (2022) Design of substrate integrated waveguide technology supported dual band mimo antenna. In 2022 international conference on electronics and renewable systems (ICEARS). IEEE
20. Passia MT, Yioultsis TV (2022) New uniplanar and broadside-coupled CSRR substrate integrated waveguides for mmWave components. In 2022 16th European conference on antennas and propagation (EuCAP). IEEE, pp 1–5

QUICLORIS: A Slow Denial-of-Service Attack on the QUIC Protocol

Anirudh Srinivas Balaji, V. Anil Kumar, P. P. Amritha, and M. Sethumadhavan

Abstract Quick UDP Internet Connections is a transport layer protocol that is being incorporated in HTTP/3. Denial-of-Service attacks exist in many variants through TCP, ICMP, and UDP protocols. Although QUIC supports 0-RTT, allows faster connection establishment and enables data encryption by default at the transport layer, there are still some security considerations in place to be addressed. This paper focuses on analyzing, implementing, and suggesting countermeasures to mitigate the Slow Denial-of-Service attack named "Slowloris attack" in the QUIC protocol. We have demonstrated the Slowloris attack in the QUIC protocol which is mentioned in RFC 9000.

Keywords QUIC · Slowloris · Denial-of-Service attack

1 Introduction to QUIC

QUIC is a relatively new general-purpose secure transport layer protocol. It creates a stateful interaction between the client and the server. In a typical client–server model, a three-way handshake process must happen in TCP (Transmission Control Protocol) before the client and the server starts exchanging the data. Since QUIC runs over UDP (User Datagram Protocol), a client can directly send a client TLS

A. S. Balaji · P. P. Amritha (✉) · M. Sethumadhavan
TIFAC-CORE in Cyber Security, Amrita School of Engineering, Amrita Vishwa Vidyapeetham, Coimbatore, India
e-mail: pp_amritha@cb.amrita.edu

A. S. Balaji
e-mail: cb.en.p2cys20006@cb.students.amrita.edu

M. Sethumadhavan
e-mail: m_sethu@cb.amrita.edu

V. Anil Kumar
CSIR Fourth Paradigm Institute, NAL Belur Campus, Bengaluru, India
e-mail: anil@csir4pi.in

A. K. Dubey et al. (eds.), *Advanced IoT Sensors, Networks and Systems*,
Lecture Notes in Electrical Engineering 1027,
https://doi.org/10.1007/978-981-99-1312-1_7

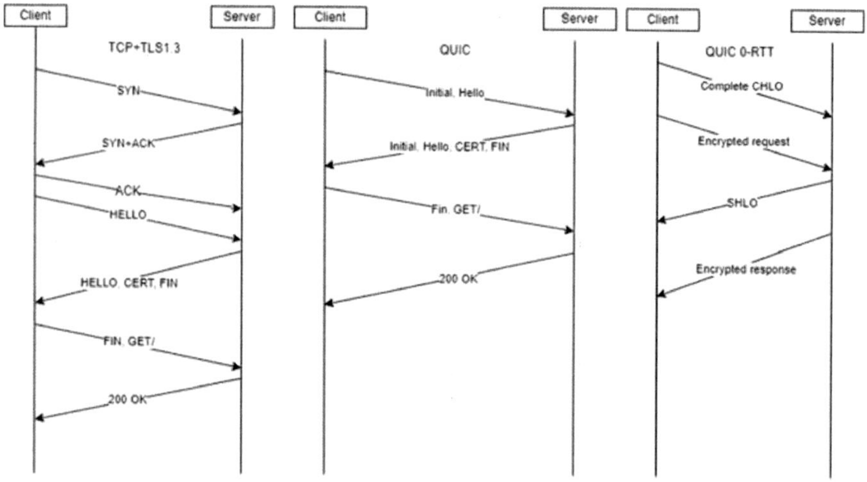

Fig. 1 Handshake processes of TCP, QUIC, and QUIC 0-RTT

(Transport Layer Security) hello message to the server and the server replies to the client with a server TLS hello message. After this, QUIC uses HTTP/3 to send the application requests (H3) to communicate with the server [1]. Unlike UDP, QUIC sends out an ACK (acknowledgement code) frame to acknowledge the data received. Ack-eliciting packets issue an ACK frame to be sent within the maximum ack delay. For improved loss detection and congestion control, all QUIC endpoints attempt to include an ACK frame if it is not sent recently. The QUIC handshake is comprised of a combination of cryptographic and transport parameters. The QUIC + TLS handshake paves way for a special feature in QUIC called "0-RTT" (Zero-Round trip time) which allows the exchange of application data as soon as possible.

To use 0-RTT, both the QUIC endpoints must enable some prior configurations to communicate. Figure 1 describes the difference in the handshake processes of TCP, QUIC, and QUIC 0-RTT.

In QUIC, communication between endpoints happens through streams which are of two types, namely, unidirectional and bidirectional streams. Unidirectional streams allow a single QUIC endpoint to send data, and bidirectional streams allow both the QUIC endpoints to transfer data. Connection identifiers are being used by QUIC endpoints for smooth network path transition. It allows QUIC endpoints to proceed after changes in the network topology or address mappings, such as those caused by Network Address Translation rebinding. Limiting the amount of data is necessary to stop the excessive consumption of memory by any malicious user. QUIC uses a limited flow control mechanism in which the receiver publicizes the total number of bytes it can accommodate for the entire connection or on a given stream. MAX_DATA and MAX_STREAM_DATA frames are sent by the receiver to the sender to publicize the limit [1]. Streams have stream IDs which are of 62-bit integer type and are unique for all connections. The client-based streams have even-numbered stream IDs and the server-based streams have odd-numbered stream IDs. Stream IDs are identified

Table 1 Types of streams IDs in QUIC

Bits	Type of streams
0 × 00	Client-based, bidirectional stream
0 × 01	Server-based, bidirectional stream
0 × 02	Client-based, unidirectional stream
0 × 03	Server-based, unidirectional stream

with the two least significant bits as shown in Table 1. STREAM frames perform encapsulation of the data that is being sent by an application. The QUIC endpoint makes use of stream ID and offset fields in STREAM frames to arrange the data. Sometimes, a QUIC endpoint can receive data at the same stream offset multiple times on a single stream. Duplicate data which has already been received can be discarded, and the offset value for a particular data must not change if it's being sent several times. However, QUIC has a few security considerations, in place which has to be taken into account when in use. One such security consideration is the applications that use QUIC protocol for communication which sends out an appropriate guidance when suspending an idle timeout. Unnecessary transmission of PING frames may have a calamitous performance impact [1].

This paper demonstrates the Slowloris attack in the QUIC protocol which is mentioned in RFC 9000 [1] and suggests countermeasures to prevent the attack.

2 Related Works

One of the design objectives in QUIC is to decrease the latencies for client–server implementations by reducing RTTs due to several, stand-alone handshakes [2]. Distributed Denial-of-Service attacks target QUIC servers. A large network has started using mitigation techniques for QUIC floods [3] but deeper study has not been done yet. Typically, a Denial-of-Service (DoS) attack affects the availability factor in the confidentiality, integrity, and availability triad by overwhelming a target with humongous number of requests until the normal requests are being processed which results in unavailability of service to the upcoming users. This attack happens with the help of a single computer to perform the attack on a chosen target. DoS attacks are classified into buffer overflow attacks and flood attacks.

A memory buffer overflow attack can consume all the available hard disk, CPU time, space, memory and hence crashing the entire system resulting in DoS attack. An attacker can oversaturate a server's capacity by flooding large number of packets eventually bringing down the service to users which results in flood-based DoS attack.

In the past, there have been some DoS attacks that disrupted the entire Internet. As the mechanisms to reduce latency are incorporated, literature [4] points out that security weaknesses in QUICs, such as lack of forward secrecy or vulnerability to replay attacks increases. HTTP/3 is intended to take advantage of QUIC's features, such as the absence of head-of-line blocking between streams [5].

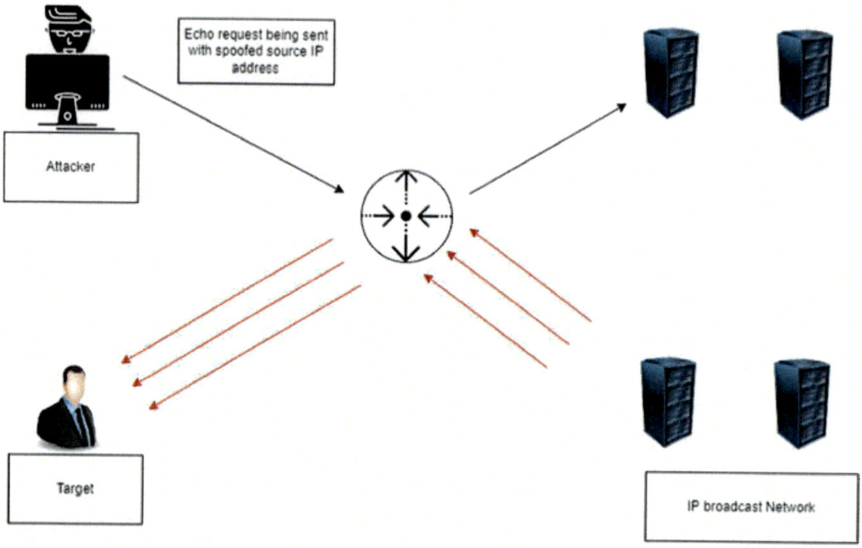

Fig. 2 Smurf DoS attack

As seen in Fig. 2, Smurf DoS attack uses the broadcast address of a vulnerable network to send large number of spoofed packets which results in flooding of a target's IP address [6].

Ping flood DoS attack [6] involves overwhelming a target by flooding with ICMP packets. The target won't be able to respond efficiently than its capacity, which eventually brings down the service. As seen in Fig. 3, Ping flood DoS can be converted to a DDoS attack by infecting other zombies in the network to bombard the target by sending ping requests.

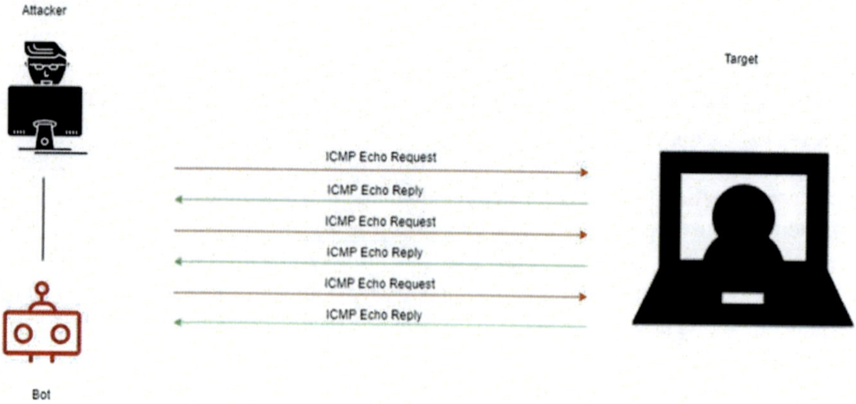

Fig. 3 Ping flood DoS attack

Apart from bombarding a target with large number of requests, the other way of performing a DoS or DDoS attack is by keeping the target alive by sending drawn-out requests. R.U.D.Y or R U Dead Yet DoS attack [7] follows this pattern of keeping a web server alive by submitting form data at a very slow pace. The victim's origin server will become unavailable to legitimate traffic in case of a successful attack.

The exploit tool sniffs form fields and exploit the form submission process so any web service that accepts form input is vulnerable to this attack.

Another variant of R.U.D.Y is the Slowloris DoS attack. As seen in Fig. 4, a malicious actor can inundate a target server by opening and sustaining many simultaneous HTTP connections. The peculiar feature of this attack is to use low amount of bandwidth to consume server's resources with requests that are slower than the normal requests. The target will only have fixed number of threads to handle the concurrent requests. The target server thread will try to stay active until the partially slow HTTP requests that are sent by the attacker is fulfilled [8]. Since this doesn't happen, the attacker exceeds the maximum limit and the server denies any further connections made, thus resulting in Slowloris denial-of-service attack.

To summarize, Slowloris happens in four steps:

(i) By sending multiple partial HTTP request headers and HTTP Keep-Alive headers, the attacker first establishes multiple connections to the targeted server.
(ii) The target creates a thread for every request that an attacker sends and closes it once the connection is finished. To be efficient, if a connection is getting prolonged, the server will timeout the connection, relieving the thread for the upcoming request.

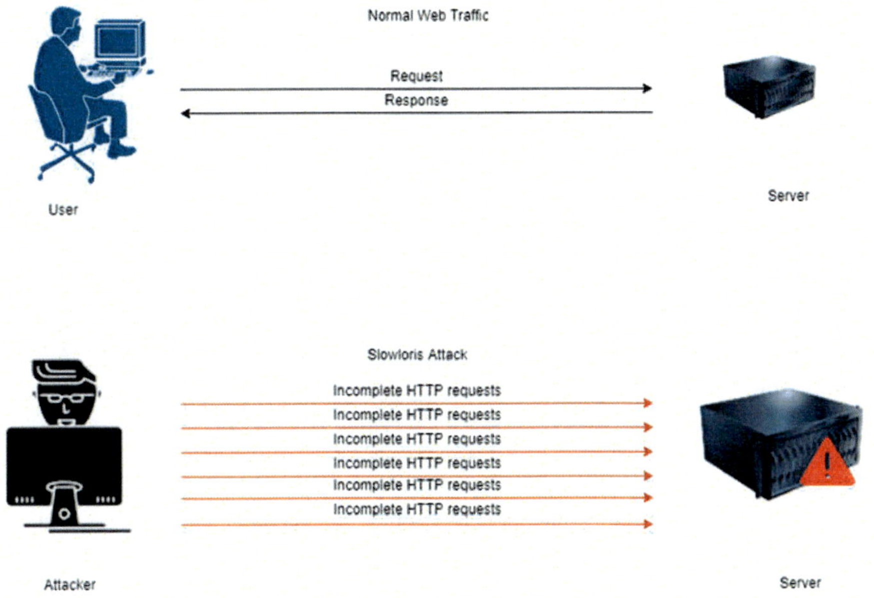

Fig. 4 Slowloris DoS attack

(iii) By sending keep-alive headers, the attacker communicates with the server to restrict it from timing out. In cases where the server times out, the attacker creates a new connection thread and keeps trying the same.
(iv) While waiting for the request to end, the target is not able to release any of the open partial connections. When all the threads are used, the server is unable to respond to additional requests, resulting in a slow denial-of-service attack.

Acharya et al. [9] quoted the DDoS attacks to be the most serious ones as it could cause severe damage to the network. Among many DDoS attacks, the UDP flood attack and the ping of death attack are regarded as the most known DDoS attacks which could cause network congestion, make the server unavailable to the users thereby disrupting the businesses. Apart from the existing strategies to mitigate any DDoS attack, an algorithm is developed to detect the two DDoS attacks and identify the attacker's node based on the number of packets being sent.

Vysakh et al. [10] addressed the DDoS attacks that are being actively exploited in IoT devices. The DDoS attacks such as the Smurf attack, Ping flood attack, Ping of death attack, and Mirai are often present in the attacker's wish list to make a service unavailable to the users. A Vulnerability scanner was developed to find the possible entry points of these attacks and notify the appropriate suggestions to the users to protect the network from these kinds of DDoS attacks.

Manoj Kumar and Vasudevan [11] introduced a DDoS-based scapy framework to generate high amount of DDoS traffic to test it against a victim based on the commands that are being received from the command-and-control center. The DDoS attack data can be generated and tested in cloud environments as well.

Samuel Black and Kim [12] addressed the common forms of application layer attacks, preventive, and detective measures to take a specific look at the HTTP flooding attacks by High Orbit Ion Cannon (HOIC) and low and slow attacks through Slowloris. Sabri et al. [13] proposed a simulative model of Slowloris DoS attack which is implemented and tested in the ActivePerl Language. The aim of the paper is to discuss about the future solution of DoS attack and propose a secure environment architecture.

Yevsieieva and Helalat [14] demonstrated slow HTTP DoS and DDoS on the virtual environments by measuring the direct impact of the attack on the target VM and the indirect impact of the attack on the neighboring VM. Sikora et al. [15] investigated about the detection and the mitigation methods for slow denial-of-service attacks. This is typically being done by a classic intrusion detection and prevention system which aims to detect, monitor, and prevent the DoS attack based on the network traffic. Once the attack is detected, it filters all the malicious traffic and blocks all the attacking IP addresses leaving the web server available to all the users.

Nawrocki et al. [16] discussed the possibility of resource exhaustion attack in the QUIC handshake process. This hypothesis is confirmed by demonstrating how this attack vector is already being used in multi-vector attacks. Unlike other denial-of-service attacks which make a service unavailable by crashing the system or by flooding the target with large number of network traffic, Slowloris is a "Slow" denial-of-service attack which relies on small traffic to affect the target. Attacks

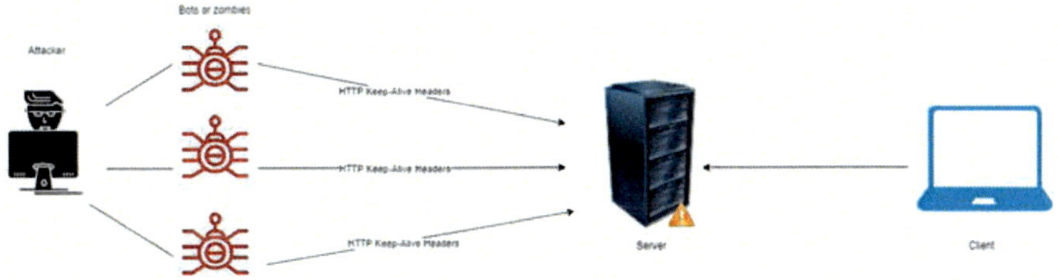

Fig. 5 HTTP slowloris attack

like Slowloris and R.U.D.Y need very little bandwidth and generate a network traffic that can go unnoticed and hence it can be tough to mitigate. As these kinds of attack don't need botnets like other DDoS attacks, it can be efficiently launched using a single system. Slowloris mainly targets threaded web servers to send slow requests in each thread which prevents normal users from accessing the service.

As seen in Fig. 5, Slowloris can be carried out by sending partial HTTP headers or HTTP Keep-Alive headers by connecting to a server. This makes the server to tie up with multiple threads that are created by the attacker to keep the connections open as long as possible affecting the availability of site to the users.

3 Proposed System

As shown in Fig. 6, we have locally configured a QUIC client–server architecture [17] with two clients named Client A, Client B, and a multi-connection server. Clients A and B send a connection request along with the CA certificate. Once the connection is established, they both begin communicating with each other, and the server responds with an ACK to client A, indicating that the client–server communication has been confirmed. From an attacker's perspective, QUIC applications that use PING frames unnecessarily can be leveraged into a Slowloris attack that can affect the performance of a server.

In our attack scenario as shown in Fig. 7, we have performed Slowloris attack in the QUIC protocol. PING frames are used to check the connectivity of a particular QUIC endpoint in the network.

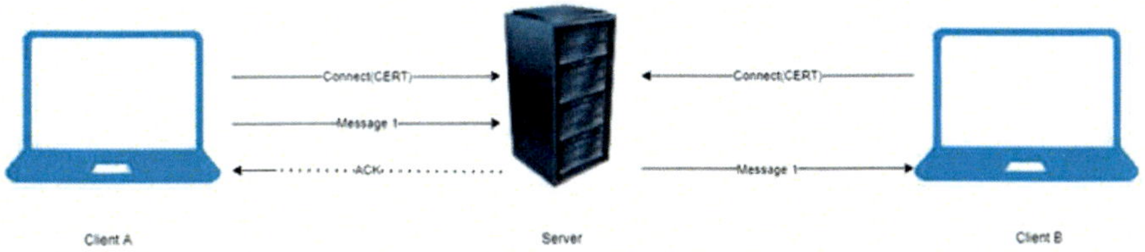

Fig. 6 Client–server establishment in QUIC protocol

Fig. 7 Slowloris attack on the QUIC endpoints

The frame format of a PING Frame as shown below indicates that it literally has no content.

PING Frame {

Type (i) = 0 × 01,

}.

The attacker can leverage this to launch a DoS attack by spawning multiple threads to send ping frames to the target QUIC endpoint, causing the connection to remain active for a longer period of time. It will eventually prevent other users from accessing the server. An attacker can also perform a Slowloris attack by opening multiple flow control windows to control the sender rate or by generating ACK frames with a high loss rate.

At first, server is run to support the multi-connection:

python3 quic_server.py –certificate certificates/ssl_cert.pem –private-key certificates/ssl_key.pem.

Then, the first client is being executed by

python3 quic_client.py –print-response –ca-certs certificates/pycacert.pem wss://localhost:4433/ws.

Finally, the other client which is the copy of the first client is executed with the same parameters:

python3 quic_client_1.py –print-response –ca-certs certificates/pycacert.pem wss://localhost:4433/ws.

In this process, the server is initiated along with the certificate so that it has the private key to decrypt the incoming messages. Client A performs a successful handshake while the server is running, and the attacker replicates the same. This procedure ensures that both clients can now communicate via the terminal with the server's assistance. During this scenario, we are using the attacker's machine to launch the QUICloris attack by using the aioquic [18] library's send_ping() API method to send PING frames via multiple threads to the server. This will cause the server to stay alive for a longer period of time and eventually the server will become unavailable to other users after attempting to send ACK frames to each of the attacker's requests.

3.1 Mitigation Techniques

In the QUIC protocol, Slowloris attack can be mitigated using two ways:

(i) Integrating security testing into the early stages of software development aids in detecting vulnerabilities in source code during the design stage when they are easy to resolve, saving expensive efforts of tackling them closer to the release date or after the release of an application. Introduce Static Application Security Testing (SAST) techniques like reviewing and limiting the unnecessary usage of PING frames that is used in the code. By this way, both developers and security engineers can collaborate to audit and fix such underlying vulnerabilities. This will encourage developers and security engineers to develop applications in QUIC which are more secure.

(ii) A server can be programmed to terminate a QUIC connection if it finds any unexpected usage of PING frames. This will avoid further traffic at the server and reduce the chances of an attacker to succeed.

4 Conclusion

Unlike other transport layer protocols, QUIC has been designed in such a way that it supports faster communication with HTTP/3. In this paper, we have implemented the Slowloris attack on the QUIC protocol and then suggested mitigation techniques for the attack. However, applications that use QUIC must be aware of the security concerns that can occur in their environment which can destruct the performance. Therefore, applications that use QUIC should now implement the respective countermeasures to defeat the denial-of-service attacks.

References

1. Iyengar J, Thomson M (2020) QUIC: a UDP-based multiplexed and secure transport. Internet Engineering Task Force, Internet-Draft draft-ietf-quic-transport-27
2. Cui Y, Li T, Liu C, Wang X, Kühlewind M (2017) Innovating transport with QUIC: design approaches and research challenges. IEEE Internet Comput 21(2):72–76
3. Cloudflare [n.d.] What is a QUIC flood DDoS attack? QUIC and UDP floods. Blog. https://www.cloudflare.com/ko-kr/learning/ddos/what-is-a-quic-flood/
4. Lychev R, Jero S, Boldyreva A, Nita-Rotaru C (2015) How secure and quick is QUIC? provable security and performance analyses. In Proceedings of IEEE symposium on security and privacy. IEEE Press, Piscataway, NJ, USA, pp 214–231
5. Kakhki AM, Jero S, Choffnes D, Nita-Rotaru C, Mislove A (2017) Taking a long look at QUIC: an approach for rigorous evaluation of rapidly evolving transport protocols. In Proceedings of the 2017 internet measurement conference, pp 290–303
6. Denial of Service. https://www.cloudflare.com/en-in/learning/ddos/glossary/denial-of-service
7. Low and Slow attack. https://www.cloudflare.com/learning/ddos/ddos-low-and-slow-attack/
8. Slowloris attack. https://www.cloudflare.com/learning/ddos/ddos-attack-tools/slowloris/

9. Acharya AA, Arpitha KM, Kumar BJS (2016) An intrusion detection system against UDP flood attack and ping of death attack (DDOS) in MANET. Int J Eng Technol 8(2)
10. Vysakh S, Binu PK (2020) IoT based mirai vulnerability scanner prototype. In 2020 third international conference on smart systems and inventive technology. IEEE, pp 97–101
11. Manoj Kumar G, Vasudevan AR (2019) D-scap: Ddos attack traffic generation using scapy framework. In Advances in Big Data and Cloud Computing. Springer, Singapore, pp 207–213
12. Black S, Kim Y (2022) An overview on detection and prevention of application layer DDoS attacks. In 2022 IEEE 12th annual computing and communication workshop and conference. IEEE, pp 0791–0800
13. Sabri S, Ismail N, Hazzim A (2021) Slowloris DoS attack based simulation. In IOP Conference series: materials science and engineering, vol 1062, no 1. IOP Publishing, p 012029
14. Yevsieieva O, Helalat SM (2017) Analysis of the impact of the slow HTTP DOS and DDOS attacks on the cloud environment. In 2017 4th international scientific-practical conference problems of infocommunications. science and technology (PIC S&T). IEEE, pp 519–523
15. Sikora M, Gerlich T, Malina L (2019) On detection and mitigation of slow rate denial-of-service attacks. In 2019 11th international congress on ultra modern telecommunications and control systems and workshops (ICUMT). IEEE, pp 1–5
16. Nawrocki M, Hiesgen R, Schmidt TC, Wahlisch M (2021) QUICs and: quantifying QUIC reconnaissance scans and DoS flooding events. In Proceedings of the 21st ACM internet measurement conference, pp 283–291
17. MaloPavol's GitHub Repository. https://github.com/MaloPavol/QUIC_chat_app
18. Aioquic GitHub Repository. https://github.com/aiortc/aioquic

Preventing Data Leakage by Trojans in Commercial and ASIC Applications Using TDM and DES Encryption and Decryption

Shashank Kumar Pathak and M. Nirmala Devi

Abstract The purpose of this work is to provide a justification to the notion of reducing the leakage of data by the data leakage Trojans in the modulation scheme of TDM (Time Division Multiplexing) by using the DES (Data Encryption Standard) encryption and decryption concept. The Data Encryption Standard (DES) is a block cipher that uses symmetric keys and can both encrypt and decode data. When it comes to producing hardware, a manufacturing company's first priority usually is to do identification and prevention of hardware Trojan assaults. Because the device Trojans can get important information about users such as encryption keys, passwords, and so on, it may be used to steal this information. Therefore, the defensive strategy used here focuses on the prevention of data leakage and not the identification of Trojan. The current system employs the idea of RECORD, which stands for randomized encryption of combinational logic for resistance to data leakage, in order to keep the data from the hardware Trojans from leaking out, despite the fact that the Trojans already know all of the information. Therefore, the suggested system of the TDM version of RECORD design is more secure than the Sequential RECORD system, and these cases of recent works have not focused on proving that TDM RECORD DES Decryption Algorithm is valid. Therefore, the proposed work of this paper will use the concept of a TDM version of RECORD with its implementation in Encryption and Decryption Algorithm with BER Testing. This method is designed in Verilog HDL and we have shown a study on this encryption and proposed decryption unit for degree of scrambling.

Keywords Data encryption standard · Trojans · RECORD · Time-division multiplexing

S. K. Pathak (✉) · M. Nirmala Devi
Department of Electronics and Communication Engineering, Amrita School of Engineering, Amrita Vishwa Vidyapeetham, Coimbatore, India
e-mail: shashank.phtk@gmail.com

M. Nirmala Devi
e-mail: m_nirmala@cb.amrita.edu

A. K. Dubey et al. (eds.), *Advanced IoT Sensors, Networks and Systems*, Lecture Notes in Electrical Engineering 1027,
https://doi.org/10.1007/978-981-99-1312-1_8

1 Introduction

The threat posed by Trojan assaults has emerged as one of the primary concerns for businesses that want to outsource the production of their hardware [1]. The danger posed by a hardware Trojan might remain latent on a fresh chip for years before causing it to become inoperable [2–4]. It is also possible for hardware Trojans to acquire passwords, keys used in encryption, or other information which may be sensitive, which would compromise upon critical security of the systems [5]. These data leakage Trojans are able to transmit confidential information over Wi-Fi networks [6] or even directly through the power signature itself. The discovery of hardware Trojans has proved to be very challenging, if not impossible. It has been shown that some Trojans can function properly with as little as five transistors in their circuitry [7]. The vast majority of the currently available defensive strategies would be useless against such an attack. This issue is not limited to the companies that develop bespoke application-specific integrated circuits (ASICs) or the companies that manufacture them. The vast majority of businesses purchase chips from commercial vendors and are not engaged in the design or manufacturing processes in any manner [8]. Protecting against Trojans on a commercially available off-the-shelf (COTS) chip is a relatively uncharted territory in terms of hardware Trojan security at the moment. This article experiments on two different aspects that may be made to the RECORD process. One, a study of RECORD based TDM encoder with DES encryption with the help of some test circuits to show how irrespective of Trojan the data is encrypted as it should be and second, a decryption unit to decrypt the data that's being sent out by the encoder which uses the randomization and the study of scrambling of data by it on wrong key input. It could be projected towards implementation on an FPGA (field-programmable gate array) or any other processor that could already be included in design, with an increased processing time of around four times as much as normal [9]. The remaining parts of this work are organized in the following approach: In Sect. 2, a brief overview of data security in hardware application and some works done on them is discussed briefly, Sect. 3 is a comprehensive overview of the TDM RECORD changed procedure is presented. The findings and implementations are investigated in Sect. 4, and the conclusions of this study will be discussed in Sect. 5.

2 Data Security in Hardware Applications

Runtime monitoring and post-manufacture testing are the two ways that are being used in the process of discovering hardware Trojans after manufacturing [10]. The identification of the changes in chip functioning that are brought about by the Trojan circuit is necessary for both monitoring during the runtime and post-manufacture testing. These approaches are dependent on the capability of tester to activate the Trojan effected circuit in order to measure the effects of the Trojan. The functionality simulation of the chip is usually not accurate enough to show the tiny changes introduced by an additional Trojan circuitry. Therefore, in order for monitoring during runtime and post-manufacture testing to be successful, a "golden chip" is typically required for comparison. This is due to the fact that the simulation of the chip's functionality cannot show the very small changes. The production method for the golden chip is a significant challenge. In addition, none of these general classifications of Trojan protection is applicable to the situation of a third-party corporation that is purchasing a COTS chip.

In general, Trojans may either be functional or parametric in nature. The former has an impact on the circuit's ability to perform its intended purpose, whereas the latter has an impact on the circuit's parameters. The functional category of Trojans includes the deletion or addition of gates or transistors to hamper the functionality of the circuit. The acts like weakening the transistor, thinning a wire, or any kind of modification with the physical parameter of the circuit are examples of the latter. An adversary would conceal the Trojan inside the circuit such that it is activated only under unusual circumstances, such as when the conditions being monitored on the internal or on the external design signals are unusually rare. This ensures that manufacturing test is passed without the revelation of the existence of the Hardware Trojan. There are now two methods available for determining whether or not Hardware Trojans are present: one uses logic test, which involves applying a test vector, and the other uses side channel analysis [11]. Only fully operational Trojans will be found using this approach. When doing a side-channel analysis, there is a possibility of additional power consumption and route latency. In light of these findings, it has been shown that this technique is effective in locating the Trojan, which is of the non-functional variety. When studying a circuit, it is unable to identify the Trojan because of the enormous process fluctuations that might occur [12]. This is another one of its drawbacks. Therefore, this is the most important thing that has to be addressed if we are looking for Trojans. It is essential that the answer to this problem include both a high-resolution strategy as well as a self-referencing approach in order to be effective. Consequently, in order to solve these problems, a novel latch-based Hardware Trojan detection methodology has been presented as a solution. In order to analyze whether or not a Trojan is present [13], this approach of self-reference, which is made on the basis of the SR latch, is used. Additionally, analysis has been carried out on both the target and reference routes for the latch's input. These are the paths that are being evaluated. The number of rounds that must be performed in this instance in order to identify the existence of a Hardware Trojan Detection is high for all possible routes [14, 15]; as a solution to this problem, this study proposes an additional MUX-based latch mechanism in addition to the latch-based approach that is already in use [10]. This approach lowers the necessary number of iterations and simplifies the needed hardware by selecting the same reference route for one or more of the target pathways. There have been several such attempts to find the presence of Trojans like the ones in [16] which uses footprints of power signal in order to accomplish its motive.

Developing the chip with security in mind from the very beginning is a more dependable technique of providing protection. Obfuscation, camouflaging of the layout, and split manufacturing are just a few of the many different concepts and approaches that fall under the umbrella of the design for security (DFS) concept. Obfuscation [17] and layout camouflaging, on the other hand, are both vulnerable to the process of reverse engineering if sufficient efforts and time are invested in the process. If numerous runs of fabrication are employed, as happens quite often in current design processes, split manufacturing is prone to vulnerability. This work is focused on a novel DFS approach which is called Randomized Encoding of Combinational Logic for Resistance to Data Leakage (RECORD), especially the TDM version of RECORD variation, that could prevent leakage of the information that could be sensitive even while the entire design is known to the attacker and multiple runs of fabrication are required. Sequential version of RECORD [18] is an extension of the original idea of the RECORD [19] that was solely defined for the combinational logic, to the sequential circuits. Sequential RECORD was developed by IBM. In order to encode incoming data bits temporarily, into a signal which would be dual-railed, hence, the approach employs the usage of two integers that are created at random. One of the railroads is represented by each of the random integers. The sequential RECORD technique is capable to successfully prevent any Trojans meant to leak data from collecting necessary data from any location on the chip by using Boolean manipulation and split manufacturing [20]. This allows the process to effectively prevent any leakage of data. This procedure does not make any attempt to identify or even prevent the presence of hardware Trojans on the chip. Any data that was obtained as a result of the assaults would be useless.

3 Comprehensive Overview of TDM RECORD

Subsequent to the design has been given to a third-party manufacture, hardware Trojans have the potential to be included at any moment. When operating from this vantage position, an adversary has access to all of the design knowledge. The currently available hardware Trojan DFS protections have been shown to be insufficient in these conditions. The three most important areas, logical obfuscation, layout camouflaging, and split manufacturing, are where you'll often find the most useful approaches. The use of RECORD in combinational logic is dissected in some detail in this article. The use of a straightforward and generic implementation that is readily adaptable to any design and does not need any circuit-specific logic or control is another advantage of RECORD [21]. A new strategy is required if there is a limited amount of space or power. TDM is an ideal alternative for minimizing the use of power and taking up less space in the overhead. The sequential RECORD procedure includes four separate but identical copies of the logic. This accomplishes two goals: first, enabling of the parallel processing of the four input vectors rapidly, and second, it assists in confusing potential attackers [22]. TDM RECORD does away with the need for duplication, while, at the same time, making it possible to implement protection measures that are much more robust. The combinational logic at the intermediate level is replicated four times and then processed in parallel. The fact

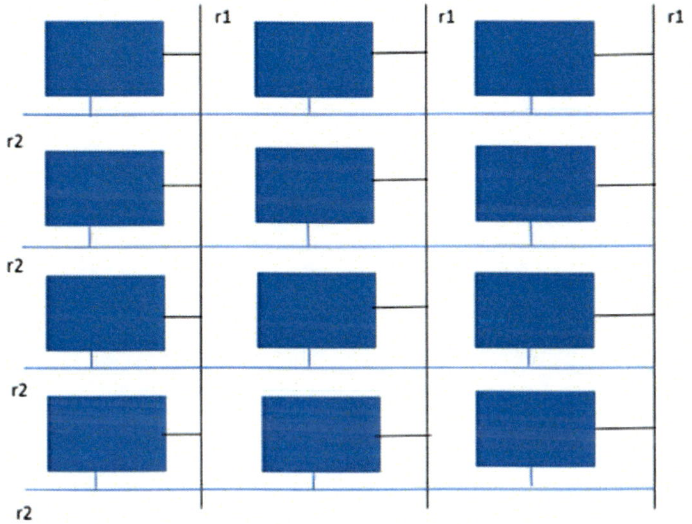

Fig. 1 Top layer framework of an array that was prefabricated

that each logic block receives a unique input vector is single feature that differentiates them from one another. The four different input variation combinations may be delivered in succession to only one of the copies of the intermediary logic when using TDM RECORD rather than having to send them to each of the four individual copies. After that, the outputs of the intermediary logic are saved in one of the four blocks of registers before being de-multiplexed to find a proper output for the final stage. In sequential RECORD, the register blocks are different from what you see above. As can be seen in Fig. 1, while sequential RECORD is being used, every register saves the de-multiplexed output of these four input vectors before they are re-indexed to the random bit on the next cycle [23].

When using TDM versioned RECORD, not all the four outputs are accessible during same cycle of the clock. Until all four input vectors have been finished, each intermediary output, vectors g1–4, are stored upon consecutive clocks. After that, the four stored outputs together with the previously mentioned two bits are demuxed and are used in choosing signals, just as previously. As can be seen in Fig. 2, the new output is re-indexed by being compared to two random bits that have been freshly produced before they are sent to the subsequent set of intermediary logic. Instead of being changed after each clock cycle, the random bits have to be held for a total of four cycles and then utilized in the generation of g1–4 [24].

Because of this constraint, control logic is required in order to determine when it will be necessary to generate fresh random bits and to exactly which block of register, the intermediary outputs, that are g1–4, will be inserted.

Figure 3 provides a visual representation of the complete procedure. At the time indicated by the clock cycle, say T, here, the dual-rail representation is obtained by using r1 and r2 through transforming data vector that's incoming. After that, the

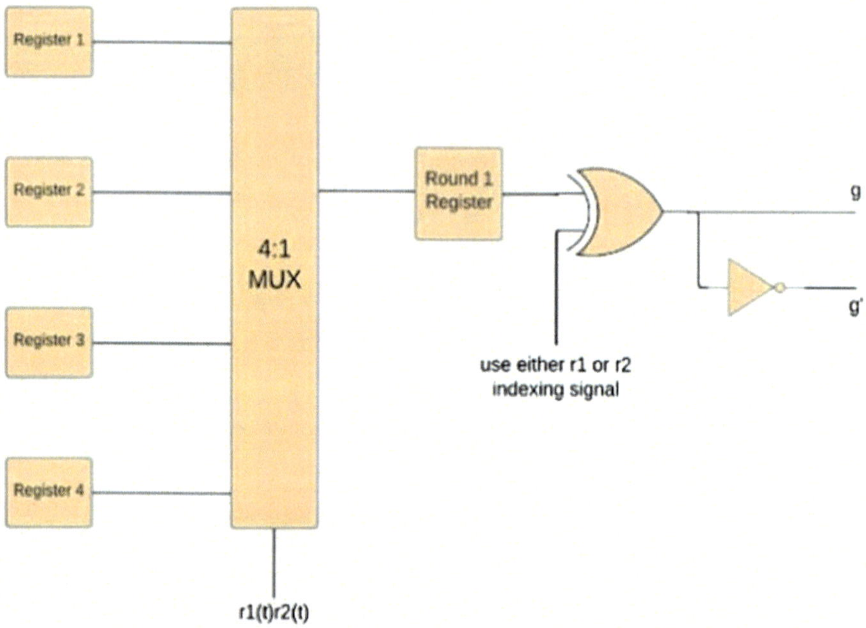

Fig. 2 Registers for every cycle of TDM RECORD and for logic re-indexing [8]

brand new bits of input, denoted by t, are received by the first block of logic. After that, it is saved in register 1, which is the next step. Register block 2 is responsible for storing the outputs. The blocks of registers are de-multiplexed at the end, during the clock cycle, T + 4, with r1 and r2 being used to serve as the select signals. In this situation, r1 and r2 are being kept constant from cycle T rather than being permitted to fluctuate from one cycle to the next. The procedure starts again at the cycle T + 5 of the clock, after which the randomized bits are permitted to get changed, the output of the block of register is modified, and then this process continues [25, 26]. Through the use of split manufacturing, the confidentiality of the random bits is preserved. The input multiplexers, control logic, registers having XOR gates, the output registers, and the final component, de-mux would all be located on the top tier of the structure. The whole intermediate logic is stored in the layer below the bottom one. There is no risk to the system's integrity caused by storage of the randomized bits for a total of four cycles of clock. An adversary still requires to wait till all four of the input data vectors had been processed prior to making any effort to deduce the randomized bit. Once this stage had been reached, the bit would vary in the same manner that it does in sequential RECORD.

There are certain obfuscation options made available by TDM RECORD that were not accessible via sequential RECORD or even RECORD. Firstly, the immunity of sequential RECORD against the attackers who attempted to deduce the randomized bits by scanning the signals that were traveling in between the top and the lower layers. Still, there was a remote possibility that the encoded data could still be deduced if any

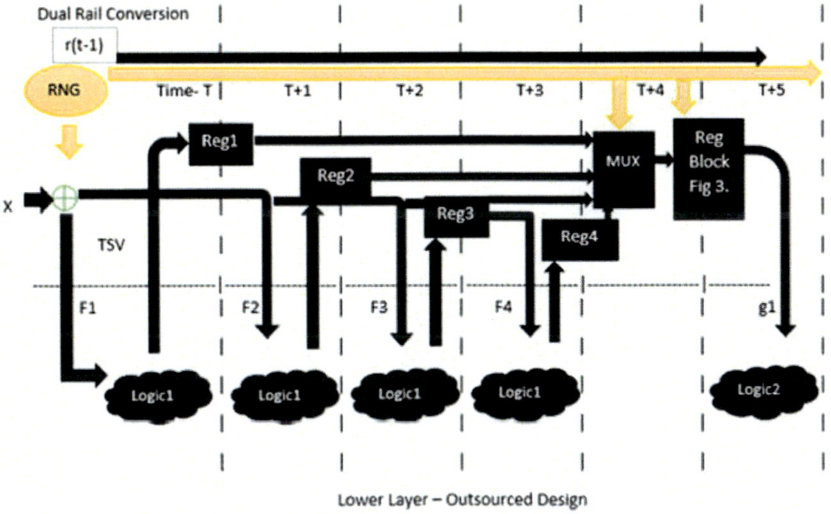

Fig. 3 The first stage of the TDM RECORD process is represented by the first cycle of data processing in the conceptual dataflow diagram [8]

adversary knew about which input vector was being received by which combinational block, and about indexing of random bits and the input bits, and also knew the indexing of the outputs of returning registers. This would allow the adversary to deduce the encoded data [27, 28]. The control logic in TDM RECORD is able to shuffle the sequence by which the vectors that are inputs get delivered to the logic by using random number generators (RNG) in the system. This allows for a more unpredictable flow of information. For instance, the input vectors might be shown in the typical order (1, 2, 3, and 4) during the first iteration, but during the second iteration, they could be presented in the order (2, 3, and 1, respectively), and so on.

Increasing the amount of random data may be done as a second measure to make it more difficult for attackers. In the past, it was not possible to add a third random bit since doing so would need an excessive amount of both space and power. Now, adding this third randomized bit will bring the total input vector number variants up to eight, though, it would do so at a very low extra cost in terms of both space and power. Just four more blocks of registers and some modifications to the controlling logic are required, but this allows for a significant increase in the number of combinations that may be achieved with regard to the top tier and the dual rail input. Time is the resource that must be spent throughout the TDM RECORD process. A minimum of 5 milliseconds, or 9 milliseconds in three random bit option's case, is added to the latency [29]. The amount of time required to create the chip also rises since the TDM RECORD idea has to be programmed into the chip from the very beginning and each design need their own unique control. In conclusion, power consumption does drop when compared to consecutive RECORD, but total energy consumption remains the same.

4 Implementation and Result Analysis

The TDM RECORD processes were developed in Verilog HDL to test their functionality and analyze their effects on the amount of space, the amount of power, and its overall performance. Here, we are using the concept of TDM RECORD (Randomized Encoding of Combinational Logic for Resistance to Data Leakage) with DES encryption [8]. It makes use of data randomization in order to prevent the hardware Trojans from performing leakage of useful information even in the scenario of the entire design being not unknown to the attacker. The encryption unit was designed and developed based upon the explanation given in Sect. 2 of this paper [8] and Fig. 3. This, as said, made use of randomized dual rail selection in order to select the select lines. The output of this unit was compared with its input and an average hamming distance (HD) of 24 out of 64 bits was found. This encryption unit was also checked with trust-HUB circuits which were having data leakage Trojans and were meant to leak the data when they knew the design as well and the output of this test is presented in Table 1. For example, if we take (AES)-T100 from trust-HUB which is being used for the purpose of leaking the AES key with a power signature with CDMA (code-division multiple access) encoding. It can be used for leaking the key being used for the DES algorithm being used here, easily. Usually, Trojans are inserted during the making of the chip at the facility of the third party, which is an untrusted one. Key data that is leaked by the Trojan will be encoded in dual-rail fashion and have indeterminate randomized bits. Trojan will have to find the random bits during runtime for converting data into single rail. To do so, it will need to information about the functional block's order and upper tier's wiring to determine the dependence of random bits. As there's a random change in the random bits on every clock cycle, every time the data processing is done it is not dependent on the previous round which makes long-term analysis of the randomized data not so effective. Therefore, in this T100 circuit, replacing AES with TDM RECORD DES circuit will help in hiding the data, Trojan will not be able to guess it.

The experimentation performed on various trust-HUB circuits consisted of same input value with same key to transfer the data. The Trojans were picked and integrated with the proposed encoder. The observation thus made suggested that this encoder is independent of any data leakage Trojan that it may encounter and continues to encode the data as it should. The Trojan produced a high capacitive output but acquiring the data that is being transferred is definitely tough for the Trojan. As we can also see in Table 1 that for the given input, the circuit produced an encoded output with a hamming distance (HD) of 38%. Though, the hamming distance will not remain the same for all the input patterns. It would vary vividly making the leakage of data tough for the Trojan.

This paper also proposes a simple TDM RECORD-based DES decryption unit which is designed to retrieve the data sent safely at the receiver's end. The entire system that was developed is illustrated in Fig. 4.

This decoder requires the key to decode the encrypted data. As the encrypted data is randomized and could be different at different clock cycles, this decoder gets those

Table 1 Calculation of hamming distance between output and input of encryption logic when used with different test circuits

Circuits	Input	Key	Output	HD%
T100	110,000,001, 010,000,100, 000,001,100, 000,010,000, 000,010,000	10,001,100,1 00,011,001, 000,110,010, 001,100,100,011	10,001,110, 010,100,010, 110,011,100,001, 001,000,100, 010,110,101	38
T1000	110,000,001, 010,000,100, 000,001,100, 000,010,000, 000,010,000	10,001,100, 100,011,001, 000,110,010, 001,100,100,011	10,001,110, 010,100,010, 110,011,100, 001,001,000, 100,010,110,101	38
T1100	110,000,001, 010,000,100, 000,001,100, 000,010,000, 000,010,000	10,001,100, 100,011,001, 000,110,010, 001,100,100,011	10,001,110, 010,100,010, 110,011,100, 001,001,000, 100,010,110,101	38
T1200	110,000,001, 010,000,100, 000,001,100, 000,010,000, 000,010,000	10,001,100, 100,011,001, 000,110,010, 001,100,100,011	10,001,110, 010,100,010, 110,011,100, 001,001,000, 100,010,110,101	38
T1300	110,000,001, 010,000,100, 000,001,100, 000,010,000, 000,010,000	10,001,100, 100,011,001, 000,110,010, 001,100,100,011	10,001,110, 010,100,010, 110,011,100, 001,001,000, 100,010,110,101	38
T1400	110,000,001, 010,000,100, 000,001,100, 000,010,000, 000,010,000	10,001,100, 100,011,001, 000,110,010, 001,100,100,011	10,001,110, 010,100,010, 110,011,100,001, 001,000,100, 010,110,101	38
T1500	110,000,001, 010,000,100,000, 001,100,000, 010,000,000, 010,000	10,001,100, 100,011,001, 000,110,010, 001,100,100,011	10,001,110, 010,100,010, 110,011,100, 001,001,000, 100,010,110,101	38
T2000	110,000,001, 010,000,100, 000,001,100, 000,010,000, 000,010,000	10,001,100,100, 011,001,000, 110,010,001, 100,100,011	10,001,110, 010,100,010, 110,011,100, 001,001,000, 100,010,110,101	38
T2100	110,000,001, 010,000,100,000, 001,100,000, 010,000,000, 010,000	10,001,100, 100,011,001, 000,110,010, 001,100,100,011	10,001,110, 010,100,010, 110,011,100, 001,001,000, 100,010,110,101	38

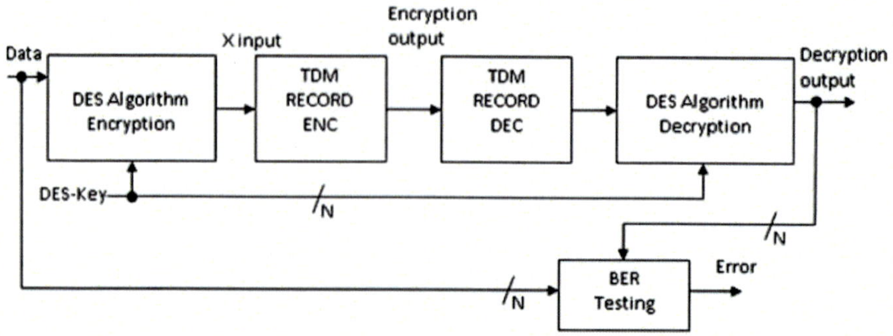

Fig. 4 Block diagram of DES TDM record method with encryption and decryption

randomized values and decrypts the data accordingly. Table 2 gives insight into how the input is randomized at the output of encryption unit and how the decoder gets it fully if the key is right and how much of the decryption output can get scrambled if the key is wrong. The measurements are given in terms of hamming distances (HDs) that are drawn between different parameters as explained after the table. Hamming distance is a parameter that gives us a measure of how different two binary data are and they can be represented in fractional values of even as percentage. Here, we have given the hamming distances in percentage.

In this table, we have considered four different hamming distance (HD) parameters:

- HD1—Between data input and encryption output.
- HD2—Between input and decryption output.
- HD3—Between encrypted values of 1 and 2, and, 3 and 4.
- HD4—Between decrypted values of 1 and 2, and, 3 and 4.

Columns 1 and 2 have same input data and columns 3 and 4 have same input data. The correct key input is same for all the data values and the wrong key input in columns 2 and 4 have got a difference of just 1 from the real key value. In this work, it is for the first time, that we have presented some data in numbers for a TDM RECORD-based DES encryption unit and analyzed the hamming distances. The hamming distance of 0 for HD2 was obtained only when the correct key bit was present. We could evidently see that the encryption unit provides a fair amount of scrambling to the given input data. Also, the output of the decryption unit that was made while carrying out this study fairly scrambles the data when the key input is wrong. The scrambling which is seen in the data here is not constant in every case. The hamming distances do vary and can have a huge range providing a good amount of security while transferring your data bits.

Table 3 was made after synthesizing the design on Xilinx XC5VLX150-2FF1759. This table shows the amount of LUTs and slice registers used by just the encryption logic and its decryption logic. Making of the decryption unit along with a BER testing occupies just a little more than 50%of the total area. The amount of lesser

Table 2 Randomization of input by the encryption logic and decryption logic on correct and wrong keys

Sl. No.	Input data to encryption unit	Output of encryption unit	Output of decryption unit	Key	HD1 (%)	HD2 (%)	HD3 (%)	HD4 (%)
1	0,000,000, 000,000,000, 000,110,000, 001,010,000, 100,000,001, 100,000,010, 000,000,010,000	0,000,000, 000,000,000, 010,001,110, 010,100,010, 110,011,100, 001,001,000, 100,010,110,100	0,000,000, 000,000,000, 000,110,000, 001,010,000, 100,000,001, 100,000,010, 000,000,010,000	0,000,000, 000,000,000, 000,000,000, 010,001,100, 100,011,001, 000,110,010, 001,100,100, 011(right)	38	0	–	–
2	0,000,000, 000,000,000, 000,110,000, 001,010,000, 100,000,001, 100,000,010, 000,000,010,000	0,000,000, 000,001,010, 100,111,000, 100,010,001, 001,000,010, 100,001,010, 001,110,101,110	0,000,000, 000,000,000, 111,110,011, 101,000,000, 101,000,101, 001,000,111, 100,011,011,001	0,000,000, 000,000,000, 000,000,000, 010,001,100, 100,011,001, 000,110,010, 001,100,100, 010(wrong)	35	32	47	30
3	0,000,000, 000,000,000, 000,000,000, 010,000,001, 000,011,000, 001,000,000, 010,100,000,111	0,000,000, 000,000,000, 101,000,010, 111,001,011, 011,111,110, 000,011,001, 111,010,011,000	0,000,000, 000,000,000, 000,000,000, 010,000,001, 000,011,000, 001,000,000, 010,100,000,111	0,000,000, 000,000,000, 000,000,000, 010,001,100, 100,011,001, 000,110,010, 001,100,100, 011(right)	40	0	–	–

(continued)

Table 2 (continued)

Sl. No.	Input data to encryption unit	Output of encryption unit	Output of decryption unit	Key	HD1 (%)	HD2 (%)	HD3 (%)	HD4 (%)
4	0,000,000, 000,000,000, 000,000,000, 010,000,001, 000,011,000, 001,000,000, 010,100,000,111	0,000,000, 000,000,000, 100,110,100, 101,110,000, 011,110,011, 101,100,101, 101,001,110,110	0,000,000, 000,000,000, 001,001,110, 010,110,001, 100,011,000, 011,100,111, 110,101,111,100	0,000,000, 000,000,000, 000,000,000, 010,001,100, 100,011,001, 000,110,010, 001,100,100, 010(wrong)	44	31	43	30

Table 3 Comparisons of TDM DES encryption and decryption method

	Comparisons of TDM-DES—synthesized on Xilinx XC5VLX150-2FF1759	
	TDM-DES encryption method	TDM-DES encryption and decryption method
Slice registers	4620	9285
Slice LUTs	2327	4579
Occupied slices	1333	2761
IOB	130	147
Delay (ns)	1.714	1.714
Power (W)	1.712	1.744

number of LUTs used confirms the fact that this RECORD-based TDM encryption and now even the proposed decryption logic utilizes very less area when compared to its previous versions [18]. Figure 5 is a snapshot from the RTL simulation of this experiment where data is the input data, decrypt_data is the output data and encry_data is the output of the encryption unit. The error bit under the BER testing group indicates if the output received at the decryption unit is right or not. We can see the error signal high in Fig. 6 when the input key is different. The analysis of these results is shown in Table 2.

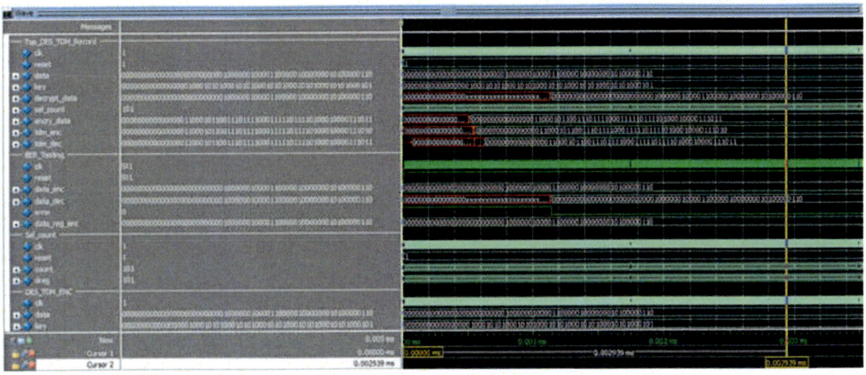

Fig. 5 Output waveform with the right key

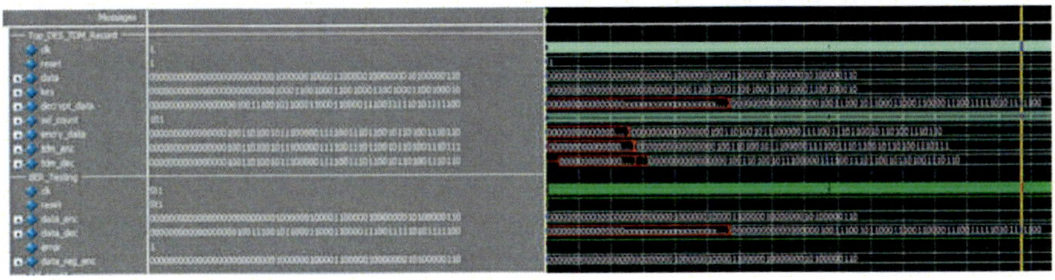

Fig. 6 Output waveform with the wrong key

5 Conclusion

The globalization of the creation of microchips creates the prospect that an adversary may plant hardware Trojans into a chip while it is being manufactured. A novel approach to protecting one's hardware against Trojans is represented by the concept of the RECORD procedure. This process does not make any attempt to identify hardware Trojans, nor does it make any attempt to prevent them. Instead, the RECORD process offers security, which can be maintained even in the presence of a running Trojan. The sequential RECORD design method has been modified in two ways. Over the original sequential RECORD technique, both those ways provide significant space and power reductions. As the input vector operation order is randomized and the random bit count is increased, TDM RECORD allows designers more freedom to prevent possible attacks. The ability of RECORD to protect sensitive data even in the scenario of the entire design being known to the attacker during production makes these design schemes more secure than conventional DFS approaches. The work carried out in this project is independent of the detection of Trojans [11, 30] as per the concept of TDM RECORD. The concept of prevention against data leakage Trojans through the encoding and decoding schemes like TDM RECORD which is used here is focused on keeping your data secured during its transfer if your hardware is infected with a data leakage Trojan. As a future scope to this work, an analysis can be drawn between the type of random number generators and their effect on this process and which random number generator is more robust. This can also be tested over other encryption techniques such as 3DES (Triple Data Encryption Standard) and AES (Advanced Encryption Standard). So far, having randomness in your encryption has been the best possible way to combat data leakage Trojans (that is, stop them from leaking any data) as the select line changes at every clock cycle making it tough for the Trojan/attacker to guess the randomized value. If you choose to apply RECORD in any way, we anticipate that you will prioritize security above anything else. Both the FPGA designer and the systems integrator have access to straightforward and efficient choices, in the form of TDM and COTS RECORD, that allow for the secure use of their designs with a reduced likelihood of data loss.

References

1. Kiamilev F, Hoover R (2008) Demonstration of hardware trojans. DEFCON 16
2. DARPA (2007) "TRUST in Integrated Circuits (TIC) - Proposer Information Pamphlet". http://www.darpa.mil/MTO/solicitations/baa07-24/index.html
3. Australian Government DoD-DSTO (2008) Towards Countering the Rise of the Silicon Trojan. https://apps.dtic.mil/sti/pdfs/ADA496426.pdf
4. Kean T (2002) Cryptographic rights management of FPGA intellectual property cores. In Proceedings of the symposium on field programmable gate arrays (FPGA'02), pp 113–118
5. Wolff F, Papachristou C, Bhunia S, Chakraborty RS (2008) Towards trojan-free trusted ICs: problem analysis and detection scheme. Proc Design Autom. Test Eur. (DATE), pp 1362–1365

6. Jin Y, Makris Y (2010) Hardware trojans in wireless cryptographic ICs. IEEE Design Test Comput 27(1):26–35
7. Saravanan K, Mohankumar N (2019) Design of logically obfuscated n-bit ALU for enhanced security. ICECA, IEEE Conference Record #45616
8. Schulze TE, Beetner DG, Shi Y (2018) Combating data leakage trojans in commercial and ASIC applications with time division multiplexing and random encoding. IEEE Trans
9. Lin L, Kasper M, Güneysu T, Paar C, Burleson W (2009) Trojan side-channels: lightweight hardware trojans through side-channel engineering. In Cryptographic hardware and embedded systems—CHES (Lecture Notes in Computer Science), vol 5747. Springer, Berlin, Germany, pp 382–395
10. Giridharan S, Ramesh SR, Nirmala Devi M (2021) A MUX based latch technique for the detection of hardware trojan using path delay analysis. In International conference on electronics
11. Huang Y, Bhunia S, Mishra P (2018) Scalable test generation for trojan detection using side channel analysis. IEEE Trans Inform Forensics Secur 13(11)
12. Sree Ranjani R, Nirmala Devi M (2018) A novel logical locking technique against key guessing attacks. Department of Electronics and Communication Engineering, IEEE
13. Wang X, Tehranipoor M, Plusquellic J (2008) Detecting malicious inclusions in secure hardware: challenges and solutions. HOST
14. Jha S, Jha SK (2008) Randomization based probabilistic approach to detect trojan circuits. In 11th IEEE high assurance systems engineering symposium
15. Smith S, Di J (2007) Detecting malicious logic through structural checking. IEEE Region 5 technical conference
16. Maruthi V, Balamurugan K, Mohankumar N (2020) Hardware trojan detection using power signal foot prints in frequency domain. In International conference on communication and signal processing. IEEE
17. Chakraborty RS, Bhunia S (2009) Security against hardware trojan through a novel application of design obfuscation. ICCAD
18. Semiconductor Industry Association (SIA) (2008) "Global billings report history (3-month moving average) 1976-March 2009" http://www.sia-online.org/galleries/Statistics/GSR1976-March09.xls
19. Defense Science Board (2005) "Task force on high performance microchip supply". http://www.acq.osd.mil/dsb/reports/200502HPMSReportFinal.pdf
20. Schulze TE, Kwiat K, Kamhoua C, Beetner D, Njilla L, Shi Y (2017) Combating data leakage trojans in sequential circuits through randomized encoding. In Proc IEEE Dependable, Autonomic Secure Comput (DASC). Orlando, FL, USA, pp 639–644
21. Schulze TE, Shi Y, Kamhoua C, Kwiat K, Chang S (2016) RECORD: temporarily randomized encoding of combinational logic for resistance to data leakage from hardware trojan. In Proc Asian HOST, Yilan, Taiwan, pp 1–6
22. Yang K, Hicks M, Dong Q, Austin T, Sylvester D (2016) A2: malicious hardware. In Proceedings of IEEE symposium on security and privacy. San Jose, CA, USA, pp 18–37
23. Kamhoua CA, Rodriguez M, Kwiat KA (2014) Testing for hardware trojans: a game-theoretic approach. In Decision and game theory for security (Lecture Notes in Computer Science), vol 8840. Springer, Cham, Switzerland, pp 360–369
24. Rostami M, Koushanfar F, Karri R (2014) A primer on hardware security: models, methods, and metrics. Proc IEEE 102(8):1283–1295
25. Tehranipoor M, Koushanfar F (2010) A survey of hardware trojan taxonomy and detection. IEEE Des Test Comput 27(1):10–25
26. Koushanfar F (2012) Provably secure active IC metering techniques for piracy avoidance and digital right management. IEEE Trans Inform Forensics Secur 7(1):5
27. Zarrinchian G, Zamani MS (2017) Latch-based structure: a high resolution and self-reference technique for hardware trojan detection. IEEE Trans Comput 66(1)
28. Li J, Lach J (2008) At-speed delay characterization for IC authentication and trojan horse detection. In Proceedings of HOST, pp 8–14

29. Nandith GS, Ramesh SR (2020) A novel approach for statistical parameter estimation and test pattern generation. In 2020 4th international conference on trends in electronics and informatics (ICOEI). Tirunelveli, India, pp 545–550
30. Saha S, Chakraborty RS, Nuthakki SS, Mukhopadhyay AD (2015) Improved test pattern generation for hardware trojan detection using genetic algorithm and boolean satisfiability. Proc Cryptographic Hardware Embedded Syst, pp 577–596

Design of Helmet-Mounted Dual-Band Conformal Antenna for Military Applications

Govind Mishra, Shuchismita Pani, and M. S. Prasad

Abstract This paper discusses the design of helmet-mounted conformal patch antenna, operating at two different frequencies, i.e., 2.45 and 4.85 GHz, which can be used for the communication purpose in military applications. This antenna is designed with substrate FR-4 of dielectric constant 4.3, thermal conductivity of 0.3 W/K/m, and having thickness of 1.60 mm. The proposed design has been done on CST studio suite software. Also, two pin diodes are connected in patch. For both diodes in ON the achieved S_{11} for the frequency 2.45 GHz is about −34.334 dB, having bandwidth of 0.08 GHz and for the frequency 4.85 GHz achieved S_{11} is −24.334 dB and having bandwidth of 0.22 GHz. The simulated results are compared and analyzed with non-conformal patch antenna and also, specific absorption rate for head has been calculated for the proposed design.

Keywords Microstrip patch · Conformal · Dual band · Pin diode · Reflection coefficient · Specific absorption rate

1 Introduction

Conformal antenna integration is required in huge number for complicated and dedicated antennas design in various sectors, like mobile communication, automotive systems, and space- and airborne applications. The phrase conformal refers to antennas that are fixed to the structural carrier's surface. These antennas can satisfy particular antenna pattern requirements via an adequate geometrical conformation.

G. Mishra (✉) · S. Pani · M. S. Prasad
Amity Institute of Space Science and Technology, Amity University Noida, Noida, UP, India
e-mail: mishragovind168@gmail.com

S. Pani
e-mail: spani@amity.edu

M. S. Prasad
e-mail: msprasad@amity.edu

A. K. Dubey et al. (eds.), *Advanced IoT Sensors, Networks and Systems*, Lecture Notes in Electrical Engineering 1027,
https://doi.org/10.1007/978-981-99-1312-1_10

A C-band conformal antenna array on cylindrical surface [1] has been designed. The operating frequency for cylindrical 8 × 2 array patch is 5.2 GHz, which gives return loss of about −14 dB with bandwidth of 4%. A conformal antenna 95 × 3 array on spherical surface having diameter 300 mm has been designed [2], and the operating frequency for this design is 9.5 GHz. The Rogers RT Duroid 5880 substrate has been used for dielectric constant 2.2 and height of substrate is 1.27 mm, and the diameter is 39 mm. The magnitude of reflection coefficient below −30 dB have been achieved. Conformal antennas with diverse substrates for antenna design have been described in [3], also construct a modeling of conformal antennas with recent improvements through reviewing early manifestations in the study. The comparative study and the effect of conformation on microstrip antenna have been studied at different angles of conformation of the patch, also the advantages of conformal antenna over simple patch antenna have been proposed [4]. A study recent study on conformal antenna and its design issues have been discussed [5, 6]. Four pairs of microstrip conformal antennas for 35 GHz are designed in study. Also, Fabrication and testing of antennas has been completed which have return loss of greater than −20 dB at 35 GHz, with first sidelobe of −16 dB, and a 60° angle between the main lobe and [7] the array plane. Monica and Jothilakshmi [8] describes a unique conformal antenna design that provides a higher bandwidth of 0.53 GHz and Gain 9 dB using a thin and elastic substrate that is ideal for airplane fuselage deployment. Also, 10.2 percent bandwidth is acquired around the working frequency of 5.2 GHz.

In this paper, a Helmet mounted dual-band conformal patch antenna of 2 × 2 array is designed operating at two different frequencies, i.e., 2.45 and 4.85 GHz. Copper (annealed) is used for ground and patch. FR-4 of dielectric constant 4.3 and thermal conductivity of 0.3 W/K/m is used as substrate and having thickness of 1.60 mm. The simulated results are compared and analyzed with non-conformal patch antenna and also SAR analysis by human head phantom has been done.

2 Antenna Design

2.1 Non-conformal Patch Antenna Design Details

Figure 1a shows the proposed structure of the patch antenna array of 2 × 2, out of which two patches are connected via pin diode, i.e., named element 1 and element 2 as shown in Fig. 1b. Figure 1b shows the dimension of the antenna, and the patch is denoted by yellow color which is of material copper having electrical conductivity 5.8×10^7 S/m, heat capacity 0.39 kJ/K/Kg, Young's modulus 120 kN/mm^2, and having thickness 0.035 mm. And the substrate is denoted by white color which is of material FR-4 having dielectric constant 4.3, electrical tand 0.025, and thermal conductivity is 0.3 W/K/m, and having thickness of 1.60 mm. The length and width of ground plane is same as dimension of substrate and the thickness of ground plane is 0.035 mm. The back view of the proposed patch antenna is shown in Fig. 1c.

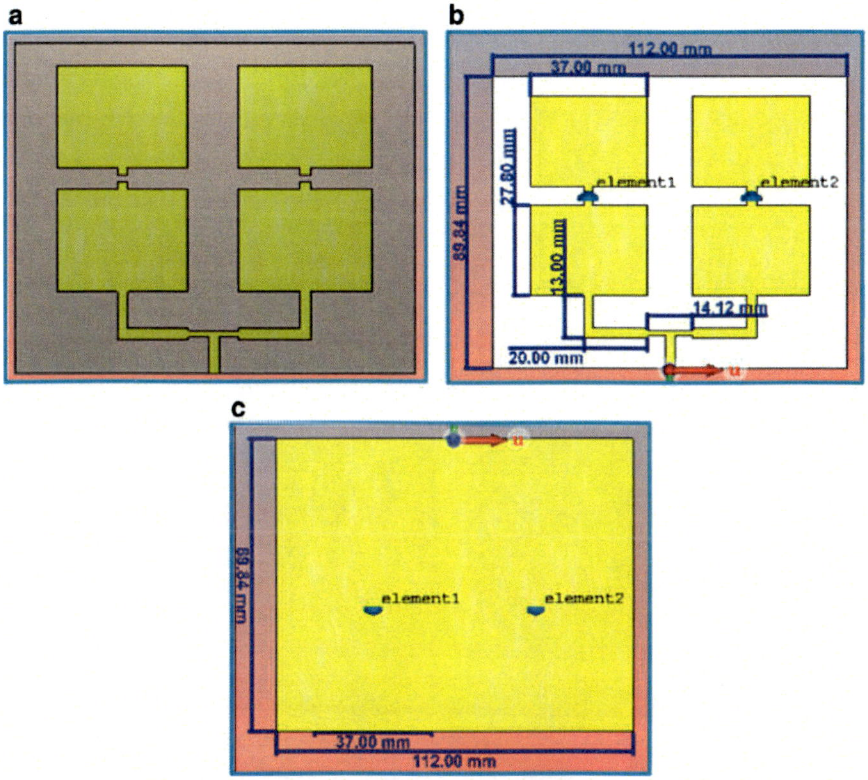

Fig. 1 **a** Front view of non-conformal 2 × 2 patch antenna **b** Dimensions of non-conformal 2 × 2 patch antenna **c** Back view of non-conformal patch antenna

2.2 *Conformal Antenna Design Details*

Figure 2a–e shows the proposed structure of the conformal (spherically conformal) patch dual band helmet-mounted antenna. The patch is conformal at 15 cm radii of sphere. In between both upper and lower patches a pin diode is connected that can be switched on or off for the flow of current. The design employs the microstrip patch feeding technology, in which the microstrip patch is directly connected to the conducting feed line, making powering up the conformal antennas very simple. The feed line impedance is approx 50Ω.

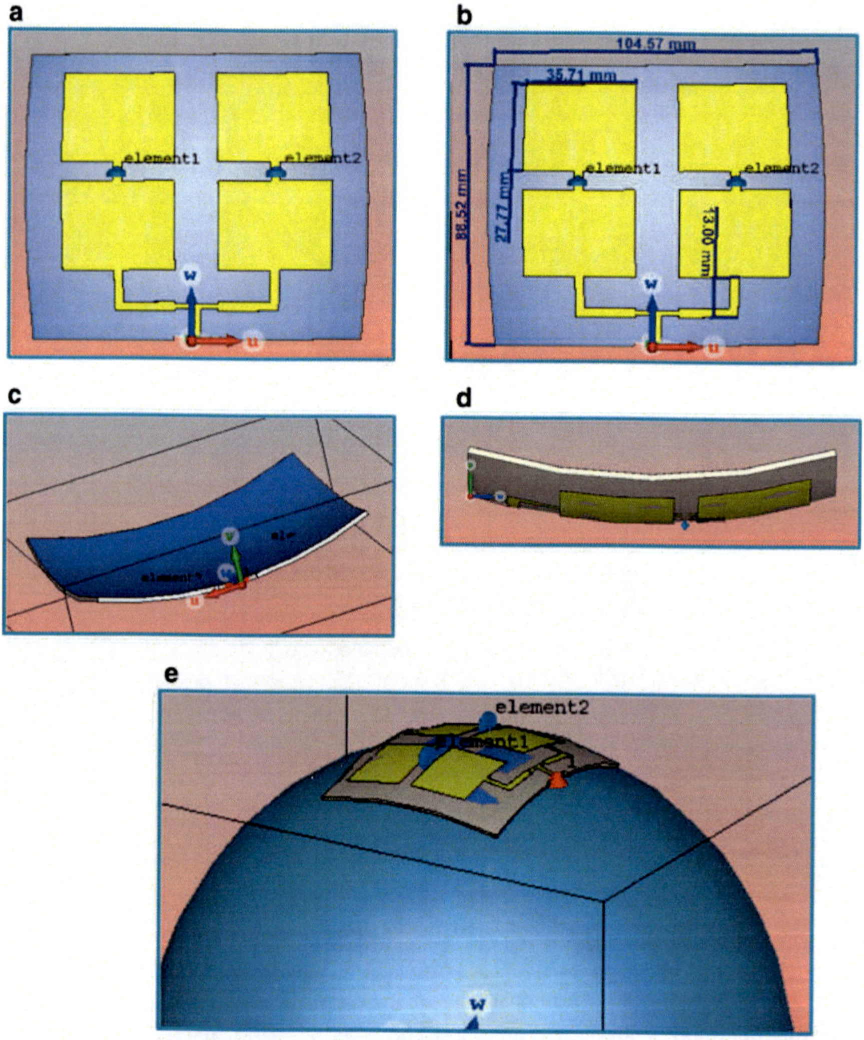

Fig. 2 **a** Front view of conformal antenna **b** Dimensions of conformal antenna **c** Conformal view of antenna **d** Side view of conformal antenna **e** Helmet Mounted view of conformal antenna

3 Simulation Results

3.1 Simulation Result of Non-conformal Patch Antenna Design

Simulation is performed on the commercial CST Studio Suite (2019). The reflection coefficient (S_{11}) is shown in Fig. 3. S_{11} plot clearly shows that the proposed antenna is radiating at two different frequencies, i.e., 2.45 GHz with −28.829 dB return loss and at frequency 4.86 GHz with −20.968 dB return loss.

Figure 4 shows farfield directivity Abs at (phi = 90°) for radiation frequency 2.45 GHz when both the pin diodes is in ON state, the magnitude of main lobe is 7.25 dBi and its direction is 43.2 deg with angular width of 63.0 deg.

Figure 5 shows farfield directivity Abs at (phi = 0°) for radiation frequency 2.45 GHz when one of the pin diodes, i.e., element1 is in ON state, and another pin diode, i.e., element2 is in OFF state. For this condition, the magnitude of main lobe is 5.74 dBi and its direction is 3.0 deg with angular width of 65.2 deg.

Figure 6 shows farfield directivity Abs at (phi = 0°) for radiation frequency 2.45 GHz when one of the pin diodes, i.e., element 2 is in ON state, and another pin diode, i.e., element 1 is in OFF state. For this condition, the magnitude of main lobe is 5.74 dBi and its direction is 3.0 deg with angular width of 65.2 deg. In this, the main lobe is slightly tilted right side at 3 deg angle. Whereas in Fig. 5, the min lobe

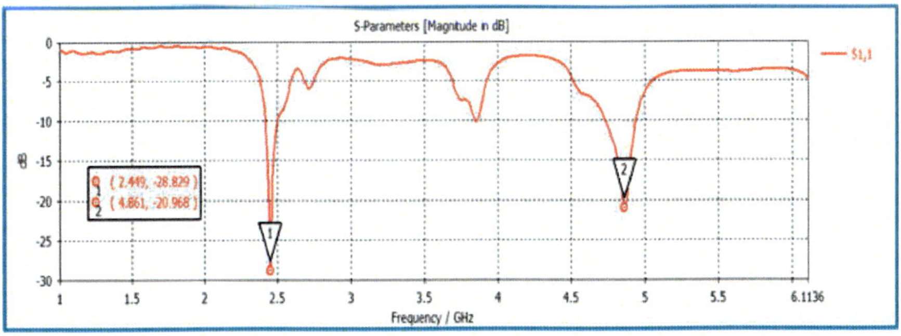

Fig. 3 S_{11} versus frequency of non-conformal patch antenna when both the pin diodes are in ON state

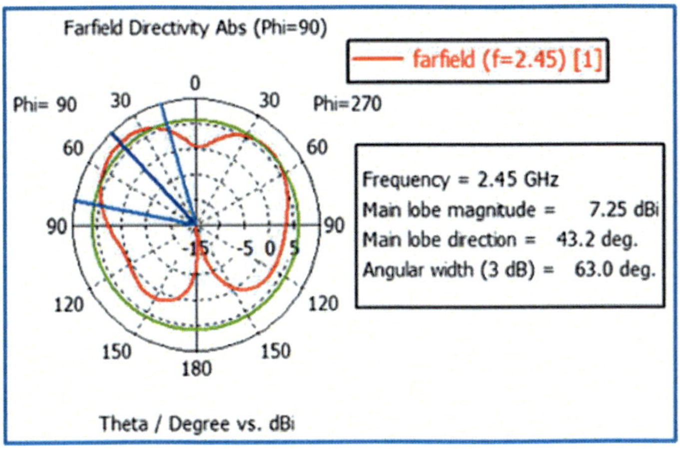

Fig. 4 Farfield at 2.45 GHz for non-conformal antenna when both the pin diodes are in ON state

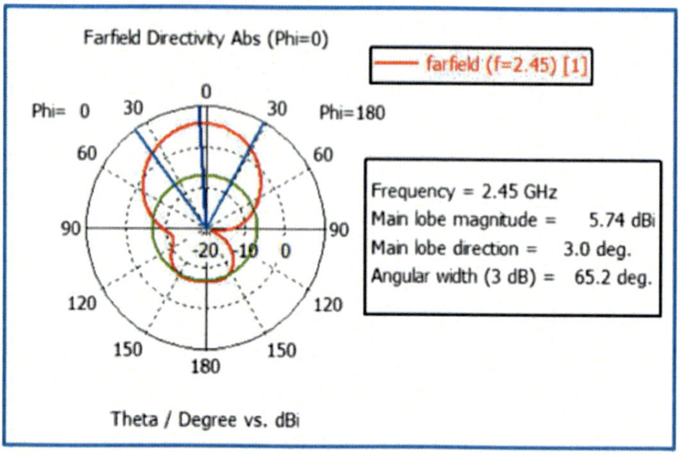

Fig. 5 Farfield at 2.45 GHz for non-conformal antenna when element 1 pin diodes are ON and element 2 pin diode is OFF

is slightly tilted left side at 3 deg angle. So, on turning pin-diode ON or OFF 3 deg angle shift in the main lobe was observed.

Figure 7 shows farfield directivity Abs at (phi = 90°) for radiation frequency 4.85 GHz when both the pin diodes are in ON state, the magnitude of main lobe is 4.61 dBi and its direction is 0 deg with angular width of 46.8 deg.

Figures 8 and 9 shows farfield directivity Abs at (phi = 0°) for radiation frequency 2.45 GHz for both conditions as mentioned in each figure. The magnitude of main lobe is 3.47 dBi and its direction is 3.0 deg with angular width of 55.2 deg. In Fig. 8, main lobe is slightly tilted right side at 3 deg angle. Whereas in Fig. 9 the min lobe is slightly tilted left side at 3 deg angle. So, on turning pin-diode ON or OFF 3 deg

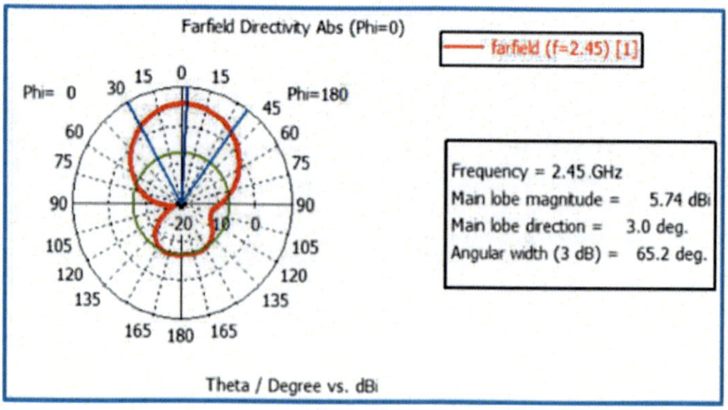

Fig. 6 Farfield at 2.45 GHz for non-conformal antenna when element 2 pin diodes are ON and element 1 pin diode is OFF

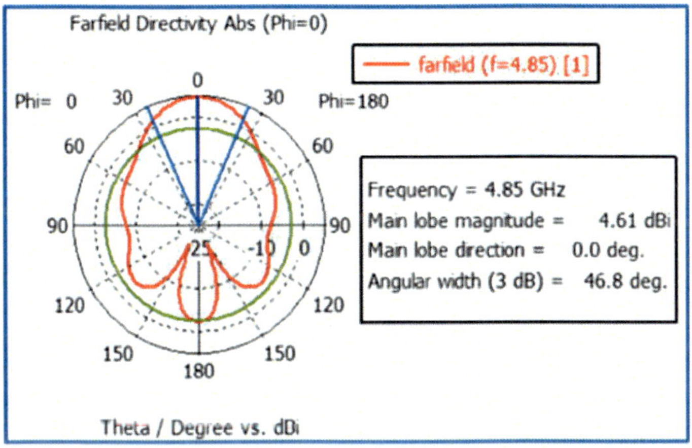

Fig. 7 Farfield at 4.85 GHz for non-conformal antenna when both the pin diodes are in ON state

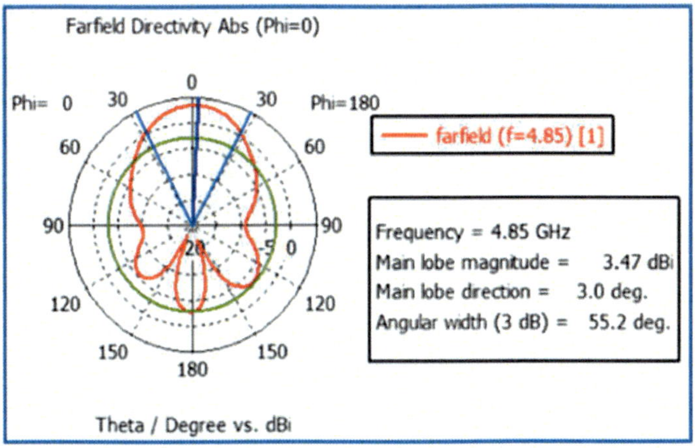

Fig. 8 Farfield at 4.85 GHz for non-conformal antenna when element 2 pin diodes are ON and element 1 pin diode is OFF

angle shift in main lobe was observed in both conditions. Whereas if both pin diodes are in ON state, direction of main lobe is 0 deg.

3.2 Simulation Result of Conformal Antenna Design

For conformal antenna, the radiating frequency is 2.45 and 4.85 GHz. The reflection coefficient (S_{11}) is shown in Fig. 10. S_{11} plot clearly shows that the proposed antenna is radiating at two different frequencies, i.e., 2.45 and 4.85 GHz. For 2.45 GHz

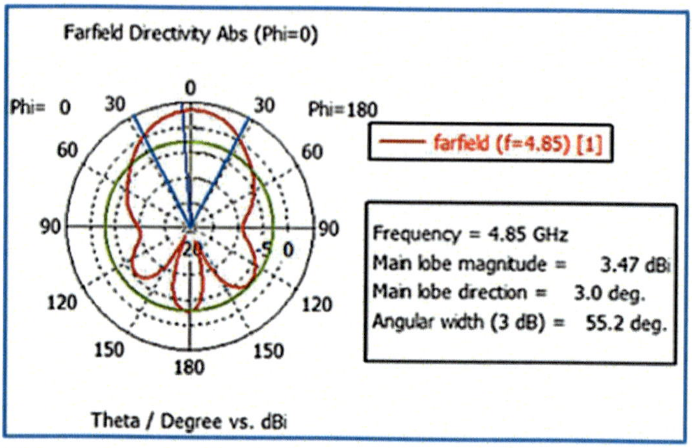

Fig. 9 Farfield at 4.85 GHz for non-conformal antenna when element 1 pin diodes are ON and element 2 pin diode is OFF

frequency return loss achieved is −34.334 dB, and having bandwidth of 0.08 GHz or 3.2 percentage bandwidth. And for 4.85 GHz frequency achieved return loss is −24.334 dB and having bandwidth of 0.22 GHz or 4.53 percentage bandwidth. As compared from non-conformal design, on making antenna curved the slight change in operating frequency has been observed, also the reflection coefficient (S_{11}) has been changed for both the frequencies. Also, some change in radiating pattern is also observed. From Fig. 11, it is clearly seen that the Power radiated or radiation efficiency is maximum at the 2.45 and 4.85 GHz frequencies.

Figure 12 shows farfield at 2.45 GHz for conformal antenna when both the pin diodes are in ON state, for this at farfield directivity Abs at (phi = 0°) have the magnitude of main lobe is 7.38 dBi and its direction is 0 deg with angular width of 58.5 deg.

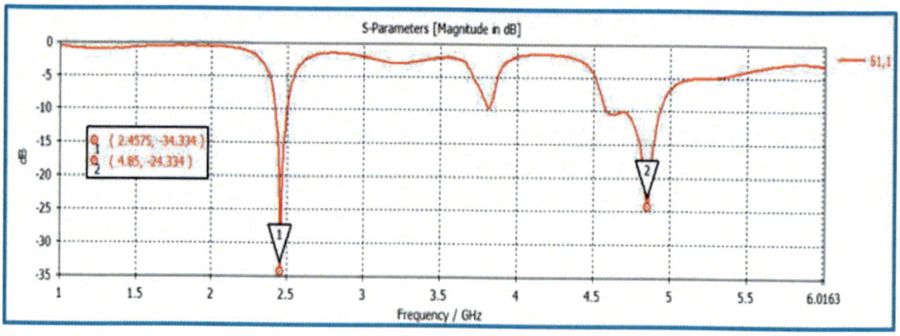

Fig. 10 S_{11} versus frequency of conformal antenna when both the pin diodes are in ON state

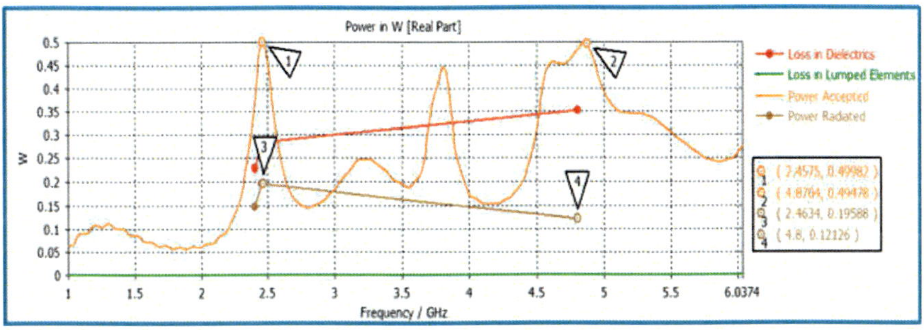

Fig. 11 Power radiated versus frequency for conformal antenna when both the pin diodes are in ON state

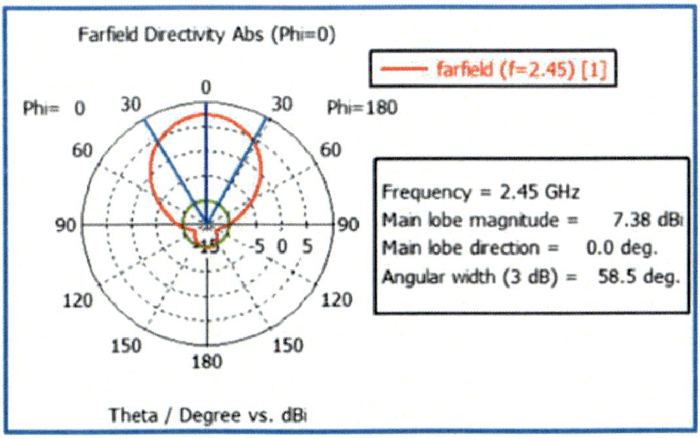

Fig. 12 Farfield at 2.45 GHz for conformal antenna when both the pin diodes are in ON state

Figures 13 and 14 shows farfield directivity Abs at (phi = 0°) for radiation frequency 2.45 GHz for both conditions. Figure 13 shows that the main lobe magnitude is 6.65dBi and main lobe is slightly tilted left side at 1 deg angle. Whereas in Figure 14, the main lobe magnitude is 6.68dBi and main lobe is slightly tilted right side at 1 deg angle. So, on turning pin-diode ON or OFF, 1 deg angle shift in main lobe was observed in both conditions. Whereas if both pin diodes are in ON state, direction of main lobe is 0 deg.

Figure 15 shows farfield at 4.85 GHz for conformal antenna when both the pin diodes are in ON state, for this farfield directivity Abs at (phi = 0°) have the magnitude of main lobe of 0.253 dBi and its direction is 180.0 deg with angular width of 28.7 deg. And Fig. 16 shows the farfield at 4.85 GHz for conformal antenna when either one diode is ON or another is OFF. In this case, farfield directivity Abs at (phi = 0°) have the magnitude of main lobe of −0.343 dBi and its direction is 7.0 deg with angular width of 89.7 deg.

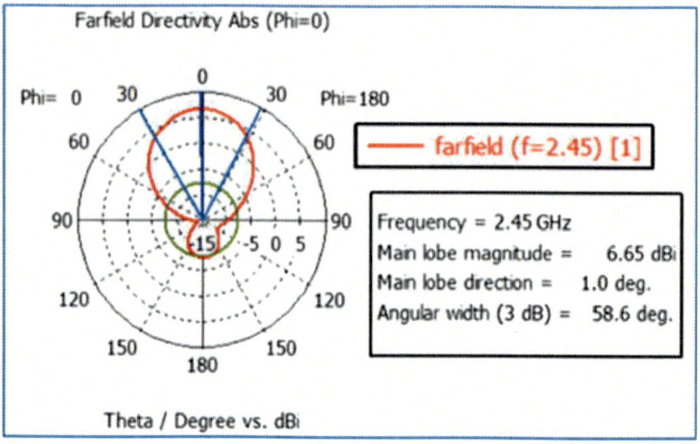

Fig. 13 Farfield at 2.45 GHz for conformal antenna when element 1 pin diodes are ON and element 2 pin diode is OFF

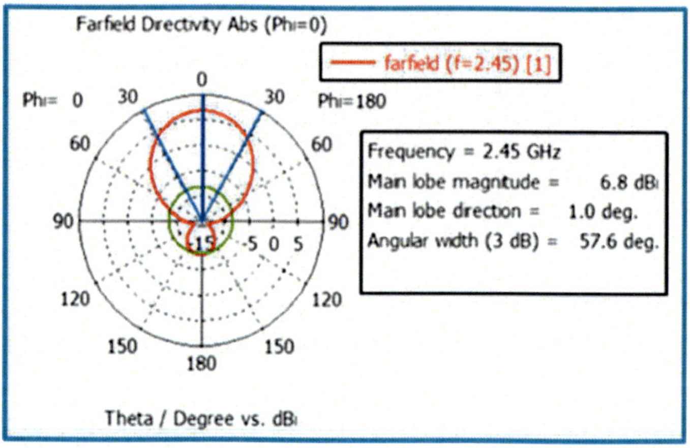

Fig. 14 Farfield at 2.45 GHz for conformal antenna when element 2 pin diodes are ON and element 1 pin diode is OFF

Figure 17 shows the result of SAR or specific absorption rate analysis of antenna at maximum power radiating frequency, i.e., 2.45 GHz. [9] The critical SAR value of human head is approx. 3.2 W/kg, and from the proposed design simulation result, the maximum SAR value obtained is 1.15876 W/kg, which is lower than the critical SAR value of the human head.

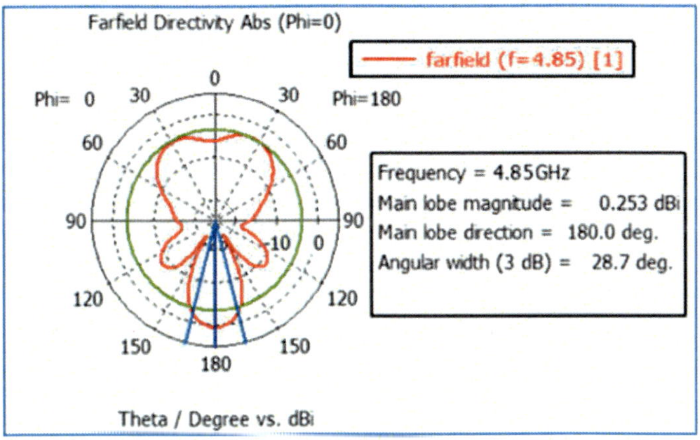

Fig. 15 Farfield at 4.85 GHz for conformal antenna when both the pin diodes are in ON state

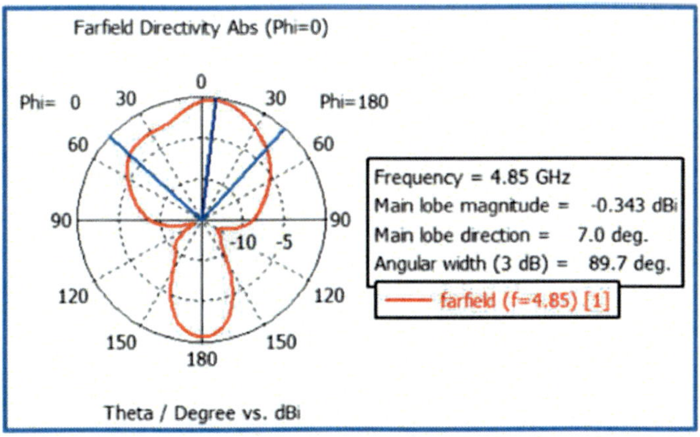

Fig. 16 Farfield at 4.85 GHz for conformal antenna when either one diodes is ON or another is OFF

4 Conclusion

The above-mentioned results suggest that the proposed helmet-mounted conformal antennas can be an excellent patch antenna substitute. The conformal antenna allows us to put the antenna on any surface without modifying the aerodynamic character or shape of the body on which it will be mounted on. Also, this helmet-mounted conformal antenna can be used for the communication purpose by the military at the dual-band operating frequency. Also, SAR (Specific absorption rate) analysis has been performed, for this antenna [9]. The critical SAR value of human head is approx. 3.2 W/kg, and from our simulation result the maximum SAR value obtained is 1.15876 W/kg, which is lower than the critical SAR value of the human head.

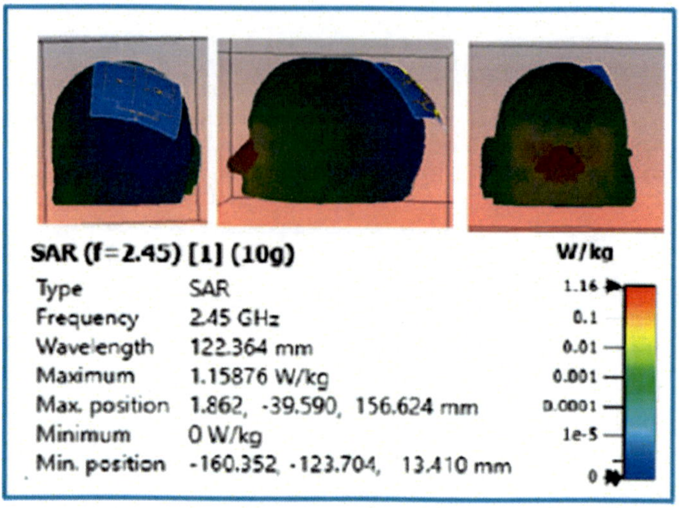

Fig. 17 SAR analysis of conformal antenna at 2.45 GHz frequency

So, it can be concluded that the proposed design cause no harm to the human and is suitable for the mounting on helmet.

5 Future Scope

The above designed antenna can be fabricated and can be tested in real-world scenario and this helmet-mounted conformal antenna can be used for the communication purpose by the military at the dual-band operating frequency. Also array size or configuration can be modified for the better performance of the antenna.

References

1. Wiesbeck W, Younis M, Loffler D (2022) Design and measurement of conformal antennae. In: Antennas and propagation society international symposium–San Antonio, TX, USA (16–21 June 2002)] IEEE Antennas and Propagation Society International Symposium (IEEE Cat. No.02CH37313)
2. Knott P (2007) Design and experimental results of a spherical antenna array for a conformal array demonstrator. In: INICA 2007 international conference on antennas—proceedings
3. Sumalatha K, Phani Kumar CR (2019) Advances in conformal antennas. Int J Eng Adv Technol (IJEAT)
4. Elleithy K, Bajwa H, Elrashidi A (2011) Conformal microstrip printed antenna. In: Conference, ASEE Northeast section conference. University of Hartford, Connecticut

5. Wu Q, Liu M, Feng Z (2008) A millimeter wave conformal phased microstrip antenna array on a cylindrical surface. In: IEEE international symposium on antennas and propagation society. pp 1–4
6. Sumalatha K, Phani Kumar CR (2019) Recent advances in conformal antennas and its design issues. Int J Innov Technol Explor Eng (IJITEE) ISSN: 2278–3075 8(9S3)
7. Wang Q, Mu N, Wang L, Safavi-Naeini S, Liu J (2017) 5G MIMO conformal microstrip antenna design. Wireless Commun Mobile Comp
8. Monica J, Jothilakshmi P (2020) A design of bandwidth-enhanced conformal antenna for aircraft applications. IETE J Res
9. Collins CM, Liu W, Wang J, Gruetter R, Vaughan JT, Ugurbil K, Smith MB (2004) Temperature and SAR calculations for a human head within volume and surface coils at 64 and 300 MHz

An Experimental Study of IEEE 802.11n Devices for Vehicular Networks with Various Propagation Loss Models

Preeti Rani and Rohit Sharma

Abstract This paper presents the experimental study of WiFi channel IEEE802.11n standard operating in the 5 GHz band under five different propagation loss models. This application is tested on two WiFi nodes with communication properties between transmitter and receiver, essential for throughput and signal strength prediction. It evaluates the throughput and signal strength of the User Datagram Protocol (UDP) traffic and its related transmission rates for communication distances. We conclude that as the distance among the WiFi nodes increases, the parameter performance decreases significantly. It has been proved that this coverage range is suitable for vehicular networks.

Keywords NS-3 · Simulation · Propagation model · Wireless channel · Vehicular network · 802.11n devices

1 Introduction

Wireless local area networks (WLAN) are broadly used because they are easy to install and can be used in many environments since they operate in an unlicensed band. The next generation of the WLAN is standardized to improve performance as IEEE 802.11n [1]. Due to the rapid development of technology, transmission and future communication entirely depend on wireless networks. Wireless networks are usually less efficient and irregular than wired networks, making quality-of-service (QoS) delivery a higher challenge for wireless communications. A wireless channel is a channel that transmits data across a specific frequency range. IEEE 802.11 WLAN, commonly referred to as WiFi, is the most broadly used broadband wireless

P. Rani (✉) · R. Sharma
Department of Electronics and Communication Engineering, SRM Institute of Science and Technology, NCR Campus, Delhi-Meerut Road, Modinagar, Ghaziabad, UP, India
e-mail: preetiresearcher1@gmail.com

R. Sharma
e-mail: rohitr@srmist.edu.in

A. K. Dubey et al. (eds.), *Advanced IoT Sensors, Networks and Systems*,
Lecture Notes in Electrical Engineering 1027,
https://doi.org/10.1007/978-981-99-1312-1_11

networking technology, having the fastest transmission rate of any standard wireless networking protocol. Depending on the IEEE standard used in the device, these frequency ranges are 2.4–2.5 GHz or 5.2–5.8 GHz. Each band is subdivided into several frequency channels [2].

There are extensive works on the throughput performance parameter of the IEEE 802.11 MAC protocol because of its popularity. Fundamental theoretical foundations have been established in this research area, concentrating on (1) flat, single-hop ad hoc networks and (2) saturated, unidirectional network traffic [3, 4, 5: p. 11, 6, 7: p. 11]. As a result, these models do not provide a valid way of analyzing the performance of flow-controlled feedback transport protocols.

The IEEE 802.11n standard was recently developed to make 802.11 faster. IEEE 802.11n utilizes a radio-frequency broadband capable of multiple-input/multiple-output (MIMO) technology. NS-3 is widely recognized and accepted by the industry and the research community as the best tool to calculate the network's performance and evaluate network efficiency. It is the most influential open-source network simulator [8]. Some validation studies attest to its accuracy with 802.11 models [9, 10, 11].

IEEE 802.11p is a wireless access standard for vehicular environments using 802.11n technology. Now that 802.11n draft-based wireless communication devices are widely available, there have not been any reports that report on the performance of these 802.11n devices [2]. Vehicle networks will likely use the following wireless technologies: IEEE 802.15.4, IEEE 802.11, and Bluetooth. The results of experiments conducted in a radio-frequency anechoic chamber are presented to explore the coexistence of IEEE 802.15.4 with IEEE 802.11 and Bluetooth. The power density is compared across the technologies using a single IEEE 802.15.4 channel [12].

IEEE 802.11p vehicular networking measurements in Northern Finland are a new standard for in-car communication [13]. Also presented are the results of a pilot testing of a developed heterogeneous networking system. Based on field measurements and pilot deployment experience, we propose a system deployment strategy with a general efficiency estimation [14]. Multiple-Node IEEE 802.11n Networks: Performance and Reliability [15] and vehicular networks used the IEEE 802.11p standard [16, 17, 13].

NS-3 is an open-source network simulator that provides a comprehensive platform for network research and studies [18]. For these reasons, we have selected NS-3 for the implementation and evaluation of this research. NS-3 provides a flexible simulation environment with various network models for varying network experiments. It is written in both C + + and Python. We present an evaluation of IEEE 802.11n devices for the vehicular network in this paper.

We created an 802.11n module in the network simulator (NS-3) platform. NS-3 is an open-source network simulator upgraded to support 802.11n (NS-3.33) in its current version.

2 Motivation

This paper presents an experimental study of UDP throughput over distance for various propagation loss models. We compare the experimental UDP throughput and received signal strength results vary the distance from 1 m up to the point where no connection is possible. In this paper, the output power of the WiFi card is fixed to 10 dBm and assumes an omnidirectional antenna with 1 dBi on both nodes.

This research looked at setting up a wireless channel in NS-3. We review the various wireless channels in NS-3 and demonstrate how to configure them in 802.11n Ad hoc WiFi simulations. None of these previous works compare the UDP throughput and signal experimental strength results for various propagation loss models based on the increased distance between the WiFi nodes. This study helps to understand the performance variations over distance. We utilized our 802.11n devices that support two spatial streams to achieve a physical data rate of 300 Mbps with a short guard interval (SGI).

3 Methodology

Wireless Channel 802.11n is represented in NS-3 by the class Yans WiFi Channel, which collaborates with the WiFi Phy class. Yans WiFi Channel Helper is a helper class in WiFi Channel. We may configure preset channels with propagation and delay models [19]. Yans WiFi Phy and Yans WiFi Channel class support propagation loss model and propagation delay model.

In NS-3, the 802.11n channel considers the following models: a constant-delay propagation model and a log-normal distribution-based propagation loss model. They are particularly named: propagation loss model and constant speed propagation delay model [19].

In this paper, we investigate five different propagation loss models:

1. Friis Propagation Loss Model.
2. Fixed Rss Loss Model.
3. Three-Log Distance Propagation Loss Model.
4. Two-Ray Ground Propagation Loss Model.
5. Nakagami Propagation Loss Model.

3.1 Propagation Loss Model

These models compute the reception power by considering the transmission power and the location of the sending and receiving antennas. The Rx signal power is calculated using propagation loss models considering the Tx signal strength and the mutual Rx and Tx antenna locations.

- *Friis Propagation Loss Model*

The Friis propagation loss model is implemented based on the equation described in Eq. (1):

$$\frac{P_r}{P_t} = \frac{A_r A_t}{d^2 \lambda^2} \tag{1}$$

For the situation of an isotropic antenna with no heat loss, use the following Eq. (2):

$$A = \frac{\lambda^2}{4\pi} \tag{2}$$

The final Eq. (3) becomes

$$\frac{P_r}{P_t} = \frac{\lambda^2}{(4\pi d)^2} \tag{3}$$

Modern extensions of the original Eq. (4) are

$$P_r = \frac{P_t G_t G_r \lambda^2}{(4\pi d)^2 L} \tag{4}$$

where P_r: reception power (W), P_t: transmission power (W), G_t : transmission gain, G_r: reception gain (unit-less), λ: wavelength (m), d: distance (m), and L: system loss (unit-less).

- *Fixed Rss Loss Model*

The Fixed Rss Loss Model maintains a consistent received power level regardless of transmission power. The received power remains constant irrespective of the transmit power; the user must configure the received power level.

- *Three-Log Distance Propagation Loss Model*

The three-log distance propagation loss model implements a log-distance path loss propagation model with three distance fields. It's the same concept as Log Distance but considers three distances (near, medium, and distant) with different exponents.

The reception power inside each field is determined using the long-distance propagation Eq. (5):

$$L = L_0 + 10 \times n_0 \log_{10}\left(\frac{d}{d_0}\right) \tag{5}$$

The transmission power is returned when the path loss is requested at a distance smaller than the reference distance d_0. The reference distance is set to 1 m by default, and the reference loss is set to 5.15 GHz in Friis Propagation Loss Model, resulting in $L_0 = 46.67$ dB. The only difference between this model and ns3:: Log Distance Propagation Loss Model is that it contains three distance variables with distinct exponents: close, medium, and distant.

- *Two-Ray Ground Propagation Loss Model*

It uses a loss propagation model with a direct ray in line of sight and a reflected ray to Earth. The formula is used in the two-ray ground reflection model mentioned in Eq. (6).

$$P_r = \frac{P_t \times G_t \times G_r \times \left(H_t^2 \times H_r^2\right)}{d^4 \times L} \tag{6}$$

The original equation by [20] assumes L = 1. L is included here to be consistent with the free-space equation. H_r and H_t are set at the appropriate node's z-position, together with a model parameter specified using Set Height Above Z.

Due to the oscillation generated by the constructive and destructive combination of the two beams, the two-ray model does not produce excellent results across short distances. For short distances, the Friis free-space model is employed instead.

Friis is utilized below the crossover distance, which is determined in Eq. (7):

$$d_{cross} = \frac{4 \times \pi \times H_t \times H_r}{\lambda} \tag{7}$$

- *Nakagami Propagation Loss Model*

The Nakagami model implements the Nakagami-m fast fading model, accounting for signal intensity changes caused by multipath fading. Because the model does not account for the path loss caused by the signal's distance traveled, it is advised to be used in conjunction with other models [19].

4 Result Analysis and Discussion

The WiFi simulation with IEEE802.11n standard operating in the 5 GHz band. In this simulation, consider two WiFi nodes, and the distance between both the nodes is an important parameter. The output of this simulation observed a variety of distances from 1 m up to the point where no connection is possible. UDP traffic with an application layer packet size of 1450 bytes is used for the performance of the present scenario are measured based on the two different parameters (Table 1).

A. **UDP Throughput**

Throughput is the rate at which data can be transported by the mechanism and quantified data sent/received by a unit of time. Throughput is measured in Mbit/Sec as per Eq. (8):

Table 1 Default parameters

Propagation model	Default parameters
Friis propagation loss model	Wave length: 58.25 mm Minimum distance: 0.5 m System loss: 1
Fixed RSS loss model	Rx signal strength: −150 dBm
Three-log distance propagation loss model	Reference loss at 1 m: 46.67 dB Distances: 1, 200, 500 m Exponents: 1.9, 3.8, 3.8
Two-ray ground propagation loss model	Wave length: 58.25 mm Minimum distance: 0.5 m System loss: 1 Height above Z: 0 m
Nakagami	Exponents: 1.5, 0.75, 0.75 Distances: 80, 200 m

$$UDP\ Throughput = \frac{Total\ UDP\ Packets \times\ Packet\ Size \times 8}{Simulation\ time \times 10^6} \tag{8}$$

B. **Received Signal Strength (RSS)**

The signal strength is reported by the physical layer of the receiving WiFi card. The received signal strength is measured in dBm. The following formula measures the RSS as per Eq. (9):

$$\text{RSS} = P_t - PL(d_0) - 10 \propto \log_{10}\frac{d}{d_0} + \text{X}_{\sigma_{RSS}} \tag{9}$$

where $PL(d_0)$ is the power loss in dBm and $\text{X}_{\sigma_{RSS}}$ is a random variable representing the noise in the measured RSS in dBm. It is often expected to be a zero-mean Gaussian random variable with variance RSS.

Figures 1 and 2 depict the UDP throughput based on the distance (m) between the WiFi node. As the distance between the WiFi node increases from 1 to 50 m, throughput is significantly reduced. As per Figs. 1 and 2, UDP throughput over distance did not differ. All five propagation loss model shows a similar value of UDP throughput over distance.

Like UDP throughput, received signal strength (RSS) varies with varying distances. It is seen in Figs. 3 and 4 for the five different propagation models. The distance is shown on the x-axis and the signal strength is shown on the y-axis.

Figures 5 and 6 show the results of the indoor communication performance tests, respectively. Applying the well-known power loss model shows how average throughput and signal strength decrease as distance increases. At 1 m distance has an average throughput of 240 Mbps and a maximum rate of 250 Mbps. It is possible to explain the higher values at 35 m due to reflection in the test environment. In further tests, we determined the maximum throughput be approximately 90 Mbps at

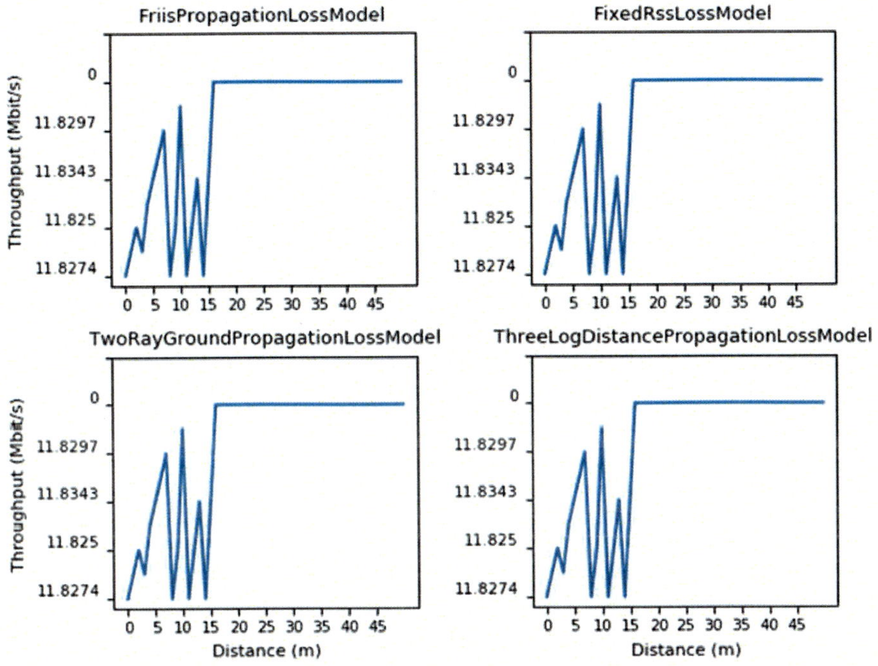

Fig. 1 UDP throughput (Mbit/Sec) based on various propagation loss models' different distances (m)

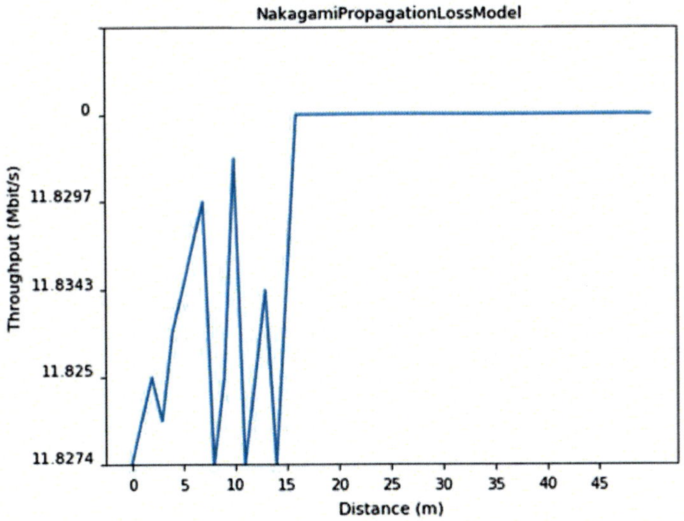

Fig. 2 UDP throughput (Mbit/Sec) based on the different distances (m) for the Nakagami propagation loss model

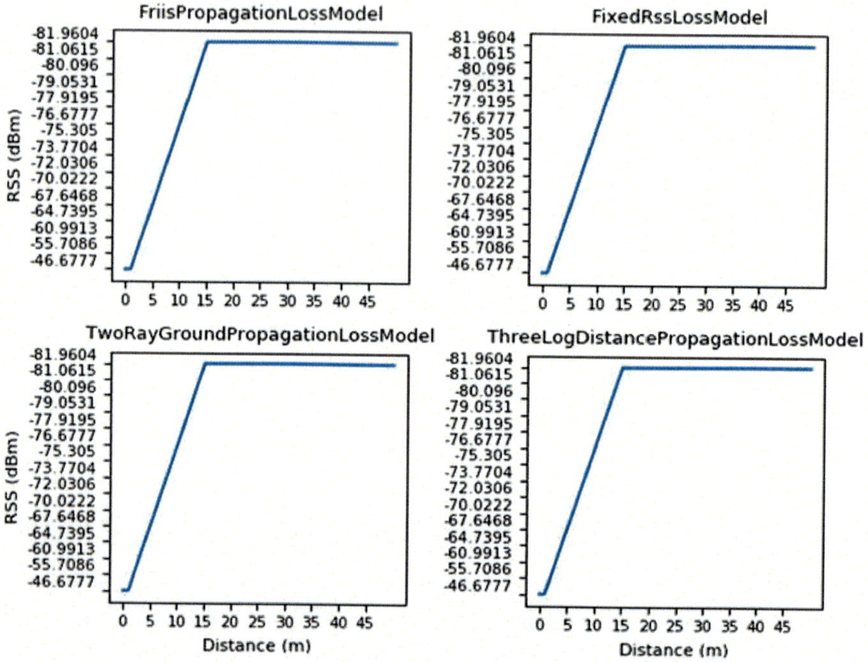

Fig. 3 RSS (dBm) based on various propagation loss models' different distances (m)

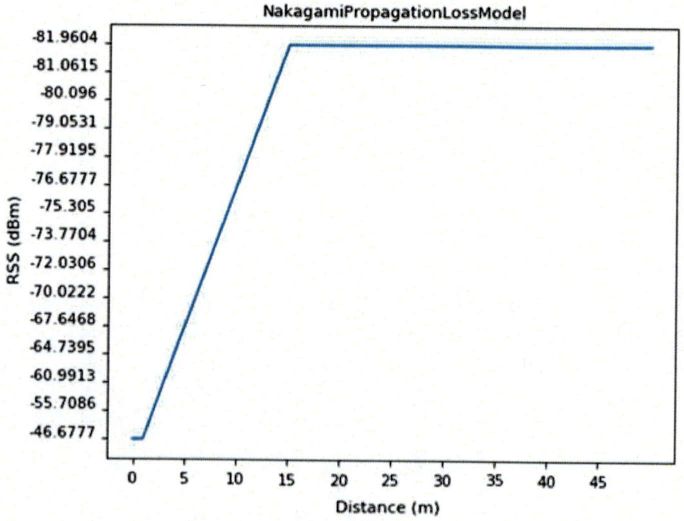

Fig. 4 RSS (dBm) based on the different distances (m) for the Nakagami propagation loss model

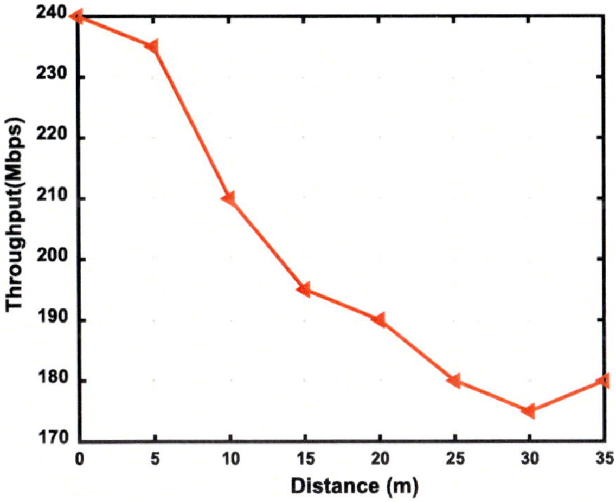

Fig. 5 Average throughput (Mbps) performance versus distance (m) for indoor communication

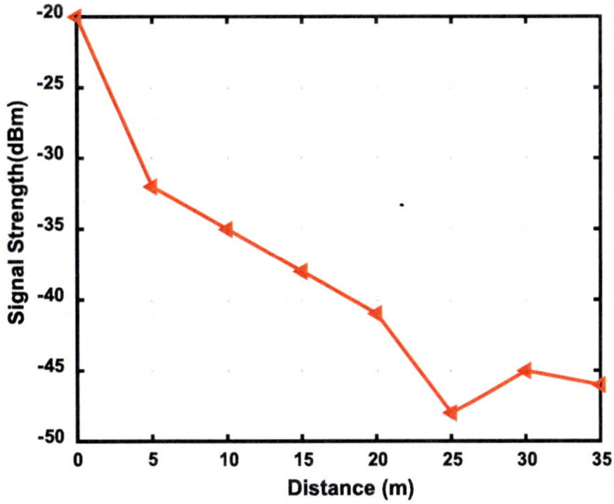

Fig. 6 Average signal strength (dBm) performance versus distance (m) for indoor communication

a distance of 100 m. It indicates that 802.11n devices can communicate indoors with high-throughput rates over long distances.

5 Conclusion

While these studies are preliminary, they indicate that even under controlled and almost ideal conditions, measured UDP throughput and signal strength of 802.11n tend to oscillate. The UDP throughput and signal strength for 802.11n are our analysis. Our measurements extend previous work on UDP traffic performance over various propagation loss models in many directions. We demonstrated that the coverage range is suitable in a vehicular network scenario.

References

1. Obayashi S (2006) Standardization and technical proposals of IEEE 802.11 n. IEICE Tech Rep 106(43):25–30
2. Matsumoto A, Yoshimura K, Aust S, Ito T, Kondo Y (2009) Performance evaluation of IEEE 802.11 n devices for vehicular networks. In: 2009 IEEE 34th conference on local computer networks. pp 669–670
3. Bianchi G (2000) Performance analysis of the IEEE 802.11 distributed coordination function. IEEE J Select Areas Commun 18(3):535–547. https://doi.org/10.1109/49.840210
4. Bruno R, Conti M, Gregori E (2002) Optimization of efficiency and energy consumption in p-persistent CSMA-based wireless LANs. IEEE Trans Mob Comput 1(1):10–31. https://doi.org/10.1109/TMC.2002.1011056
5. Cali F, Conti M, Gregori E (2000) Dynamic tuning of the IEEE 802.11 protocol to achieve a theoretical throughput limit. IEEE/ACM Trans Netw 8(6):785–799. https://doi.org/10.1109/90.893874
6. Kumar A, Altman E, Miorandi D, Goyal M (2005) New insights from a fixed-point analysis of single cell IEEE 802.11 WLANs. In: Proceedings IEEE 24th annual joint conference of the IEEE computer and communications societies, vol 3. pp 1550–1561. https://doi.org/10.1109/INFCOM.2005.1498438
7. Tay YC, Chua KC (2001) A capacity analysis for the IEEE 802.11 MAC protocol. Wireless Netw 7(2):159–171
8. Khan AR, Bilal SM, Othman M (2012) A performance comparison of open source network simulators for wireless networks. In: 2012 ieee international conference on control system, computing and engineering. pp 34–38. https://doi.org/10.1109/ICCSCE.2012.6487111
9. Jonsson A, Akerman D, Fitzgerald E, Nyberg C, Priyanto BE, Agardh K (2016) Modeling, implementation and evaluation of IEEE 802.11ac in NS-3 for enterprise networks. In: 2016 Wireless Days (WD). pp 1–6. https://doi.org/10.1109/WD.2016.7461452
10. Khosroshahy M (2006) Study and implementation of IEEE 802.11 physical channel model in yans (ns3 prototype) network simulator. INRIA Report 1–61
11. Ravindranath NS, Singh I, Prasad A, Rao VS (2016) Performance evaluation of IEEE 802.11 ac and 802.11 n using NS3. Indian J Sci Technol 9(26):1–9
12. Zacharias S, Newe T, O'Keeffe S, Lewis E (2012) Coexistence measurements and analysis of IEEE 802.15.4 with WiFi and bluetooth for vehicle networks. In: 2012 12th International conference on ITS telecommunications. pp 785–790. https://doi.org/10.1109/ITST.2012.6425289
13. Rani P, Hussain N, Khan RAH, Sharma Y, Shukla PK (2021) Vehicular intelligence system: time-based vehicle next location prediction in software-defined internet of vehicles (SDN-IOV) for the smart cities. In: Al-Turjman F, Nayyar A, Devi A, Shukla PK (eds) Intelligence of Things: AI-IoT based critical-applications and innovations. Springer International Publishing, pp 35–54. https://doi.org/10.1007/978-3-030-82800-4_2
14. Sukuvaara T, Ylitalo R, Katz M (2013) IEEE 802.11p based vehicular networking operational pilot field measurement. IEEE J Select Areas Commun 31(9):409–417. https://doi.org/10.1109/JSAC.2013.SUP.0513037
15. Lee S, Lim A (2012) Reliability and performance of IEEE 802.11n for vehicle networks with multiple nodes. In: 2012 International conference on computing, networking and communications (ICNC). pp 252–256. https://doi.org/10.1109/ICCNC.2012.6167422
16. Hussain N, Rani P (2020) Comparative studied based on attack resilient and efficient protocol with intrusion detection system based on deep neural network for vehicular system security. In: Distributed artificial intelligence. CRC Press, pp 217–236

17. Hussain N, Rani P, Chouhan H, Gaur US (2022) Cyber security and privacy of connected and automated vehicles (CAVs)-based federated learning: challenges, opportunities, and open issues. In: Federated learning for IoT applications. Springer, pp 169–183
18. Narayan S, Jayawardena C, Wang J, Ma W, Geetu G (2015) Performance test of IEEE 802.11ac wireless devices. In: 2015 International conference on computer communication and informatics (ICCCI). pp 1–6. https://doi.org/10.1109/ICCCI.2015.7218076
19. Propagation Loss Models NS-3 (2016) (n.d.). https://www.nsnam.org/docs/models/html/propagation.html
20. Rappaport TS (2002) Wireless communications—principles and practice, (The Book End). Microw J 45(12):128–129

Cloud Computing, 5G and IoT

Performance Evaluation of Genetic Algorithm and Flower Pollination Algorithm for Scheduling Tasks in Cloud Computing

Pardeep Singh, Gourav Bathla, Deepak Panwar, Alankrita Aggarwal, and Shivani Gaba

Abstract Resource management during the cloud computing process is one of the prominent issues currently under research. The reason behind this is to procure an optimal utilization of these resources with the least cost and time. Allocation and scheduling of users' tasks over the servers is an NP-hard problem, which can be solved by applying various heuristic and meta-heuristic approaches. With this motive, the paper has analyzed the performance of two meta-heuristic approaches, i.e., Genetic Algorithm (GA) and Flower Pollination Algorithm (FPA) in scheduling the tasks over virtual machines (VM). Tasks to VMs allotment is mainly responsible for the utilization of virtual resources which are actually running over physical machines (PM) or servers. The prime parameter considered in the research is makespan which indirectly measures the consumption of resources for performing a set of tasks. Simulation for the research is performed for a defined set of tasks that need to be allocated over a particular number of servers using the cloudsim tool. Results are measured and analyzed for both the algorithms applied in the scheduling process and also a default approach of first come first serve (FCFS) is compared with the overall results. Analyses prove the GA approach overperforms the FPA in reducing the makespan in various configurations considered.

P. Singh
Computer Science & Engineering, Graphic Era Hill University, Dehradun 248001, India

G. Bathla (✉)
School of Computer Science, University of Petroleum and Energy Studies, Dehradun, Uttrakhand 248001, India
e-mail: gouravbathla@gmail.com

D. Panwar
Computer Science & Engineering, Manipal University Jaipur, Jaipur, India

S. Gaba
Depatment of Computer Science, PIET, Panipat, Haryana, India

A. Aggarwal
Department of AML, Chandigarh University, Mohali, India

A. K. Dubey et al. (eds.), *Advanced IoT Sensors, Networks and Systems*,
Lecture Notes in Electrical Engineering 1027,
https://doi.org/10.1007/978-981-99-1312-1_12

Keywords Task scheduling · Cloud computing · Genetic algorithm · Flower pollination

1 Introduction

Emerging as a novel intelligence computing technique, the technology of Cloud Computing is in high demand for the software industry cause of its high usefulness and applicability in multiple sectors nowadays such as telecommunication, medical, education, engineering, services, manufacturing, and research. For example, educational clouds can provide education services virtually through cloud applications to a variety of learners living in urban or rural areas [1]. Not only the novel startups in the IT industry are getting benefits from the services of this sophisticated computing methodology but also the old regimes gradually adopting this online available infrastructure for availing a better service to their users of various fields which is more reliable, scalable and economical than the traditional offline structure. Due to this, data centers comprising the physical infrastructure for the cloud environment are expanding exponentially which mainly consists of huge numbers of server stacks handling the computation load of numerous clients. Further, these servers are virtually divided into various virtual machines which can be created, modified, and destroyed as per the user's real-time requirements [2]. Hence, the number of VMs required is depending upon the number of tasks at any instant and also the methodology of assigning these tasks over these available VMs. Thus, the task scheduling approach plays a prominent role in the optimal management of virtual resources by efficiently assigning tasks to them.

Scheduling a given number of tasks over available machines is an NP-hard problem, in which it is difficult to find an exact solution, however, meta-heuristic approaches have been very useful in providing nearly optimal solutions to the scheduling problem. In this research work, two meta-heuristic algorithms, i.e., Flower Pollination Algorithm (FPA) and Genetic Algorithm (GA) have been introduced to allocate resources to available tasks. Both of these algorithms are nature-inspired algorithms, extended for multi-objective optimization by calculating the fitness value. The purpose of the task scheduling process using FPA and GA is to find an optimized mapping between tasks and resources so that makespan could be minimized. The makespan in a cloud environment means the time taken by the virtual machines to process all the tasks (cloudlets) assigned to them. The assignment of the tasks in form of cloudlets to VMs is another field of research considered by a researcher to achieve different objectives [3]. We will be implementing the above-mentioned algorithms to optimally allocate the cloudlets to VMs. One of the key challenges of cloud computing is scheduling tasks over suitable resources such that the resources can be efficiently utilized. To ensure efficient management of resources in the cloud, a good task scheduling algorithm must be deployed in the task scheduler. An optimal task scheduling in the cloud refers to the process of allocating the available resources to the user's task in an efficient manner that tends to achieve the goal. Various task scheduling algorithms exist in the literature that varies depending upon the type of objective parameter that they try to optimize.

The paper has been organized into six sections, after the introduction, it defines the basic system model of cloud computing that assist in understating the problem of task scheduling in the field. The next section briefs about GA and FPA approaches as applied to finding the optimal solution. Details about the simulation environment and experimentation are defined in the further section followed by results analysis and finally conclusion and future scope at the last.

2 System Model

Figure 1 is defining the cloud architecture with various layers, which includes the resource layer interacting with the layer of consumer objects on one end and layers of hosts or physical machines (PMs) on another. Users availing the cloud services, though the gadgets and other computational devices produce the tasks for computation over the cloud which in form of a cloudlet passes to a particular data center. The data center broker is an agent handling these cloudlets and assigns them further to a particular VM for computation depending upon the scheduling policy applied. This policy is responsible for managing the balanced distribution of dynamically varying workloads among the available VMs with the goal of achieving various objectives prioritize by the policy. On the upper layer, Virtual Machines are dynamically configured by the hypervisor or Virtual Machine Manager (VMM) using the virtualization concept as per the configuration required for handling a workload. In this phenomenon, not only a PM can be segregated into multiple VMs but also multiple PMs can be aggregated into a VM. Figure 1 is only representing the segregation concept of dividing a host into many VMs.

However, the work has focused mainly on the task scheduling process, as the allotment of VMs over hosts is managed by another concept of VM consolidation which is a separate research area [2]. Considering the problem of scheduling as NP-hard, an optimal solution can find out using various heuristic or meta-heuristic approaches. A solution for the scheduling approach basically, defines the mapping among the VMs and Tasks (active) at any instant. So, considering a solution for the problem according to any meta-heuristic or nature-inspired approach and defining an objective function for minimizing or maximizing the objective can lead to an optimal solution. Following this methodology, two nature-inspired algorithms are applied to find the optimized VMs-to-tasks mapping in order to reduce the makespan. The next two sections are detailing about the approaches.

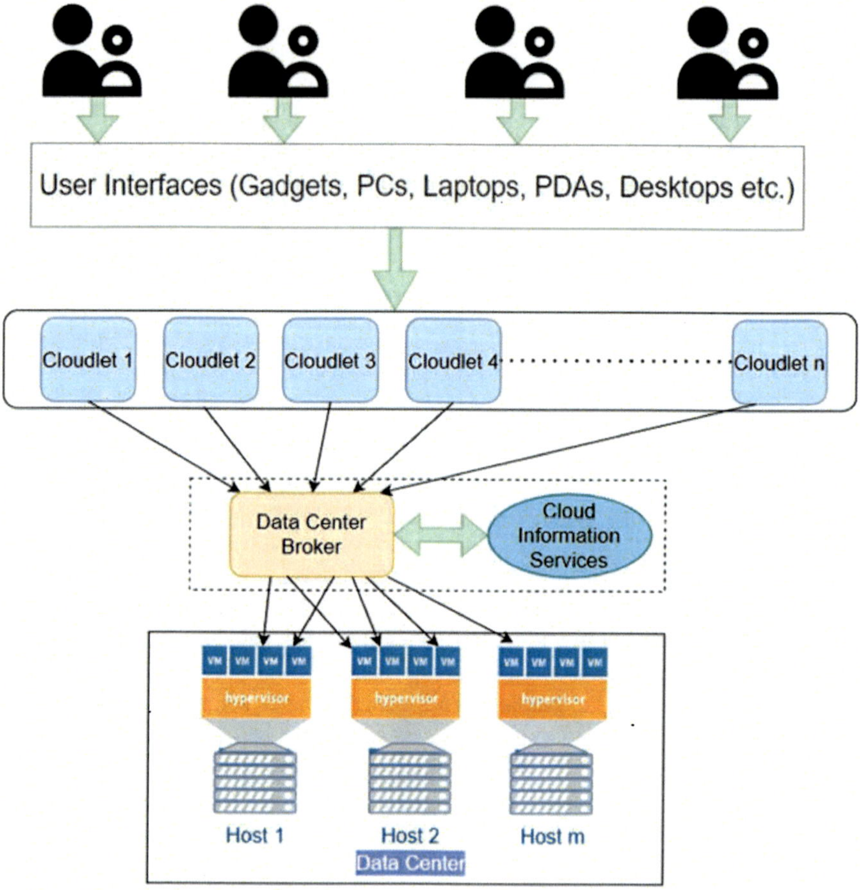

Fig. 1 Cloud computing system architecture

2.1 *Flower Pollination Algorithm (FPA)*

FPA algorithm is inspired by the process of pollination in plants and flowers, which can be divided into two types mainly, i.e., biotic or cross-pollination and abiotic or self-pollination. The first kind is basically, dependent upon biotic pollinators such as birds and insects where pollinators travel long distances having levy flight properties leading to global pollination [4, 5]. The latter type is local pollination which doesn't require external pollinators and primarily wind is responsible for making this short distant pollination [6]. This natural process has inspired the developers to process any computational problem by finding the solution based on a novel evolutionary approach. Initially, an algorithm based on flower pollination by bees named FPAB was developed by [7]. Further, FPA based on flower reproduction is developed by [4], that compared the approach with GA and PSO approaches. As a swarm-based algorithm, it can be applied to various real-world optimization problems in different fields such as wireless and wired networking, electrical and mechanical engineering problems, signal and image processing, meteorology, etc. [6].

Figure 2 is describing the basic flow chart of FPA. The process of FPA starts with initializing the controlling parameters, i.e., size of the population, scaling parameter, and switching probability. Switching probability is responsible for deciding the selection of global or local pollination operations for individual pollen in the population. Further, a set of solutions or pollens called initial population with the defined size is created for the given problem. Every solution in the generation must match the format where operators of FPA can be applied. The whole process loops in a predefined number of iterations that terminated with defined stopping criteria. In each iteration, local or global pollination is performed for every solution that existed in the population. In local pollination, decision parameters of the current solution lead towards two random solutions extracted from two pollens of the population, whereas, in global pollination, it leads towards the best global solution.

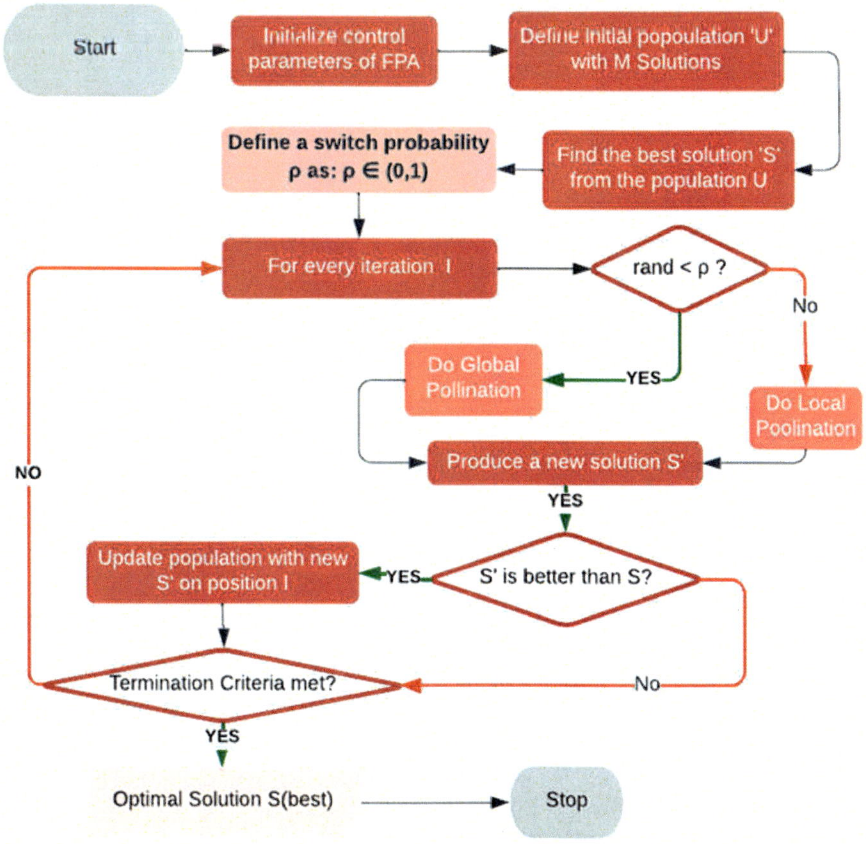

Fig. 2 Flowchart of flower pollination algorithm

2.2 Genetic Algorithm

Define in Fig. 3, GA is applied to a problem by defining an encoded solution for the problem called a chromosome. Initially, a set of such chromosomes called the initial population is generated. Another main requirement to apply GA is defining a fitness function for the same problem, which is able to evaluate and provided a fitness value for a chromosome. This fitness function is directly related to the objectives required to be achieved while solving the problem, e.g., for a traveling salesman problem a shorted path as an encoded solution can be evaluated for minimum value by a fitness function [8]. Similarly, in task scheduling problem, a mapping of tasks and VMs as chromosome is evaluated for minimum makespan [9], cost, and energy [10, 11]. After selecting the suitable solutions based on different selection techniques, the crossover operator is applied to the chromosomes for generating the next generation. The mutation is an optional operation applied to chromosomes after crossover. This cycle of fitness check, selection, and crossover operations further on new generation continues till the convergence or termination criteria are not met.

3 Literature Review

Many research works have been proposed and developed for task scheduling by various researchers. These researches can be categorized mainly into three categories, i.e., heuristic approaches, meta-heuristic approaches, and hybrid approaches. In the heuristic approach, some investigative concept is applied to the problem for finding the solution, in the meta-heuristic application any nature-inspired algorithm

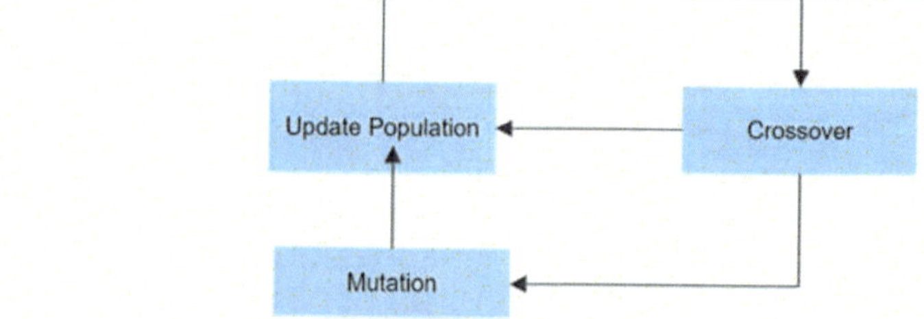

Fig. 3 Process of genetic algorithm

or evolutionary approach is applied to the problem and hybrid application is a combination of both first two types. This work focuses mainly on analyzing two primary meta-heuristic approaches, i.e., GA and FPA, and analyses their application for task scheduling in a common environment. So, some researches performed in field in these approaches are reviewed further.

Authors in [12], considered the task scheduling problem as a Directed Acyclic Graph (DAG) and developed an approach of representing the schedule accordingly for applying GA. The work has tried to solve the static task scheduling problem by implementing the standard GA and then modifying the fitness function two times to define CPGA and TDGA. CPGA focused on reducing the scheduling length and prove to perform better than MCP [13], whereas TDGA has fitness function to reduce communication delay by task duplication.

Burak Tagtekin et al. [14] applied the GA on job scheduling process to schedule jobs in a pipeline. It aims to reduce makespan and load balancing by automatically prioritizing the tasks using GA and accordingly modifying the schedule and allocating jobs to processor. Experiments performed in three cases have justified the proposed approach. Kaur et al. developed a modified GA by adding outputs of LCFP (Longest Cloudlet to Fastest Processor) and SCFP (Shortest Cloudlet to Fastest Processor) approaches with some random solution in initial population. Evolution is made by choosing solutions with minimum makespan and minimum cost for next generation and total 30 iterations are executed in java-based simulators. However, it represents a static scheduling and not considering the dynamic mapping of tasks in the schedule. Also, it not considered any priority parameter for multiple users for the tasks during the allocation [9].

Sahraei et al. also modified the GA approach to reduce makespan and cost. The authors encoded the jobs-machine mapping for GA and developed a fitness function considering both makspan and cost in different scenarios. Compared the results with traditional MAX–MIN and MIN–MIN approaches and justified the results obtained for the considered goals [15].

Kaur et al. [16] proposed a new approach for Task Scheduling using Flower Pollination Algorithm (TSFPA) to allocate resources to task in cloud environment. The results were compared with that of genetic algorithm, FCFS, and round-robin approach of scheduling using cloudsim toolkit. Gupta et al. [17] implemented the Flower Pollination Algorithm for task scheduling problem to find an optimal solution of mapping between VMs and tasks, and compared various parameters including makespan and resource utilization with other nature-inspired approaches, i.e., GA and Particle Swarm Optimization (PSO).

Bezdan et al. [18] also considered Task Scheduling (TS) as an NP-hard problem and accordingly applied a nature-inspired meta-heuristic approach of FPA with some enhancements to find an optimal solution with minimum makespan. Results of FPA application in TS for makespan reduction are compared with other similar meta-heuristic approaches, e.g., ACO, PBACO, Min–Min, and FCFS; however, more similar approaches can also be compared to find the best out of them.

Malti et al. [19] modified the objective function of FPA for three different objectives, i.e., makespan, cost, and reliability of task execution. The authors implemented the proposed algorithm on cloudsim with a specific configuration and compared it with Pareto-based TOPSIS and aggregation approach for all the three objectives.

4 Problem Statement

Scheduling of tasks or cloudlets over the available virtual machines to achieve various objectives is considered as an optimization problem. The problem is to minimize or maximize a particular parameter as an objective without affecting other parameters, e.g., reducing the makespan of tasks without affecting load balancing. This work has analyzed the implementation of novel approaches for reducing makespan while leaving the effect analysis over other parameters as a future perspective. Time to execute all the tasks by available computing resources is considered as makespan and it directly represents the performance of cloud services. The phenomenon behind reducing the makespan is directly proportional to the extent to which resources are utilized. More unutilized resources more the waiting time for tasks and hence more the makespan. So, makespan can be defined as inversely proportional to the rate of utilization of virtual machines. An optimal solution for the problem tends to increase this rate of utilization in order to decrease the makespan. The problem of mapping tasks over VMs can be more clarified mathematically as

- Let $V = \{V_1, V_2, V_3 \ldots\ldots V_m\}$ $\; i.e. m\, number\, of\, virtual\, machines$
- Let $T = \{T_1, T_2, T_3 \ldots\ldots V_n\}$ $\; i.e. n\, number\, of\, tasks\, and$ $n\, is\, integer\, multiple\, of\, m.$
- Let $E = \{E_1, E_2, E_3 \ldots\ldots E_n\}$ where E_i is the execution time of ith task including the intial setup time and processing time with the condition $E_1 \leq E_2 \leq E_3 \leq \cdots \leq E_n$.
- Also, it is assumed that one VM can execute one task at a time.
- Two VMs can't execute the same task.
- All the tasks are of equal priority.
- Aim is to minimize the makespan $M = \sum_{j=1}^{n} E_j$.

5 Methodology

The goal of the work is to implement and compare GA and FPA approaches to the problem of task scheduling in the cloud computing environment. The methodology used for achieving the goal is to implement both approaches in a cloud environment with a common configuration. For this, a cloud simulator named cloudsim [20] is used and modifications in the simulator code are performed one by one for both the algorithms. Details about the configurations applied in the experiments are mentioned in the further section and below is the pseudocode for both algorithms as per the cloudsim simulator:

Table 1 Variables used in Psudeocode for GA and FPA

Variable	Purpose	Variable	Purpose
Vmsize	Total number of VMs	cloudletId	Identity of a cloudlet
cloudletsize	Total number of cloudlets	cloudletList	List of available cloudlets
vmcapacity	Threshold capacity of each VM	vmExecutionTime	Time taken by a VM to execute a cloudlet
solcount	Total number of solutions in the population	3d arraylist	Data structure to store initial generated population

5.1 Flower Pollination Algorithm Psudeocode

```
Step 1: for i=0 to solcount-1
        Initialize a new row to 3d arraylist.
        for j=0 to cloudletsize-1
                get clouletId for given index from cloudletList
                generate a random bin between 0 to m=vmsize-1
                if(size of each vm<=vmcapacity)
                        assign the given cloudlet to selected vm.
                        Increase the size the selected vm by1.
                else generate other vm and repeat the process
        if(i==solcount)
                stop the algorithm.
       Create a 2d array "fitness" to store solution no and fitness of that solution.
Step 2: Calculate the fitness of each solution in the population.
        for i=0 to solcount-1
                fitness[i][0]=i;
                for j=0 to vmsize-1
                        get mips for particular vm
                        for k=0 to size of jth vm-1
                                calculate total cloudlet length for jth vm.
                        Duration for jth vm=mips of jth vm/total cloudlet length
                fitness[i][1]=max(duration for each vm of solution)
Step 3. Find the best solution from given population
        for i=0 to solcount-1
                find the solution with best fitness.
        store the best solution from population into a 2d arraylist.
```

Step 4. Generate new solutions

```
    Create a new 2d arraylist "newsol" to store new solutions.
    for i=0 to solcount-1
        Generate a random number "x" between 0 and 1.
        if(x<=0.8)
            generate a random number "sol2" between 0 and solcount-1;
            if(sol2!=i)
                perform local pollination between ith sol and sol2 of population
                generate a random number "vmno" between 0 and vmsize-1;
                for j=0 to vmsize-1
                    if(j<=vmno)
                        store the elements from ith solution to newsol
                    else
                        store the elements from sol2 to newsol
        else
            perform global pollination between ith solution and best solution
            generate a random number "vmno" between 0 and vmsize-1;
            for j=0 to vmsize-1
            if(j<=vmno)
                store the elements from ith solution to newsol
            else
                store the elements from best solution to newsol
    repair newsol using step5.
    if(new solution fitness<fitness of ith solution)
        replace ith solution with new solution and update its fitness
    if(new solution fitness<best solution fitness)
        replace best solution with new solution and update the best solution no.
```

Step 5. Repair newsol.

```
    Create a 2d arraylist "cloudletassign" to check cloudlet assignment to vms
    for i=0 to vmsize-1
        for j=0 to size of given vm-1
            add the vm number of cloudlets to the list.
    Check the number of vms on which a particular cloudlet is assigned.
    for i=0 to cloudletsize-1
        find the size "s" of the cloudletassign list at position i;
        if(s==2)
            delete the cloudlet i from one of the 2 vms using step6.
    for i=0 to cloudletsize-1
        find the size "s" of the cloudletassign list at position i;
        if(s==0)
            assign the cloudlet i again to a vm using step7.
```

Step 6. Delete the cloudlet assigned to 2 vms from one of the 2 vms.

```
    Calculate the vmExecutionTime for each vm.
    if(vmExecutionTime of 1st vm >=vmExecTime of 2nd vm)
        delete the cloudlet from 1st vm.
    else if(vmExecutionTime of 2nd vm >=vmExecTime of 1st vm)
        delete the cloudlet from 2nd vm.
```

Step 7. Assign the unassigned cloudlet to a vm

```
    Calculate the vmExecutionTime for each vm and sort the list in ascending order.
    for j=0 to vmsize-1
        if the vm having least exec time is not overflowed
        assign the cloudlet to that vm.
        break;
```

Step 8. Run the program for multiple iterations and find the final solution

5.2 Genetic Algorithm Psudeocode

```
Step 1. for i=0 to solcount-1
        Initialize a new row to 3d arraylist.
        for j=0 to cloudletsize-1
                get clouletId for given index from cloudletList
                generate a random bin between 0 to m=vmsize-1
                if(size of each vm<=vmcapacity)
                        assign the given cloudlet to selected vm.
                        Increase the size the selected vm by1.
                else generate other vm and repeat the process
        if(i==solcount)
                stop the algorithm.
Create a 2d array "fitness" to store solution no and fitness of that solution.
Step 2. Calculate the fitness of each solution in the population.
        for i=0 to solcount-1
                fitness[i][0]=i;
                for j=0 to vmsize-1
                        get mips for particular vm
                        for k=0 to size of jth vm-1
                                calculate total cloudlet length for jth vm.
                        Duration for jth vm=mips of jth vm/total cloudlet length
                fitness[i][1]=max(duration for each vm of solution)
Step 3. Generate new solutions by crossover
        Create 2 new 2d arraylist "newsol1" and "newsol2" to store new solutions.
        for i=0 to solcount-1
                for j=i+1 to solcount-1
                        perform crossover between ith sol and jth sol of population
                        generate a random number "vmno" between 0 and vmsize-1;
                        for k=0 to vmsize-1
                                if(j<=vmno)
                                        store the elements from ith solution to newsol1
                                        store the elements from jth solution to newsol2
                                else
                                        store the elements from jth to newsol1
                                        store the elements from ith solution to newsol2
        repair both solutions using step4 and add them to initial population.
Step 4. Repair newsol.
        Create a 2d arraylist "cloudletassign" to check cloudlet assignment to vms
        for i=0 to vmsize-1
                for j=0 to size of given vm-1
                        add the vm number of cloudlets to the list.
        Check the number of vms on which a particular cloudlet is assigned.
        for i=0 to cloudletsize-1
                find the size "s" of the cloudletassign list at position i;
                if(s==2)
                        delete the cloudlet i from one of the 2 vms using step5.
        for i=0 to cloudletsize-1
                find the size "s" of the cloudletassign list at position i;
                if(s==0)
                        assign the cloudlet i again to a vm using step 6.
```

```
Step 5. Delete the cloudlet assigned to 2 vms from one of the 2 vms.
        Calculate the vmExecutionTime for each vm.
        if(vmExecutionTime of 1st vm >=vmExecTime of 2nd vm)
                delete the cloudlet from 1st vm.
        else if(vmExecutionTime of 2nd vm >=vmExecTime of 1st vm)
                delete the cloudlet from 2nd vm.
Step 6. Assign the unassigned cloudlet to a vm
        Calculate the vmExecutionTime for each vm and sort the list in ascending order.
        for j=0 to vmsize-1
                if the vm having least exec time is not overflowed
                        assign the cloudlet to that vm.
                        break;
Step 7. Modify existing solutions by mutation.
        for i=0 to Population.size-1
                generate a random number "x" between 0 and 1.
                if(x>0.5)
                        generate two random numbers a and b between 0 and vmsize-1
                        if(a!=b)
                                generate a random number c between 0 and size of ath vm in solution.
                                generate a random number d between 0 and size of bth vm in solution.
                                Interchange cth and dth cloudlets in the solution.
                        else continue.
Step 8. Copy the best solcount number of solutions from initial population into a 3d arraylist.
        Clear the initial population.
        Copy back the best solcount solutions from temporary arraylist to initial population.
Step 9. Run the program for specific number of times and print the best solution at end.
```

6 Experimentation and Results

6.1 Experiment Setup

The basic examples provided in the cloudsim simulator library provide a specific simulation environment for cloud computing where data centers holding physical machines (PMs) with a defined configuration are virtually divided into VMs and tasks in the forms of cloudlets allocated over these available VMs. The configuration of PMs, VMs, and cloudlets can be modified by the experimenter according to the needs of the experiment but within a specified limit of the system in use. Different files in the cloudsim core library imitate various components of data centers, e.g., data center, broker, VM Scheduler, cloudlet Scheduler, Host Allocator, VM Allocator, etc. This facilitates a programmer to modify a specific phenomenon of a component, execute it and test its outcomes in various conditions. The work has modified the cloudlet scheduling algorithm defined by the broker component and implemented two meta-heuristic approaches, i.e., FPA and GA one by one under a similar configuration. Experiments are performed to evaluate the makespan under two types of circumstances, first keeping the VMs fixed and analyzing the effect of an increasing number of cloudlets, and another by keeping the cloudlets fixed and analyzing the makespan under increasing order of VMs. In the first case, 12 VMs have been configured seven times, i.e., with 20, 40, 60, 80, 100, 500, and 1000 cloudlets. Later, the cloudlets are fixed at the count of 100, and VMs' count is increased five times, i.e., 10, 20, 30, 40, and 50. Both the cases of experiments are not only executed with the default scheme of First Come First Serve (FCFS) but also revised two times for

Table 2 Makspan obtained in the first phase of experiments

No of cloudlets	No of Vms	FCFS	FPA	Genetic algorithm
20	12	2.78	2.68	2.64
40	12	5.48	5.276	5.046
60	12	8.12	7.78	7.558
80	12	11.55	10.95	10.61
100	12	15.28	14.44	14.12
500	12	178.34	166.99	158.47
1000	12	598.45	560.10	540.11

FPA and GA approaches to obtain the value of makespan by every simulation. The obtained results are shown in Tables 2 and 3 for both experiment types.

6.2 Resuts and Analysis

Table 2 displays the value of makespan obtained in all the seven configurations under the three approaches with fixed VMs and varying cloudlets. Results are clearly showing the difference among all the three approaches and this can be analyzed from the table that the value of makespan is directly proportional to the number of cloudlets in all the cases cause more cloudlets will increase the burst time of task to be completed. Also, FPA is optimizing the makespan in comparison to the FCFS approach, however, GA is outperforming both the FPA and FCFS in all the configurations by making a less makespan value for similar conditions.

The line graph shown in Fig. 4 is mainly comparing the performance of applying FPA and GA approaches with FCFS for task scheduling graphically in all the simulations performed, where the red line represents the performance difference between GA and FCFS, and the blue line is representing the same among FPA and FCFS. The horizontal axis in the graph corresponds to the number of cloudlets and the vertical axis represents the percentage difference in the value of makespan. This is evitable that GA has outperformed the FPA gradually in increasing order of cloudlets, i.e., GA has not only reduced the makespan more but also performed better for a high number of cloudlets as compared to FPA.

Table 3 explores the results for the second set of experiments where cloudlets are set fixed at 100, and VMs are increased progressively in five different simulations with a similar workload. It can be analyzed from the table that makespan is reduced linearly with increasing no. of VMs in all the simulations executed for all the approaches tested. However, GA has again successfully overtaken the FPA in this phase too, by achieving less makespan in all the five simulations.

The graph in Fig. 5 is picturizing the performance of applied approaches in the second phase of experiments with the same color coding used in Fig. 4. This can be analyzed that performance difference of GA and FPA is increasing with enhancing

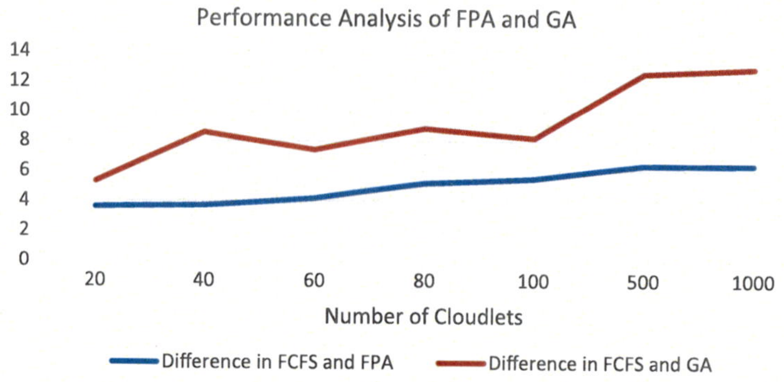

Fig. 4 Comparison of GA and FPA performances for Ist phase

Table 3 Makspan obtained in IInd phase of experiments

No of cloudlets	No of Vms	FCFS	FPA	Genetic algorithm
100	10	18.47	17.19	16.92
100	20	9.85	8.97	8.52
100	30	7.32	6.3	5.85
100	40	5.28	4.89	4.55
100	50	4.88	4.11	3.73

the number of VMs, i.e., GA is more suitable for simulations with high configuration than applying FPA approach.

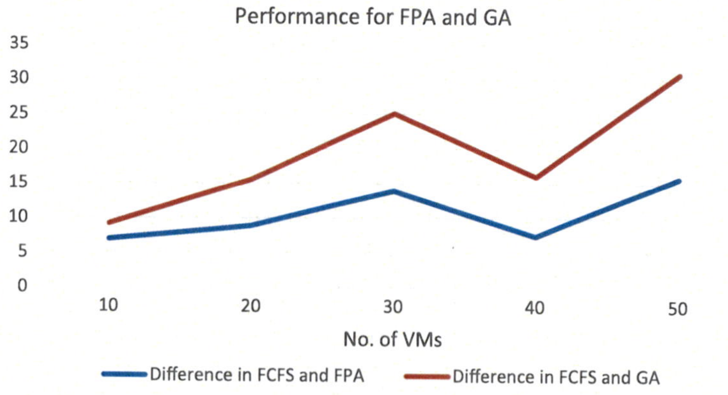

Fig. 5 Comparison of GA and FPA performances for IInd phase

7 Conclusion and Future Work

This paper aims to analyze the application of two meta-heuristic approaches, i.e., GA and FPA for the task scheduling process in a cloud computing environment. In this regard, a simulator tool named cloudsim is used to form a cloud environment and modifications are performed in the code of cloudlet scheduler algorithms available in the tool to implement the required approaches. Results mainly focused on the makespan value obtained for the execution of various simulations with different configurations. Experiments are performed in two phases to find two prospectives of effects of GA and FPA approaches, i.e., when cloudlets or tasks are increased and when VMs are increased keeping the workload constant. Analyses of results conclude that GA has performed better than the FPA in every simulation of both phases. FPA has successfully reduced the makespan to the FCFS approach, however, couldn't succeed from GA for the same. The reason for the low performance of FPA is more complications in calculating the solutions than GA. More research in the field can be conducted by implementing other heuristic and meta-heuristic approaches [21] e.g., branch and bound [22], crow–penguin optimizer [23], crow search optimization [24], firefly optimization [25], etc. Other than task scheduling, these or some other meta-heuristic approaches can also be applied in VM scheduling and tested for optimizing server consolidation [2, 11].

References

1. Wang B, Xing H (2011) The application of cloud computing in education informatization. In 2011 International conference on computer science and service system (CSSS). Nanjing, China
2. Singh P, Sengupta J, Suri P (2019) A novel approach of virtual machine consolidation for energy efficiency and reducing sla violation in data centers. Int J Innov Technol Explor Eng 8(5):547–555
3. Singh P, Prakash V, Bathla G, Singh RK (2022) QoS aware task consolidation approach for maintaining SLA violations in cloud computing. Comp Electr Eng 99:107789. Elsevier
4. Yang X-S (2012) Flower pollination algorithm for global optimization. In International conference on unconventional computing and natural computation. Springer, pp 240–249
5. Pavlyukevich I (2007) Levy flights, non-local search and simulated annealing. J Comput Phys 226(2):1830–1844
6. Khursheed M-U-N, Nadeem MF, Khalil A, Sajjad I, Raza A, Iqbal MQ, Bo R, Rehman WU (2020) Review of flower pollination algorithm: applications and variants. In: 2020 International Conference on Engineering and Emerging Technologies (ICEET)
7. Kazemian M, Ramezani Y, Lucas C, Moshiri B (2006) Swarm clustering based on flowers pollination by artificial bees. In: Swarm intelligence in data mining. Springer, pp. 191–202
8. Singh P, Singh RK, Joshi D, Bathla G (2022) Knowledge application to crossover operators in genetic algorithm for solving the traveling salesman problem. Int J Softw Innov (IJSI) 10(1):1–20
9. Kaur S, Verma A (2012) An efficient approach to genetic algorithm for task scheduling in cloud computing environment. Int J Inform Technol Comp Sci (IJITCS) 4(10):74
10. Tiwari R, Sille R, Salankar N, Singh P (2022) Utilization and energy consumption optimization for cloud computing environment. In: Cyber security and digital forensics. Springer, Gurgaon, pp 609–619

11. Sengupta J, Singh P, Suri P (2020) Energy aware next fit allocation approach for placement of VMs in cloud computing environment. In: Future of information and communication conference. San Francisco
12. Omara FA, Arafa MM (2010) Genetic algorithms for task scheduling problem. J Parallel Distrib Comp 70(1):13–22
13. Wu M-Y, Shu W (1997) On parallelization of static scheduling algorithms. IEEE Trans Software Eng 23(8):517–528
14. Tagtekin B, Ozturk MU, Sezer MK (2021) A case study: using genetic algorithm for job scheduling problem. arXiv:2106.04854v1
15. Sahraei SH, Kashani MMR, Rezazadeh J, Farahbakhsh R (2019) Efficient job scheduling in cloud computing based on genetic algorithm. Int J Commun Netw Distrib Syst 22(4):447–467
16. Kaur J, Sidhu BK (2017) A new flower pollination based task scheduling algorithm in cloud environment. In: 2017 4th international conference on signal processing, computing and control (ISPCC). New York
17. Gupta I, Kaswan A, Jana PK (2017) A flower pollination algorithm based task scheduling in cloud computing. In: International conference on computational intelligence, communications, and business analytics. Springer, pp 97–107
18. Bezdan T, Zivkovic M, Antonijevic M, Zivkovic T, Bacanin N (2021) Enhanced flower pollination algorithm for task scheduling in cloud computing environment. In: Machine learning for predictive analysis. Springer, pp 163–171
19. Malti AN, Benmammar B, Hakem M (2022) QoS based task scheduling algorithm in cloud computing. In E3S web of conferences. EDP Sciences, p 01014
20. Calheiros RN, Ranjan R, Beloglazov A, De Rose CA, Buyya R (2011) CloudSim: a toolkit for modeling and simulation of cloud computing environments and evaluation of resource provisioning algorithms. Softw Pract Exp 41(1):23–50
21. Singh H, Tyagi S, Kumar P, Gill SS, Buyya R (2021) Metaheuristics for scheduling of heterogeneous tasks in cloud computing environments: analysis, performance evaluation, and future directions. Simul Model Pract Theory 111:102353
22. Singh P (2021) Scheduling tasks based on branch and bound algorithm in cloud computing environment. In: 8th International conference on signal processing and integrated networks (SPIN). Noida
23. Singh H, Tyagi S, Kumar P (2020) Crow–penguin optimizer for multiobjective task scheduling strategy in cloud computing. Int J Commun Syst 33(14):e4467
24. Singh H, Tyagi S, Kumar P (2021) Cloud resource mapping through crow search inspired metaheuristic load balancing technique. Comput Electr Eng 93:107221
25. Pankwar D, Saini G, Agarwal P, Singh P (2022) Firefly optimization technique for software quality prediction. In: Lecture notes in networks and systems, vol 425. Springer, Singapore, pp 263–273

Malicious Node Detection in Heterogeneous Internet of Things

Supriya Goel, Somya, Swapnil Sutar, and Priyanka Mekala

Abstract In recent years, security has been gaining popularity in resource-constrained IoT ecosystems as it has less capability to run conventional encryption standards. IoT comprises several communication protocols like Bluetooth Low Energy, Wi-Fi, and Zigbee for different applications, including home automation, smart city, etc. With such a heterogeneous system, it becomes complex to secure it as with every different protocol comes more vulnerabilities in the network. Anomaly-based detection methods have increasingly attracted the scientific community in recent years. Deep packet inspection evaluates the network traffic and forms a set of informative features that formalizes the system's normal and anomalous nature. We classify a normal or abnormal activity using machine learning algorithms and present the results of our detection system implemented on a heterogeneous IoT testbed. This system is applicable for companies, offices, government organizations, or secret agencies to increase network security to protect their systems.

Keywords IoT · Security · Anomaly detection · BLE · Wi-Fi · Communication protocol

Supported by organization C. R. Rao Advanced Institute of Mathematics Statistics and Computer Science.

S. Goel · Somya · S. Sutar (✉) · P. Mekala
C. R. Rao Advanced Institute of Mathematics Statistics and Computer Science Hyderabad Central University Campus, Gachibowli, Hyderabad, India
e-mail: swapnilsutar04@gmail.com

S. Goel
e-mail: supriya@crraoaimscs.res.in

Somya
e-mail: 201915004@daiict.ac.in

P. Mekala
e-mail: mekalapriyanka@gmail.com

A. K. Dubey et al. (eds.), *Advanced IoT Sensors, Networks and Systems*,
Lecture Notes in Electrical Engineering 1027,
https://doi.org/10.1007/978-981-99-1312-1_13

1 Introduction

IoT or Internet of Things is a concept referring to the billions of everyday physical devices around the world embedded with sensors, software, and other technologies, all exchanging data via the internet. The number of these IoT devices has reached 9.5 billion in 2019 [1] from 1.7 billion in 2015 [2] and is further expected to grow at a much faster rate. With the growing number of devices, there are increasing numbers of challenges in IoT like security, powering billions of sensors, e-waste, and among that security has turned out to be a significant concern. According to SonicWall Research labs, there's a 30% increase in IoT malware attacks (i.e., a total of 32.4 million) worldwide in the third Quarter of 2020 [3]. 2016 Dyn cyberattack is an example of increased concern in which Mirai malware used IoT endpoints as botnets and used them to launch DDoS attacks against well-known websites. Today, IoT devices are embedded all around us in seemingly everything, e.g., wearables, cars, homes, and some of that IoT communication will contain data that must be protected or otherwise shielded. Compromising a single component or communication channel in IoT-based systems can paralyze the part or complete Internet network. Although there is no standard architecture for IoT, it can be divided into four layers, first is physical or perception consisting of sensors interacting with the physical world, second is the network layer (gateways and routers), it aggregates data from various sensors, and third is processing layer which has data processing algorithms and stores and processes the data sent from gateways to the cloud, and fourth is the application layer that acts as an interface and provides a view of analytics [4]. For example, in healthcare, body sensors will form the physical layer, gateways will include the network layer, and the application layer will display readings such as medical intakes, physical movements. The given architecture makes it easy for us to identify the vulnerabilities.

Remaining paper is organized as follows: Sect. 2 describes the motivation for this paper, whereas Sects. 3, 4, and 5 cover the Background and related work which gives an insight into the malicious node detection. Section 6 discusses the methodology and framework used. The experiment and hardware details are presented in Sect. 7. Finally, Sect. 8 contains a discussion of the results and their importance and uses. The thesis ends with a conclusion in Sect. 9 which summarizes and concludes the paper.

1.1 Limitations and Threats

IoT devices are small in size and constrained in power, memory, computational capability, network bandwidth and all these factors pose several limitations in the areas of security and privacy. IoT cannot adopt the same security system as used till now because of the constraints mentioned before and due to IoT's heterogeneous nature in terms of protocols like BLE and IEEE 802.11 employed in the single system.

IoT deals with vast amounts of data which makes it more prone to flooding attacks. As the IoT system expands, attack methods have also become more complex. Today's networks are vulnerable to

- Denial of Service—In DoS, an attacker makes a machine or network resource unavailable for users.
- Spoofing—Spoofing occurs when an attacker impersonates the identity of a genuine node and sends malicious traffic to disrupt the network's normal operation.
- MITM (Man in the Middle)—Spoofing attack can lead the attacker in the "middle" of the conversation between two parties where this illegitimate node can either eavesdrop or insert malicious data.
- Botnet attacks—by forming a network of hacker-controlled IoT nodes which are further used to carry out mass attacks.
- Physical attacks—Such attacks risk is also there as sensors are deployed in open spaces.

The traditional security approach to cope up with attacks includes encryption, authentication, firewall, etc. Encryption is an effective countermeasure but due to the lack of computing power it's not so easy to encrypt all the data and hence we require a second layer of security. Thus we have propose a novel and next-generation anomaly detection system that works with IoT devices and can detect behavioral anomalies in the heterogeneous network. Such an intrusion detection system is independent of the type of IoT devices, communication protocols, and network structure.

The goal of our research is to make an anomaly detection system for heterogeneous IoT infrastructure. Most of the work done is concentrated on specific communication protocol or has not been implemented and little goes for spoofing detection in heterogeneous IoT networks along with the implementation. Spoofing attacks can lead to other types of attacks, via this, an attacker can disrupt the network's normal operation, take control of devices, and even remove some devices from the network. The main goals of this report are to develop the system with the following features:

- Raspberry Pi-based system that detects anomalies in the network (that points to an attack) irrespective of IoT Communication protocols
- Acceptable detection rate, lower false positives, and negatives without adding any overhead to the IoT devices and network.
- Our system should be suitable for small businesses or offices who want to improve network security in order to secure their systems but cannot afford to invest in an enterprise-grade solution.

While achieving this certain assumptions are made, such as the adversary cannot fully control the target device physically or remotely, detection infrastructure is secure and can't be compromised by the attacker. The attacker doesn't use jamming, and it is an indoor environment with static sensors.

1.2 Malicious Node Detection

Any node which follows the abnormal activity to attack other nodes in network, such node is said to exhibit malicious behavior. It can disturb the operation of the entire network. We will discuss several security solutions to detect such malicious behavior in this section. Firewalls and anti-malware software alone is not enough to protect an entire network from attack. The identity of a malicious node can be verified through cryptography algorithms but authentication is not always possible because of additional infrastructural overhead due to key handling and management. Many papers have applied an Intrusion Detection System (IDS) for increased security. Intrusion detection is a monitoring system that refers to the detection of malicious activity and gives immediate alerts. An intrusion is different from the normal behavior of the system, and hence anomaly detection techniques are applied in the intrusion detection domain. An intrusion detection system works by the theory that there is a difference in the network flow between legitimate users and malicious users. It can be designed in various ways, some are shown in the block diagram of Fig. 1. To classify malicious activity, an IDS employs two major ways such as Statistical anomaly detection and Rule-based anomaly detection.

- Rule-based detection system in which a set of pre-defined rules (requires expert knowledge) are used to detect any malicious activity. In this, network traffic is examined for known attack patterns (known as signatures) and matches the patterns which point to an attack. It fails to detect unknown or modified attacks and those attacks that exploit the recently discovered vulnerabilities, i.e., Zero-day attacks.

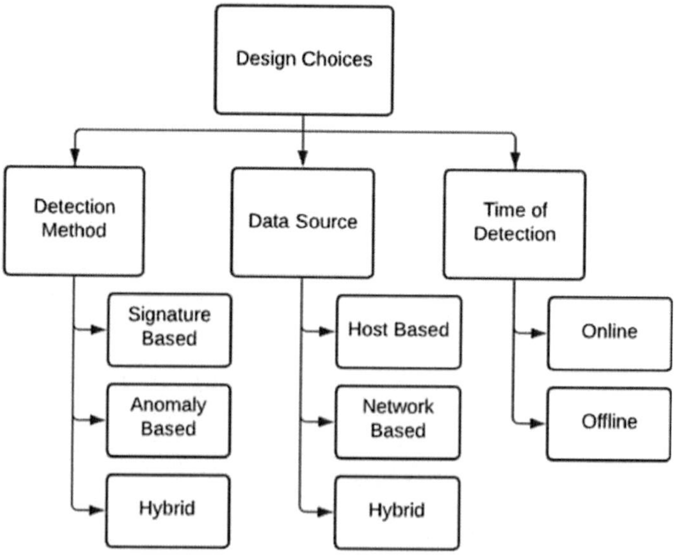

Fig. 1 Design choices

Since they require to save huge amounts of rules (expert knowledge), they can be resource demanding and therefore, it is not possible to use a pattern based approach for IoT networks.

- Anomaly-based detection systems compare a normal recorded behavior with current input and use a type of statistical calculation to determine anomalies in the network traffic. The advantage is that a new attack for which a signature doesn't exist can be detected. Furthermore, IoT technologies do not stay the same indefinitely. Instead, they continue to improve and add new features, e.g., BLE, Wi-Fi, and hence new vulnerabilities with new attack signatures. The difficulty with these systems is that when unusual actions are recognized, they generate a large number of unnecessary false alarms and are not flexible enough for a complex system.
- Hybrid detection—They start with anomaly detection, then try to relate it with the corresponding signature, e.g., Kalis.

An IDS can be placed in several locations throughout the network. When placed in the physical network it is known as a network-based IDS and if on a host system, it is known as a host-based IDS. In classical settings detection system can be installed on hosts directly on a network device such as switch, router, or host system, or it can be installed as a standalone device but such host-based approaches cause issues, due to resource constraints of devices processing and deep inspection analysis is only possible for selective hosts in a system. The best placement for an IDS is determined by the environment it will protect. And instead of capturing packets at the network layer and inspecting their payloads, deep inspection of header data is done as packet payloads are often encrypted. It will send an alert if it detects any suspicious data packets. The alert is saved in a log file that administrators can access.

2 Literature Survey

A lot of methods are used for the objective of malicious activity detection. To secure the network [5] presents lightweight encryption algorithms for low-resource devices for IoT environments. Satam et al. [6], An et al. [7] use an n-gram-based approach to characterize the normal behavior of the Bluetooth and HTTP protocol, respectively. Several machine learning algorithms [4, 8, 9] as well as deep learning are being explored to improve the accuracy, reduce the false positives and the overhead to the system. Du et al. [10] follow a method to distribute an anomaly detection scheme across several end-devices. In [11], the authors focused on the security of IoT devices in the smart city and suggested the AD-IoT system (Anomaly Detection-loT), which is based on a Random Forest ML-based architecture. Using machine learning-based dataset assessment, the suggested technique can efficiently identify any type of suspicious behavior occurring at distributed fog nodes. But experiments on datasets are challenging as labeled attack samples are difficult to get, and this information is frequently out of date due to new breaches. The NSL-KDD and DARPA datasets are the most widely used by researchers when creating new IDS. Because both datasets were developed more than a decade ago, neither can accurately reflect the network behaviors of today's IoT networks.

Zarpelão et al. [12] analyzed IoT intrusion detection research and classed it by detection method, security threat, placement strategy, and validation strategy. The authors' key point is that IoT intrusion detection techniques are still in the early stages of development. They specifically pointed out that the recommended solutions do not address a wide range of attacks or IoT devices. Furthermore, many of the systems currently available have never been extensively examined or validated.

SNORT [13] is the most widely used Signature-based Intrusion Detection/ Prevention system. SNORT detects intrusion attempts by analyzing network traffic in real time. In SNORT, the signature compromises of the header that consists of the source address, destination address, and ports and its options that include payload and metadata which are used to determine whether the network traffic corresponds to a known signature or not. But Snort becomes too resource demanding for IoT hence Alessandro et al. propose RPiDS [14], an IDS architecture customized for IoT environments that includes Raspberry Pi equipped with Snort. Results show that the Raspberry Pi can host Snort, hence providing portability and security on demand (i.e., everywhere, anytime).

Gajewski et al. focus on attacks on Home automation systems and have proposed two step method to improve attack detection accuracy [15]. First, packet traffic data is collected on the Home gateway and with the help of ML algorithm (NB model) classifies the data records as normal or anomalous. Second, Across ISP data center behavioral similarities are found between observations from Home gateways (this correlation process requires more computational resources), if any similar anomalies are found that might point to an attack [16].

Kalis is one of the few methods that do not target individual protocols. It follows a Hybrid approach and proposes that processing network events and traffic through all the detection techniques requires a high amount of system resources and can cause delays in attack reaction but we can restrict our set of attack detection techniques by knowing what attacks are possible in that network. From collected observations, Kalis finds out characteristics of the network (whether mobile or static, powerful or constrained) and then finds out possible selective attacks and hence narrows down the detection set [17].

Due to various vulnerabilities in BLE, spoofing attacks are a threat to BLE devices. All spoofing attacks will result in certain anomalous features in the advertisement packets that will contain BLE's identity. Jianliang et al. have proposed BlueShield to obtain certain features like advertising interval, RF signal frequency offset Received Signal Strength Indicator (RSSI) and determine those anomalies.

Summerville et al. have proposed a lightweight packet anomaly detection approach that is feasible to run on resource-constrained IoT end nodes using pattern matching to detect signs of undesirable activities. The authors propose a technique called bit pattern matching for small IoT devices to perform feature selection. As these devices use few and relatively simple protocols and hence are highly similar in network payloads.

3 IoT Communication Protocols

Out of several communication protocols in IoT like Bluetooth, Wifi, Zigbee, Lora, Nb-IoT, we have considered the first three due to their popularity and compatibility with the upcoming devices. The main difference between the three is shown in table.

	Bluetooth Low Energy (BLE)	Wi-Fi	Zigbee
IEEE spec	802.15.1	802.11	802.15.4
Max. data rate	1 Mbps	54 Mbps	250 kbps
Power consumption	0.01–0.5 W	High	30 mW
Frequency	2.4 GHz	2.4 Ghz and 5 GHz	2.4 GHz, 900 MHz and 868 MHz
Applications	Smart wearables	Broadband internet access	Home automation (monitoring and control)

3.1 *Bluetooth Low Energy*

Bluetooth Low Energy (BLE) [18] was started as Bluetooth 4.0 with the goal of low-power consumption. The modulation technique used is GFSK, and the theoretical upper limit for the data rate is 1 Mbps. Its throughput is around hence other technologies like Wi-Fi still have maintained their position. The standard uses a technique called FHSS frequency hopping spread spectrum, in which the radio hops between channels on each connection event. Since Wi-Fi and classic Bluetooth also work in the 2.4 GHz band, this technique avoids interference. BLE has two types of packets

1. Advertising—It broadcasts data for applications that do not require full connection establishment or to discover slaves. This type of packet consists of up to 31 bytes of payload along with the basic header information such as Bluetooth device address. These packets are broadcasted blindly without scanning any nearby device, and the state of the device is said to be in advertising mode.
2. Data packets—These packets carry general user data between master and slave and the device is said to be in a connected state (transitions between states are represented in Fig. 2).

Around 80% of BLE devices lack secure pairing and communicate data without employing any form of secure authentication. As cyberattacks get more sophisticated, it is necessary to equip BLE devices with advanced security measures while also applying security patches to address vulnerabilities. However, such firmware upgrades and security patches solutions may not be widely implemented by various vendors, especially for BLE devices with limited or no I/O capabilities, and may even fail to provide backward compatibility with legacy BLE devices. As a result of

Fig. 2 State diagram of BLE

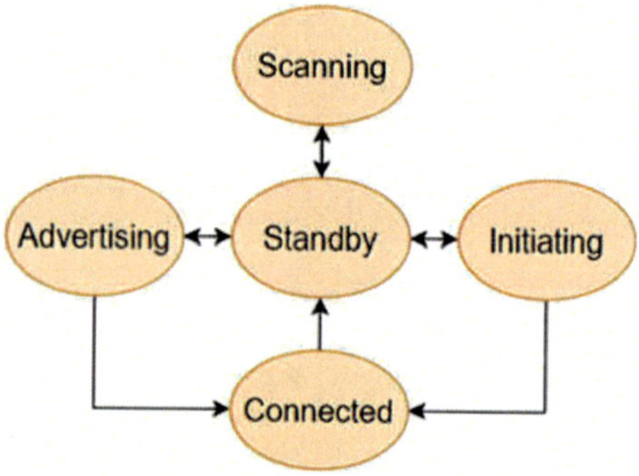

the limitations of these solutions, a more straightforward approach to reducing the threat of spoofing attacks is required.

3.2 Wi-Fi

Wi-Fi (Wireless Fidelity) [19] is used for in-building network connectivity. IEEE 802.11 specifies the MAC and Physical layer protocols for implementing Wi-Fi communication. The Wi-Fi protocol uses the 2.4 GHz and 5 GHz bands to communicate. Wi-Fi networks are still prone to data link layer attacks. Devices on all Wi-Fi networks can be successfully attacked by the deauthentication attack. Man-in-the-middle (MITM) and denial-of-service (DoS) attacks can both be carried out with these methods. An intrusion detection system that can detect attacks on the Wi-Fi protocol with a minimal number of false positives and negative alarms is required.

3.3 Zigbee

Zigbee [19] popular for low-power mesh network speed of 250 kbit/s, 128-bit AES encryption, and numerous power-saving features such as sleep with scheduled wake-ups. IoT works with different protocols on different levels and Zigbee is one among them It is based on IEEE 802.15.4 that defines its Physical and MAC layers. Zigbee's main objective is to save energy such that sensors can survive years with an AA battery. Applications include a home automation system, smart lighting system.

Zigbee consists of three main components: Coordinator, Router, and End Device. The only device that can start a Zigbee network is a coordinator, and each Zigbee network must have only one coordinator. Zigbee routers serve as intermediary devices, allowing data to flow back and forth between them and other devices. A Zigbee End Device provides only basic functionality and cannot send or receive directly with other devices [20].

Despite the fact that the protocol has network-wide and pairwise encryption and authentication, it still has a number of weaknesses that have been uncovered and addressed in the past. One of Zigbee's flaws is the unencrypted initial handshake between the network coordinator and a joining device. Zigbee security is necessary as our door lock has the same security as our lights [21].

4 Methodology

Spoofing attacks, in which an attacker impersonates a benign BLE device and feeds harmful data to its consumers, are common on IoT devices. One important type of attack that leads to several others (e.g., DoS, MITM, jamming, energy depletion, and various exploits at the protocol level) and is easy to launch is spoofing [22]. We aim to detect these spoofing attacks and detect that malicious node with the experiment setup of Fig. 3. Broadly our work phase can be said to be in three steps. First, node identification, then, analyze and extract the features and lastly classify the abnormal data (also shown in Fig. 4). Our focus will be on the parameters that are hard to predict by the attacker. Each advertising packet comprises a device's unique identifier as well as information about the device's services.

The most important step in developing threat detection models is feature selection. The set of attributes or features regarded to be the most effective attributes is retrieved during this step in order to develop appropriate detection algorithms (detectors). Defining a large feature space can help the system detect a variety of attacks, but redundant features that just add overhead and delay to the attack detection and prevention process should be avoided. Our focus will be on the parameters that are hard to predict by the attacker.

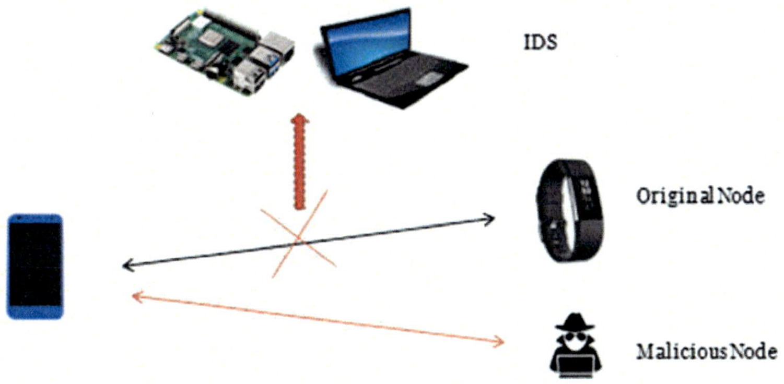

Fig. 3 Experiment setup

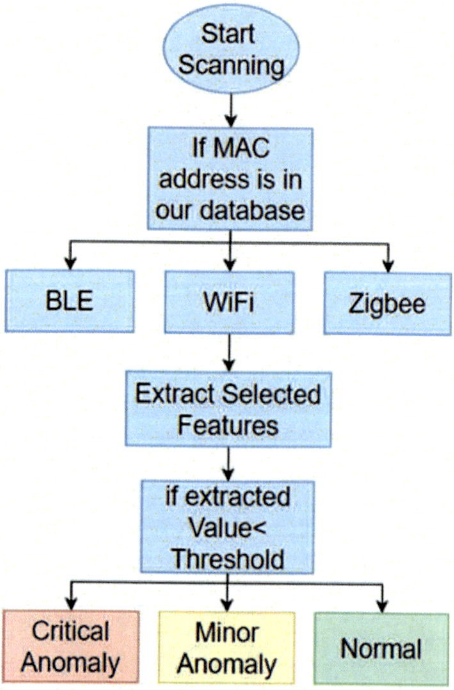

Fig. 4 Flow chart of method

4.1 RSSI

The authentic value of the physical features (e.g., RF signal's strength) of the advertising packets would be different at different channel paths, this mechanism ensures that even an attacker with the capability to mimic all the physical features, cannot trick correctly imitated values at the same time.

The detection system monitors all network node RSS values and examines them for indicators of spoofing attempts [23]. When an attack is identified, the detection module notifies the victim, notifying them of the attacker's existence and identity. If R does not fall within the Rmin < R < Rmax range, the program comes to a halt and raises the alert, signaling the presence of a spoofing attempt. This experiment will use a static threshold as the dynamic threshold means more computations and adds a small delay for queuing and grouping of frames.

4.2 Advertising Interval and Traffic Features

The real-time advertising interval can be lower than the predicted INT value if both the BLE device and the attacker are broadcasting advertising packets, and so can be used to detect spoofing attempts. Even while connected, some BLE devices may continue to advertise, but most (intermittent) will not advertise after connection and those devices can be kept under state supervision to detect advertisement on their behalf by a malicious node even after the legitimate device is connected. Numerous researchers are trying to identify the characteristics of traffic generated as a product of IoT device communication. The traffic characteristics (such as number of packets, last active time duration) generated by individual IoT devices can be a key factor in researching the relationships of generated traffic to certain processes in the communications network. Such features are used to identify IoT devices in the network, detect unauthorized devices in the network, and detect network traffic anomalies.

4.3 Techniques for Detection

- In general, any density estimation approach can be applied to model the normal classes. These rely either on the distance from the neighboring point or relative data density [24]. Density around an anomalous point should be significantly lower than the normal ones. Some well-known approaches are kNN and LoF. Because normal data is far more abundant than anomalous data, the one-class classification algorithm can be used to identify a single class by learning from a training set [25].
- Statistical Methods: Anomalies are detected using statistical approaches that rely on a set threshold, mean and standard deviation, and probability. It follows the rule that normal data will fall under high-probability regions and outliers will fall under the low-probability area. Probability threshold will divide the mentioned two regions.
- Isolation Forest is designed to support anomaly detection algorithm which is based on decision trees approach. It works by randomly selecting features and isolates each point with the assumption that anomalous points will be isolated first. The contamination parameter in this represents the percentage of outliers in the dataset and can be found by trial and error [7].
- Clustering-based techniques can also be used where the normal belong to clusters, whereas anomalous ones do not.

5 Experiment Design

5.1 Hardware Description

The software and measuring instruments used in the studies will be demonstrated and described in this part. The commands for controlling the software and tools will also be demonstrated and explained.

We have used Raspberry Pi 4 Model B running with the operating system Kali Linux. It was chosen due to its compactness and portability. Michael Ossmann's

Ubertooth One is an open-source Bluetooth test tool. The device's main function is to decode Bluetooth packets, it also has to figure out the master piconet MAC address, get the packet's SNR (Signal to Noise ratio), figure out the device's hopping sequence, and use the spectrum analyzer to reveal any Wi-Fi or Bluetooth activity in the 2.4 GHz band (Fig. 5).

Our system has an internal Wifi card of Broadcom BCM 4321 that supports wireless monitor mode. When set up in monitor mode (default is managed mode), it enables a device having a wireless network interface controller to monitor all wireless network traffic. In a nutshell, it aids in the "monitoring" of packets received without any filtering and without linking any access point. The tool used to change the mode to monitor mode is Aircracking and with the following commands:

```
sudo airmon-ng start wlan0
sudo airmon-ng check kill
sudo airodump-ng wlan0mon
```

The CC2531 USB dongle is a fully functional USB device that connects to IEEE802.15.4 / ZigBee applications through a PC. The dongle can be used as a Zigbee packet sniffer by plugging it straight into the PC/Raspberry. Commands used are with the zigbee2mqtt tool [19]:

```
cd /opt/zigbee2mqtt
npm start
```

5.2 *Software Tools*

- Spooftooph: Spooftooph is an application that automates the spoofing or cloning of Bluetooth data (Address, Class, and Name). We spoofed the BLE earphones and started advertising packets as the same, shown in Fig. 6 in the name of OP Headset and 98:09:CF:D2:70:A8 MAC (Media Access Control) address discarding its original name and MAC address.
- Python: Scapy, Bluepy are python libraries used that can handle Wifi and Bluetooth scanning.

Fig. 5 Raspberry Pi 4 model B

Fig. 6 Malicious node

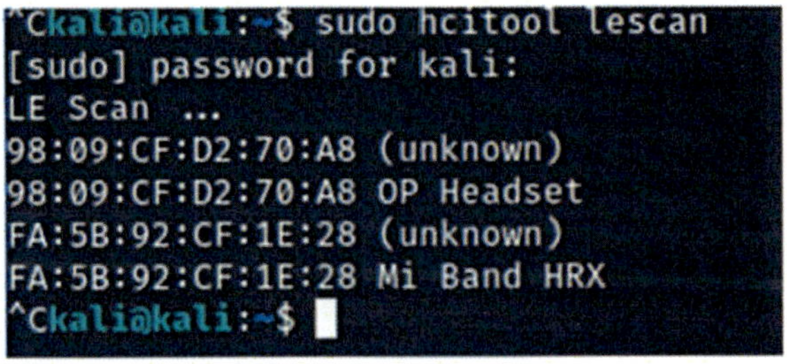

- Wireshark: Wireshark is a free and open-source packet analyzer. It helps in deep packet inspection of protocols, live inspection, and offline analysis.

5.3 Testbed

We used a heterogeneous network testbed to evaluate the performance of the suggested spoofing detection system. The testbed was set up in an office building within the CR Rao AIMSCS Institute's IoT lab in Hyderabad. There were seven nodes in the network: one victim, one attacker, and five legitimate nodes. We utilized six different IoT devices as shown in Fig. 8, these cover the mainstream IoT applications (e.g., Smart Plug, Watch, Camera, earphones, Voice Assistant) and popular manufacturers like Xiaomi, Qubo, OnePlus, Tuya. Labeled dataset for the internet of things (IoT) is very rare and difficult to find. So with this testbed, our database was created and stored. IoT devices were arranged as shown in Fig. 7 with a static environment.

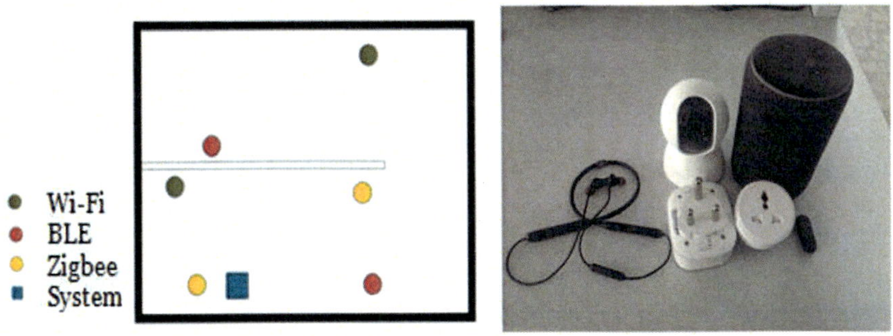

(a) Overview of IoT lab (b) IoT devices

In practise, we want to locate our device first before making any connection (Node Scanning). The system will scan the data packets in the nearby area with the help of given sniffers. Most BLE devices broadcast on three channels (37, 38, and 39) on a regular basis in order to be discovered. We have used Raspberry Pi 4 as our system, Ubertooth one as BLE sniffer and the packets sniffed are being analyzed in

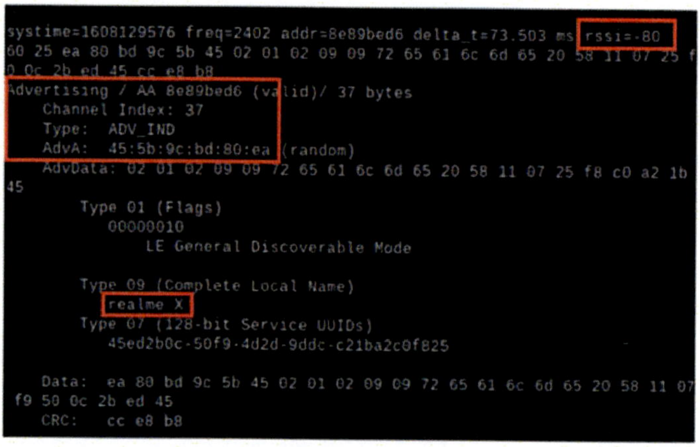

Fig. 7 Packet capture

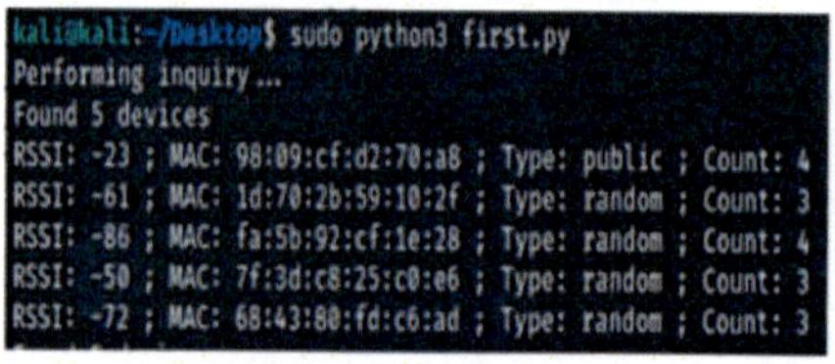

(a) Scanning of Wi-Fi

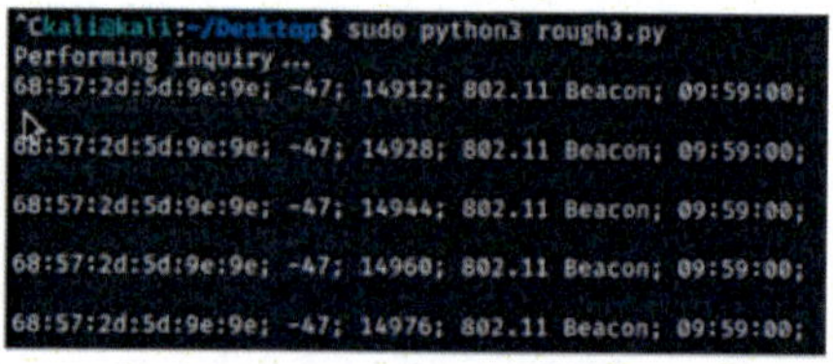

(b) Scanning of BLE

Fig. 8 Scanning snaps

the Wireshark tool. Wireshark is a free and open-source packet analyzer. It is capable of capturing, inspecting, analyzing, and visualizing packets or frames.

The Bluetooth Low Energy Link Layer Format (btle) is captured by Wireshark as the packet's outermost layer. It includes an Access Address, a PDU, and a CRC calculated from the Payload content. The most significant portion of the BLE packet is the advertisement data. It's content usually consists of flags, a device's short name, and manufacturer-specific data. All these parameters are visible in Fig. 7 where the packet dissection of a Realme mobile is shown. The following commands are needed, the first one turns Bluetooth on and the second one is to start capturing BLE packets.

```
sudo hciconfig hcio up
ubertooth-btle -f -c /tmp/pipe
```

Figure 8 shows how the packet captured can be seen and its MAC address needs to be extracted to identify whether the node belongs to our network. Following the collection of network traffic, the next step is to extract features in order to identify the devices. We further develop Python codes using the libraries bluepy and scapy to extract the details.

It retrieves the device name, MAC address, and the advertising data INT value, which equals subtracting the current advertising packet's time-of-arrival from the previous advertising packet's time-of-arrival. The observed INT is equal to a pre-

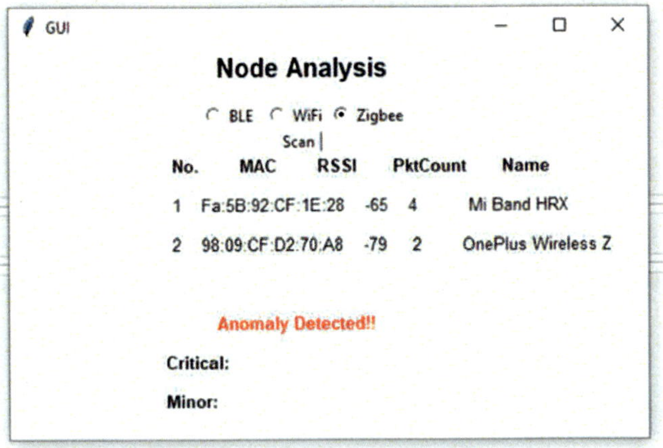

Fig. 9 Traffic capture of BLE

determined advertisement time with a delay of 0 to 10 milliseconds, as described by the BLE specification (p. 2750 in [26]). It also saves the BLE device's determined attributes as well as an associated device identifier. Examine the features of each device after extracting the advertising packets and their features. Any interval between two advertising packets must always be greater than INT's lower bound. If it doesn't, the monitor flags it as an abnormality and sounds an alarm [27].

Representation of the system was formed with the help of GUI library in Python (Fig. 9), and the user interface is simplified for the user to check if any anomaly present.

6 Performance Evaluation

The experiment was performed with data collected over the testbed for a period of a few hours. The advertising node was placed at different distances from the R-Pi. When a few RSSI values were plotted over different distances, we could observe there was a sudden shift in the RSSI values. Hence RSSI could prove the most decisive factor to be considered with respect to the distance factor and such abrupt change in RSSI values of advertising packets can detect an attack. Figure 10a shows the variations and we can easily distinguish among three different distances [19].

Collection of such values was observed to have close to Gaussian Distribution (Fig. 10b) having probability density function $f(x) = \frac{1}{\sigma\sqrt{2\pi}} \exp\left(-\frac{1}{2}\left(\frac{x-\mu}{\sigma}\right)^2\right)$. This statistical distribution method was used to predict the anomalies along with others. Other unsupervised algorithms were also used as the amount of abnormal data or the anomalies is few or rare, supervised algorithms could not be used. Unsupervised

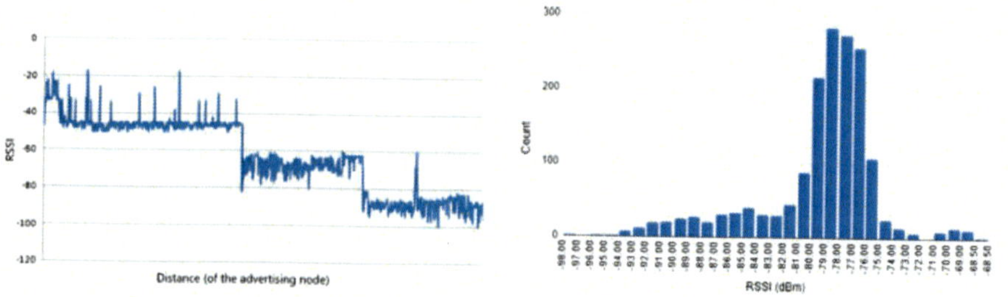

(a) RSSI with varying distance (b) Distribution of RSSI

Fig. 10 RSSI graphs

Table 1 Algorithms Used and its TP & FN

Algorithm	TP	FN
Statistical	0.98	0
Isolation forest	0.83	0.19
Local outlier factor	0.98	0.05

would analyze the data and form boundaries according to normal data only, anything that does not fit the data falls in the anomaly category.

Metrics considered for evaluating the algorithms are TP (True Positive) and FN (False Negative). Abnormal RSS values can be due to malicious activity or environmental changes and hence can also cause false alarms. The statistical algorithm has proved to be more accurate and with the least false negatives but others seem to have large FP which is the major issue in anomaly detection problems as shown in Table 1.

7 Conclusion and Future Work

The detection of malicious nodes is a critical concern in the Internet of Things. Several outlier detection algorithms for IoT have been proposed in recent years. The proposed anomaly detection system based on portable raspberry pi detects anomalies in the network (that points to an attack) irrespective of IoT Communication protocols. The statistical approach has an acceptable detection rate, and the lowest false positives and negatives. It does not add any overhead to the IoT devices and network. The application for this system is to increase network security in small companies or offices in order to protect their systems against increasingly complex threats in such environments. When there is no spoofing, the sequence of RSS sample vectors for each MAC address is close to each other and fluctuates around a mean vector. A spoofing attack, on the other hand, involves many nodes claiming the network from various physical locations.

Future work includes developing a classification model of IoT devices in a smart home environment. And integrating our spoofing detector into a real-time localization system that can both detect the spoofing attacks, as well as localize the adversaries in our IoT network. The experiment can be further modified by considering dynamic systems and hence using dynamic thresholds that will constantly be trained during implementation and change itself according to prevailing scenarios. Such can be a case in the Smart City network. Further, more unsupervised algorithms can be considered and comparison can be done. As more and more IoT protocols are taking the lead like LoRa, cellular IoT, considering these systems can be truly heterogeneous.

References

1. IC Insights, accessed 24 December 2020 https://www.icinsights.com/news/bulletins/Internet-Of-Things-Market-To-Nearly-Double-By-2019-/
2. IoT Analytics, Lueth K, accessed 18 December 2020, https://iot-analytics.com/state-of-the-iot-2020-12-billion-iot-connections-surpassing-non-iot-for-the-first-time/
3. SonicWall, "Mid-Year Update 2020 SonicWall Cyber Threat Report", accessed 18 December 2020, https://www.sonicwall.com/2020-cyber-threat-report/
4. Chaabouni N, Mosbah M, Zemmari A, Sauvignac C, Faruki P (2019) Network intrusion detection for IoT security based on learning techniques. IEEE Commun Surv Tutor 21:2671–2701
5. Singh S, Sharma PK, Moon SY, Park JH (2017) Advanced lightweight encryption algorithms for IoT devices: survey, challenges and solutions. J Ambient Intell Humanized Comput 1–18
6. Satam P, Satam S, Hariri S (2018) Bluetooth intrusion detection system (BIDS). In: Proceedings of IEEE/ACS 15th international conference on computer systems and applications (AICCSA), pp 1–7
7. An Y, Yu FR, Li J, Chen J, Leung VCM (2021) Edge intelligence (EI)-enabled HTTP anomaly detection framework for the Internet of Things (IoT). IEEE Internet Things J 8(5):3554–3566
8. Hasan M, Islam MM, Zarif MII, Hashem MMA (2019) Attack and anomaly detection in IoT sensors in IoT sites using machine learning approaches. Internet Things, vol 7, no 100059
9. Shanthamallu US, Spanias A, Tepedelenlioglu C, Stanley M (2018) A brief survey of machine learning methods and their sensor and IoT applications. In Proceedings of IEEE conference on information, intelligence, systems and applications, pp 1–8
10. Du Q, Wei Y, Mao Y (2019) Distributed deployment of anomaly detection scheme in resource-limited IoT devices. In: 2019 IEEE 19th international conference on communication technology (ICCT)
11. Alrashdi I, Alqazzaz A, Aloufi E, Alharthi R, Zohdy M, Ming H (2019) Ad-iot: anomaly detection of IoT cyberattacks in smart city using machine learning. In: IEEE 9th annual computing and communication workshop and conference (CCWC). IEEE 2019, pp 0305–0310
12. Zarpelo BB et al (2017) A survey of intrusion detection in Internet of Things. J Netw Comput Appl 84:25–37
13. Roesch M (1999) Snort—lightweight intrusion detection for networks. In: Proceedings of USENIX LISA'99
14. Sforzin A, Mármol FG, Conti M, Bohli JM (2016) Raspberry Pi IDS: a fruitful intrusion detection system for IoT. In: Proceedings of 13th IEEE international conference on advanced and trusted computing (ATC), pp 440–448
15. Gajewski M, Mongay Batalla J, Mastorakis G, Mavromoustakis CX (2020) Anomaly traffic detection and correlation in smart home automation IoT systems. Trans Emerg Telecommun Technol

16. Alrashdi I, Alqazzaz A, Aloufi E, Alharthi R, Zohdy M, Ming H (2019) AD-IoT: anomaly detection of IoT cyberattacks in smart city using machine learning. In: Proceedings of the 2019 IEEE 9th annual CCWC, pp 305–310
17. Midi D et al (2017) Kalis-a system for knowledge-driven adaptable intrusion detection for the Internet of Things. In: Proceedings of 2017 IEEE 37th international conference on distributed computing systems (ICDCS 17)
18. Wu J, Nan Y, Kumar V, Payer M, Xu D (2020) BlueShield: detecting spoofing attacks in bluetooth low energy (BLE) networks. In: 23nd international symposium on research in attacks, intrusions and defenses (RAID 2020)
19. Hussain F, Hussain R, Hassan SA, Hossain E (2019) Machine learning in IoT security: current solutions and future challenges. ArXiv, vol abs/1904.05735
20. Stelte B, Rodosek GD (2013) Thwarting attacks on ZigBee-Removal of the KillerBee stinger. In: Proceedings of network service management (CNSM), pp 219–226
21. Chen Y, Trappe W, Martin RP (2007) Detecting and localizing wireless spoofing attacks. In: Proceedings of IEEE conference on sensor, mesh and Ad Hoc communications and networks
22. Cerar G, Yetgin H, Fortuna C (2020) Learning to detect anomalous wireless links in IoT networks. in: IEEE Access, pp 363–368
23. Tang Z, Zhao Y, Yang L, Qi S, Fang D, Chen X, Gong X, Wang Z (2017) Exploiting wireless received signal strength indicators to detect evil-twin attacks in smart homes. Mob Inf Syst 4:1–14
24. Siby S, Maiti RR, Tippenhauer N (2017) IoTScanner: detecting and classifying privacy threats in IoT neighborhoods
25. Hussain S, Rahman MS (2009) Using received signal strength indicator to detect node replacement and replication attacks in wireless sensor networks. In: SPIE proceedings on data mining, intrusion detection, information assurance, and data networks security, vol 7344. Orlando, USA
26. Bluetooth, "Bluetooth Smart or Version 4.0+ of the Bluetooth specification", accessed 18 December 2020
27. Jokar P, Arianpoo N, Leung VCM (2013) Spoofing detection in IEEE 802.15.4 networks based on received signal strength. Ad Hoc Netw 11(8):2648–2660

Performance Evaluation of HSRP and GLBP Over OSPF and RIP Routing Protocols

Aayush Agarwal, Shilpi Sharma, and Evin Tom Xavier

Abstract After the rise of Covid-19 in the whole world, most of the organizations were forced to shift their business and services online for the users. Thereby, for the businesses and services to run smoothly, full-time availability of the network is a must. But there may arise situations in which availability might not be present. This problem is solved by using redundancy network protocol in network infrastructure. Among the routing and redundancy protocols used, it is prudent to evaluate which routing protocol gives best performance with which redundancy protocol. This paper aims to evaluate the performance of the most widely used routing protocols OSRP and Rip with Cisco Proprietary redundancy protocols HSRP and GLBP. The result of this study indicates that OSPF and RIP work best with HSRP and GLBP, respectively.

Keywords HSRP · GLBP · OSPF · RIP v2 · Interior Gateway Protocol · Internal routing protocols

1 Introduction

Leading rapidly developing communication technology demands to increase the application speed and convergence time if there arises a redundancy. The availability of the network is very necessary in the current information technology era. Various organizations and companies need a network to protect the operation of the system damage business, data loss/damage, and data processing errors.

A. Agarwal (✉) · S. Sharma · E. T. Xavier
Department of Computer Science and Engineering, Amity School of Engineering and Technology, Amity University Noida, Noida, UP, India
e-mail: aayush.agarwal.97@protonmail.com

S. Sharma
e-mail: ssharma22@amity.edu

E. T. Xavier
e-mail: evintomg@gmail.com

A. K. Dubey et al. (eds.), *Advanced IoT Sensors, Networks and Systems*, Lecture Notes in Electrical Engineering 1027,
https://doi.org/10.1007/978-981-99-1312-1_14

While building a network infrastructure, one should keep in mind that how the network going to tackle with failures and problems. Routers and switches work continuously alike, and any kind of disruption during their run time can lead to the possibility of them not working properly, or not working at all. Due to this the network won't be available for an certain time, until the technician properly fixes the issue. This, of course, will take time, causing some activities to be delayed from all departments in a company [1].

In the current times, computer networks have become an essential part of operation in majority of companies and institutions, which enables system to give different services and required accesses to distribute the shared resources among connected and related devices across various groups which are called Enterprise Networks [2]. Redundancy can be connected to various layers of network stack, and different protocols were created likewise. Besides, load balancing to increase the efficiency becomes the straightforward extension of the redundancy schemes [3].

In the current times, after the spread of COVID-19, in almost all the areas of the world, everything feasible is made to operate from an online source. Grocery, medicines, and food, everything is being ordered online. University and schools have begun taking their classes in online mode via video chat, making networks of vital importance. Thereby, redundancy of such networks has increased, as no enterprise will like the loss of availability to their resources or services.

The aim of this paper is to evaluate the performance of Hot Standby Router Protocol (HSRP) and Gateway Load Balancing Protocol (GLBP) redundancy protocols over Open Shortest Path First (OSPF) and Routing Information Protocol (RIP) routing protocols. HSRP and GLBP are both Cisco Proprietary protocols and can work over Cisco devices only. OSPF and RIP are open standard protocols and can be used on any device irrespective of the manufacturer. The network topology takes up each combination one by one and then gives a detailed analysis of the packets sent and received, which help in evaluating their performance.

1.1 A Subsection Sample

Open Shortest Path First (OSPF) is a routing protocol used to determine the correct route for packets within IP networks. It is an interior routing protocol and uses link state algorithm for communicating with an autonomous system. OSPF can detect changes inside the topology, such as link failures, and converges on a new loop-free routing structure within seconds [4]. Being the first widely implemented routing protocol that could converge a network in seconds and ensure loop-free paths, OSPF has many features that allow the imposition of policies on route propagation that may be appropriate for keeping local, for load sharing, and for selective route importing over IS-IS.

A hierarchical structure is imposed by the OSPF areas on the flow of data through the network. A network using OSPF always has at least one zone, if more than one area exist in the topology, let's say 2 area or 3 areas, then one of the areas should be elected as the backbone area. OSPF has only 2 levels in its hierarchy, the backbone and all other areas attached to it. OSPF has the facility to divide the devices (routers), into distinguished groups that exchange routing information within the area, but summarize that routing information by advertising external routes. A large OSPF network looks something like a large bubble (the spine area) with many smaller bubbles (heel areas) attached directly to it. Edge of the area Routers (ABR) are used to connect the areas. Each area will choose a designated router (DR) and a backup designated router (BDR) to aid in flooding of Link State advertisements (LSAs) throughout the area. OSPF message addresses are determined by the type of network to which the OSPF interface is connected. One of the following OSPF network types must be selected when configuring an interface on an OSPF router, it can be either broadcast, point to point or non-broadcast multiple access.

1.2 Background Knowledge of RIPv2

The Routing Information Protocol (RIP) is a distance vector Interior Gateway Routing Protocol (IGP) used by routers to exchange routing information. RIP uses hop count as a routing metric. RIP avoids routing loops by implementing a limit on the number of hops allowed on a route from source to destination. The maximum number of hops allowed for RIP is 15. However, this hop limit also limits the size of networks that RIP can support. RIP version 2 (RIPv2) was developed due to the shortcomings of the original RIP. With RIP, a router sends its complete routing table to all other connected routers every 30 s. Triggered updates can also occur if a router stops working before the 30-s timer expires. RIP performs "routing by rumor" and is more prone to loops than other routing protocols. That's because one RIP router sends its complete routing table to all other routers. All other routers do the same, and because there is no actual neighbor relationship or route calculation, the routers have little first-hand knowledge of available networks. Based on RFC 1388, 1723, and 2453, RIP is a distance vector routing protocol. Its main limitation is that it cannot support a network that has more than 15 hops. RIP assumes that something more than 15 hops is infinite, so it considers that the route is not valid. Despite this limitation, RIP works very well for basic path communications between devices.

OSPF (Open Shortest Path First) is a router protocol used in large networks of autonomous systems, Routing Information Protocol (RIP), an older routing protocol that is installed on many corporate networks works for non-autonomous systems. Like RIP, OSPF is referred to by the Internet-Engineering-Task-Force (IETF) as one of several Gateway Protocols (IGPs). Using OSPF, a host that obtains a change in a routing table or detects a change in the network immediately and multicasts the information to all other hosts on the network so that they all have the same routing table information. Unlike RIP where the whole routing table is sent, the host using OSPF only sends the part that has changed. With RIP, the routing table is sent to a neighboring host every 30 s. OSPF broadcasts updated information only when a change has occurred. Instead of just counting the number of hops, OSPF bases its route descriptions on "link states" that they support account for additional network information. OSPF also allows the user assigns cost metrics to a given host router so that some roads are privileged. OSPF supports a variable network subnet mask to be able to subdivide a network. RIP is OSPF Compliant for Router-to-End Station Communication. Since many networks that already use RIP in use, router manufacturers tend to include RIP support inside a router designed primarily for OSPF.

1.3 Background Knowledge of HSRP

The Hot-Standby-Router-Protocol (HSRP) provides a mechanism that is structured to support non-disruptive IP traffic failover in certain circumstances. In particular, protocol protects against first hop router failure. When the source host device cannot reach, the IP address of the first jump router dynamically. The protocol is structured for use on LAN with multi-access, multicast, or broadcast capability (e.g., Ethernet). HSRP is not designed as a replacement for dynamic router discovery mechanisms and those protocols should be used instead whenever it is possible. Great classes of legacy host implementations that don't support dynamics intended to have the capabilities of configuring a default packet transferring router. HSRP provides a failover service to those hosts. All routers participating in HSRP are supposed to run IP routing protocols and have a compliant set of routes. The discussion of what protocols are appropriate and if-routing is far beyond the range of this journal. Using HSRP, a group of routers work together to illustrate the illusion of a single virtual router to hosts in the LAN topology. This group is known as an HSRP group or a standby group. A single router device from the group is chosen from the group and is responsible for forwarding the packets that the hosts send to the virtual router. This router is known as an active router. Another router is chosen as the backup router. In the case that the active router fails, standby takes over the packet forwarding functions of the active router. Although any number of routers can run HSRP, only the active router has the responsibility of packets sent to the virtual router. To minimize network traffic, only assets and standby routers are programmed to send periodic HSRP messages once the protocol has completed the election process. If the asset router fails, standby router becomes active router. In the case that standby router fails or becomes the active router, another router if available in the network and configured for HSRP is chosen as the standby router.

1.4 Background Knowledge of GLBP

The main reason for GLPB improvement was to connect a requirement that existed with HSRP, i.e., the simple use of load adjustment. The problem that exists is that an isolated switch within a group is constant sending of activity at a given time weather situation. To the point where only a loner the switch sends an information, a lot of sessions without movement, the transmission capacity is left sitting out of standby interface switches. GLPB works a bit unique compared to two other L2 redundancy protocols. How? 'Or' What GLPB redundancy functions may be limited confusing, especially if the architect has been involved with HSRP or potentially VRRP. With the GLPB there are two unique forms of proofreading: AVG excess and repetition of FAV. AVG Replenishment is similar to VRRP and HSRP.

The remainder of paper is structured as follows. Next section presents the review of the literature used. In Section 3, we apply the proposed approach to a network for all the combinations one by one. Section 4 represents the comparison of all the combinations on various factors-based efficiency. Finally, conclusions and future work are presented in Sect. 5.

2 Literature Review

See Table 1.

Table 1 Literature review

Serial no.	Title	Methodology	Limitations
1	Application of Vlan and HSRP technology in the dual core campus network [5]	Proposed incorporating VLAN on switches with HSRP redundancy protocol. The system used level 3 switches and was tested for a dual core campus network and was tested for feasibility and efficiency at max run time	The network proposed here in applies to small networks and not to large networks. Further there is no criterion in order to calculate the packets lost and time to switch to another router
2	Network disaster recovery design using hot standby router protocol (HSRP) [6]	Proposed a network redundancy topology in which the redundant link is placed physically on a different geological location. This prevents any loss of availability of a network, if one building comes into destruction due to a natural disaster	The network proposed herein doesn't take into consideration any of the QoS factors while switching of the primary router takes place
3	First-hop redundancy protocols in omnet ++ [7]	Proposed a network which uses OMNET ++ protocol, over which a redundancy protocol was installed. The paper only checked whether the redundancy protocol was working or not	OMNET ++ is not a widely used protocol. Further, the paper only checks whether redundancy link works or not. The paper didn't check any QoS values related with it

(continued)

Table 1 (continued)

Serial no.	Title	Methodology	Limitations
4	Performance analysis of network protocol attacks using evil foca [8]	The network is attacked at layer 2 using Evil Foca tools. Performance of remedial tools is evaluated for their effectiveness in rectifying the attack	Network herein used doesn't take into account any attacks which might go to another router which is attached to the router being attacked via redundancy means
5	Hot standby router protocol for a Private University in Malaysia [9]	A network is designed for a university campus and the redundancy protocol is checked for effectiveness	QoS values related to switching of primary router with routing protocol wasn't measured
6	Network-level redundancy for campus LAN [10]	A network is designed for a university campus LAN and the redundancy protocol is checked for effectiveness	QoS values related to switching of primary router with routing protocol wasn't measured
7	Performance comparison of first-hop redundancy protocols [11]	All the FHRP were evaluated for their performance on a single routing protocol. The evaluation criterion had the following factors: convergence time, CPU utilization, Bandwidth consumption, Traffic flow	Combinations of FHRP wasn't taken with different routing protocols
8	Review on high availability and faster convergence in Ip networks [12]	Network availability was evaluated using different routing protocols and HSRP	Some QoS values were overlooked and not all the FHRP were included
9	Automatic retrieval and analysis of high availability scenarios from system execution traces: A case study on hot standby router protocol [13]	A case study on HSRP which included performance evaluation of the redundancy protocol in consideration with different availability scenarios	Different FHRP weren't included like VRRP and GLBP
10	Designing a reliable and redundant network for multiple VLANs with Spanning Tree Protocol (STP) and Fast Hop Redundancy Protocol (FHRP) [14]	A network was proposed which used STP on layer 2 and FHRP on layer 3. The network was then tested for availability and effectiveness	Different routing protocols weren't used on layer 3 while testing different FHRP
11	Medium size enterprise, don't they also need network redundancy? [15]	The paper evaluates the availability needs of a medium-sized enterprise and came to conclusion that their network needs FHRP	

3 Research Methodology

3.1 Problem Formulation

In an enterprise network, redundancy is of vital importance and is a necessity to ensure 100% availability of the network to the customer and workers. Routers, work all time during day and night to facilitate routing over the networks. During this process, any activity may happen which may cause the router to boot again. If this situation arises, the network won't be available to anyone until the router completely boots up. This can be harmful to the business and network. To ensure 100% run time and availability, a secondary back up router is installed with the same configurations as the master router, so that if any case like this arises, the backup router can immediately switch over and the control until the initial router boots up. The router which is initially routing the requests is called 'Active Router,' and the router which is in backup is called 'Standby router'. The rules by which the router switches over from standby to active in case of a problem is called 'redundancy protocol'. There are, namely, three redundancy protocols present in the market, out of which two are Cisco proprietary. They are Hot Standby Redundancy Protocol (HSRP), Virtual Router Redundancy Protocol (VRRP), and Gateway Load Balancing Protocol (GBLP). In this research, we will be using the two Cisco Proprietary redundancy protocols, HSRP and GLBP.

3.2 Conceptual Model

At this stage, a network topology is suggested and created. This will be done using the software GNS3, which is also one of the tools for network simulation. The following are used for the simulation of OSPF and RIP routing protocol, each with HSRP and GLBP:

2 Cisco 3720 routers
2 Cisco 3720 routers to act as server routers
4 Cisco IOS switches
2 End point PCs (Fig. 1).

3.3 OSPF on Router

OSPF areas are used to impose a hierarchical structure data flow over the network. In network using OSPF there is always at least one zone, and if there are more than one area, one of the two areas should be the spinal area. In OSPF there are only two levels in the hierarchy: spine and all others. They are connected to it. Areas are used to group router to a manageable group that exchanges routing information locally, However, to summarize the routing information, External route is used. A standard

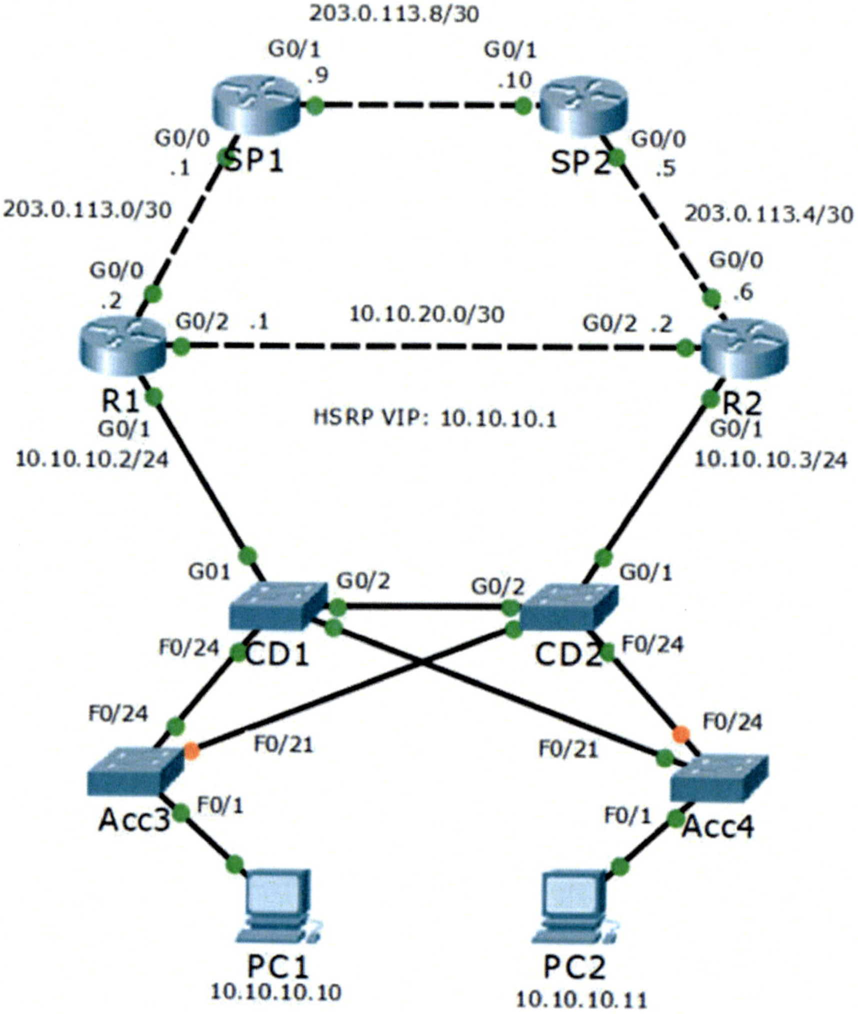

Fig. 1 Network topology

OSPF network looks like lots of small bubbles like big bubbles (spine area) Bubbles (heel part) directly attached to it (Fig. 2).

Figure 2 shows how loopback interfaces are configured on every router with the id 192.168.0.x, where x is the number of the router from 1 to 4. For the topology

```
R1(config)#interface loopback0
R1(config-if)#ip address 192.168.0.1 255.255.255.255
```

Fig. 2 Assigning loopback address to R1 router

```
R1(config)#router ospf 1
R1(config-router)#network 10.0.0.0 0.255.255.255 area 0
R1(config-router)#network 192.168.0.0 0.0.0.255 area 0
```

Fig. 3 Command to configure OSPF on routers

explained above, R1 and R2 have the same numbers as written. Routers S1 and S2 have been allotted the numbers 3 and 4, respectively.

Figure 3 shows the way OSPF is configured on router R1. Same configuration has been repeated on all the routers R1 and R2. However, the area was changed to 2 in case of S1 and S2.

3.4 RIP on Router

Routing information protocol or RIP comes in two version. Version 1 and version 2. Both the versions are compatible with cisco iOS routers. Version 1 is pretty much outdated; hence version 2 is used more nowadays. Rip sends the current router's information to the neighbor routers and keeps sending hello message on a specific interval of time to keep its IP address table updated (Fig. 4).

3.5 Redundancy Protocols on Router

The redundancy protocols on routers provides a mechanism which is made to support non-disruptive failover of IP traffic in certain circumstances. In particular, the protocols protect against the failure of the active router when the source device cannot learn the IP address of the virtual router dynamically.

```
R1(config)#router rip
R1(config-router)#version 2
R1(config-router)#network 10.0.0.0
R1(config-router)#network 172.16.0.0
```

Fig. 4 Command to configure RIP

4 Result and Discussion

4.1 Scenario 1 OSPF + HSRP

The smallest jitter value in the network was found to be 41.1 ms, while the smallest value was found to be 0.2 ms. During the experiment, the packet loss was constant during a certain interval and wasn't spaced over all the pings requests. The packet loss during the whole experiment was of 14 counts, for all 1060 pings. The entire experiment was conducted for 1420 s and the total downtime came out to be 19.1% for the whole. The smallest throughput value during the whole experiment was found to be 2 Bytes/sec. The maximum value of throughput came out to be 1046 bytes/sec.

4.2 Scenario 2 OSPF + GLBP

The largest jitter value in the network was found to 42 ms, while the smallest value was found to be 0.1 ms. During the experiment, the packet loss was constant during a certain interval and wasn't spaced over all the pings requests. The packet loss during the whole experiment was of 35 counts, for all 1060 pings. The entire experiment was conducted for 1420 s and the total downtime came out to be 19.35% for the whole. The smallest throughput value during the whole experiment was found to be 2 Bytes/sec. The maximum value of throughput came out to be 1928 bytes/sec.

4.3 Scenario 3 RIP + HSRP

The largest jitter value in the network was found to 95 ms, while the smallest value was found to be 0.5 ms. During the experiment, the packet loss was constant during a certain interval and wasn't spaced over all the pings requests. The packet loss during the whole experiment was of 24 counts, for all 1060 pings. The entire experiment was conducted for 1420 s and the total downtime came out to be 24.88% for the whole. The smallest throughput value during the whole experiment was found to be 2 Bytes/sec. The maximum value of throughput came out to be 992 bytes/sec.

4.4 Scenario 4s RIP + GLBP

The largest jitter value in the network was found to 41.5 ms, while the smallest value was found to be 0.1 ms. During the experiment, the packet loss was constant during a certain interval and wasn't spaced over all the pings requests. The packet loss during the whole experiment was of 10 counts, for all 1060 pings. The entire experiment was conducted for 1420 s and the total downtime came out to be 19.1% for the whole. The smallest throughput value during the whole experiment was found to be 2 Bytes/sec. The maximum value of throughput came out to be 1702 bytes/sec.

4.5 Evaluation

Figure 5 shows the comparison of average jitter time for all the scenarios consisting of OSPF + HSRP, OSPF + GLBP, RIP + HSRP, and RIP + GLBP. The smaller the jitter of a network, the faster the data transmission process and the better. OSPF + HSRP has the smallest average jitter time value and is the fastest network among all the networks considered. RIP + HSRP has the largest average jitter time value and is the slowest network if average jitter time is to be considered. OSPF + GLBP and RIP + GLBP had somewhat close average jitter time value and don't have considerable difference in their speed. Meanwhile OSPF has approximately close values with both HSRP and GLBP, however, RIP has different values with HSRP and GLBP, meaning that RIP shouldn't be used along with HSRP redundancy protocol.

Figure 6 shows comparison of packet loss for all the scenarios consisting of OSPF + HSRP, OSPF + GLBP, RIP + HSRP, and RIP + GLBP. The smaller the number of packet loss of a network, the faster the data transmission process and the better. RIP + GLBP has the lowest packet loss number and is the fastest network type if packet loss is considered. OSPF + GLBP has the biggest number of packet loss and is the slowest network if packet loss is considered. OSPF + HSRP and RIP + HSRP have values in between the extremes, with OSPF + HSRP value for packet loss being lower than the latter. OSPF works fastest when incorporated with HSRP, however, RIP works fastest when incorporated with GLBP.

Figure 7 shows the comparison of average downtime time for all the scenarios consisting of OSPF + HSRP, OSPF + GLBP, RIP + HSRP, and RIP + GLBP.

Fig. 5 Jitter values in all scenarios

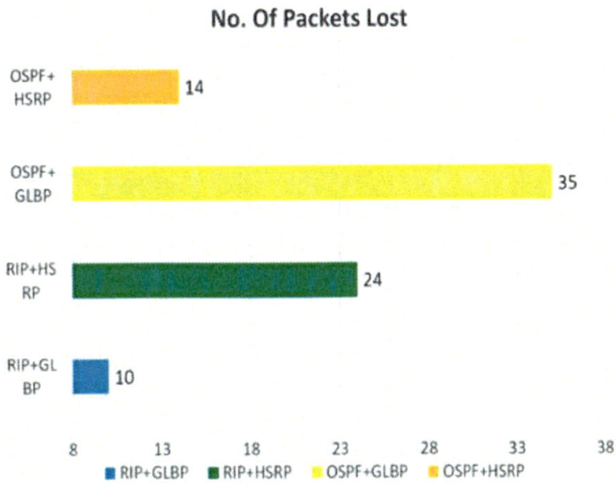

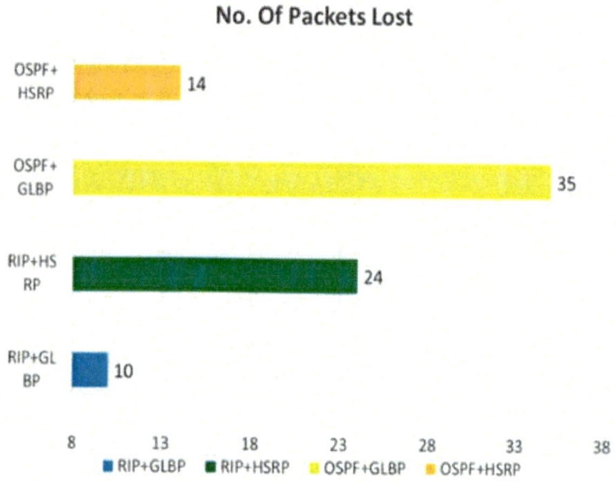

Fig. 6 Number of packets lost in all scenarios

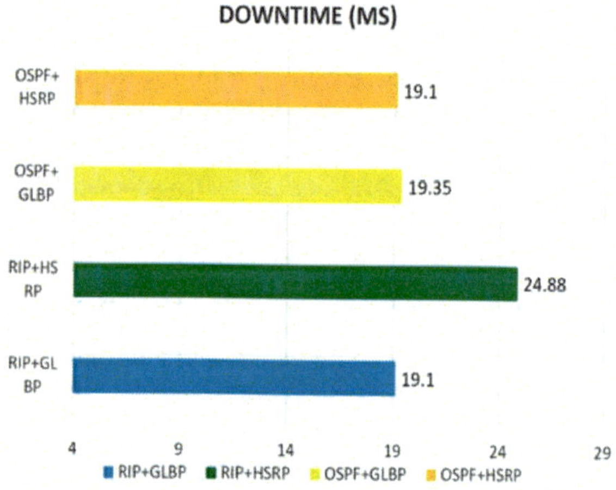

Fig. 7 Downtime values in all scenarios

The smaller the downtime of a network, the faster the data transmission process and the better. RIP + GLBP has the smallest average downtime value and is the fastest network if only downtime is considered. RIP + HSRP has the biggest average downtime value and is considered the slowest network if only downtime is considered. OSPF with HSRP and GLBP produced close results with a difference of 0.2 ms only. From this observation, they both have approximately equal speed. OSPF can be used with either HSRP or GLBP for execution; however, RIP should be used with GLBP only.

Figure 8 shows the comparison of average throughput time for all the scenarios consisting of OSPF + HSRP, OSPF + GLBP, RIP + HSRP, and RIP + GLBP. The bigger throughput of a network, the faster the data transmission process and the better. OSPF + GLBP has the highest average throughput value and is the fastest network if only throughput is considered. RIP + HSRP has the smallest average throughput

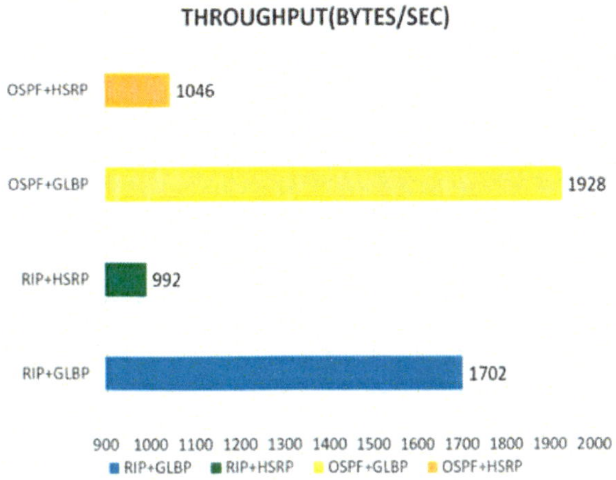

Fig. 8 Throughput values in all scenarios

value and is the slowest network if only average throughput value is considered. OSPF + HSRP has the second smallest value after RIP + HSRP and is followed by RIP + GLBP. From the above figure, it is clear that OSPF produced the best results when incorporated with GLBP, and the same goes with RIP.

5 Conclusion

Based on the results of the performance evaluation of OSPF and RIP routing protocols with each HSRP and GLBP redundancy protocols, it can be concluded that OSPF routing protocol works best when incorporated with HSRP, even though HSRP has less average throughput when compared to GLBP. It can also be concluded that RIP v2 works best when incorporated with GLBP, RIP + GLBP has outperformed RIP + HSRP in all scenarios. Therefore, when using OSPF routing protocol, HSRP redundancy protocol is preferred because it was higher QOS value when compared to GLBP. And when using RIP routing protocol, GLBP is preferred as it contains higher QOS value when compared with HSRP.

Further among OSPF + HSRP and RIP + GLBP, the latter has more QOS value, they both don't have a considerable difference when compared to average jitter, packets lost and downtime, however, RIP + GLBP has considerably higher average throughput value which makes the network fast. For a small network (15 routers) OSPF + HSRP is recommended.

There is still scope of development in this area. Further research can be done for both versions of RIP, i.e., RIP version 1 and RIP version 2, and based on the performance a suitable Redundancy Protocol can be recommended. With OSPF, performance evaluation can be done with VRRP redundancy protocol, and then it can be compared with existing results of HSRP and GLBP.

References

1. Julia RH, Suseno HB, Wardhani LK, Khairani D, Hulliyah K, Muharram AT (2020) Performance evaluation of first hop redundancy protocol (FHRP) on VRRP, HSRP, GLBP with routing protocol BGP and EIGRP. In: 2020 8th International Conference on Cyber and IT Service Management (CITSM)
2. Alam T et al (2018) Design and implementation of a secured enterprise network using dynamic multipoint VPN with HSRP protocol. In: 2018 International conference on innovations in science, engineering and technology (ICISET)
3. Channabasavaraj PM, Megha KR, Sushmitha M, Baliga S (2019) Comparative analysis of HSPR, VRRP and GLPB network redundancy protocols. Int J Res Electr Electron Commun Eng
4. Udhayamoorthi M, Karthik S, Senthilkumar C, Mohan KS, Dinesh PS (2015) Enhanced designing of network using IPv6 protocol and enabling HSRP for redundancy. In: 2015 International conference on soft-computing and networks security (ICSNS)
5. Haiyan Y (2018) Application of Vlan and HSRP technology in the dual core campus network. In: 2018 International conference on smart grid and electrical automation (ICSGEA)
6. Harahap AR (2020) Network disaster recovery design using hot standby router protocol (HSRP). Al'adzkiya Int Comp Sci Inform Technol (AIoCSIT) J 1(2)
7. Veselý V, Holuša J, Ryšavý O (2017) First-hop redundancy protocols in omnet. In: Proceedings of the 7th international conference on simulation and mo deling methodologies, technologies and applications. SIMULTECH
8. Zaw ZM, Kyaw T (2020) Performance analysis of network protocol attacks using evil foca
9. Mohamed AMA, Soon JNP, Wan WS, Yuen PK, Heng LE (2018) Hot standby router protocol for a Private University in Malaysi. Int J Sci Eng Technol
10. Wai KK (2019) Network-level redundancy for campus LAN. Int J Trend Sci Res Dev 3(5)
11. Shakshuki T, Mansur TSM (2019) Performance comparison of first hop redundancy protocols. Int Conf Techn Sci
12. Singh A, Kaur EH (2017) Review on high availability and faster convergence in ip networks. Int J Adv Res Comp Sci 8(7)
13. Sheghdara M, Hassine J (2020) Automatic retrieval and analysis of high availability scenarios from system execution traces: a case study on hot standby router protocol. J Syst Softw
14. Shahriar F, Newaz S, Rashid SZ, Rahman MA, Rahman MF (2019) Designing a reliable and redundant network for multiple VLANs with Spanning Tree Protocol (STP) and Fast Hop Redundancy Protocol (FHRP)
15. Chris Foo XJ, Siew JX, JosephNg PS, MayKang CM, Phan KY, Lim JT (2017) JomNetwork: medium size enterprise, don't they also need network redundancy? J Inform Syst Res Innov

Comparative Analysis of Load Balancing Algorithm in Cloud Computing

Prakash Choubey and B. Mohapatra

Abstract Cloud computing is the delivery of computing services—including servers, storage, databases, networking, software, analytics, and intelligence over the Internet. To provide computing services to users in efficient manner, there is a need of effective utilization of resources across the VM. But in today's world there are billions of users who are dependent on cloud computing services this leads to a huge no. of traffic across data centers. While serving users' request, some data center will have a maximum number of users' task traffic and some data center will have less amount of traffic. These problems create unbalancing of loads across the data centers. To rectify this problem, we need a load balancer or load balancing technique. Load Balancing technique is a mechanism of redistributing users' task/request across the data centers so that no virtual machine should have maximum no. of users' task traffic, less no. of users' task traffic or no users' request. The effectiveness of a load balancing technique depends upon the algorithms used behind it. The cloud analyst simulator is such a tool which uses various types of algorithms, for example—throttled algorithm, round robin algorithm, etc., and different service broker policy like closest data center, and some other parameters (e.g., data center configuration), with the help of cloud analyst we can find out data center service request time and total cost for particular data center. The objective of this paper is to compare the service request time and total cost (VM cost and total data transfer cost) of a data center by using existing load balancing algorithms with different service broker policies.

Keywords Cloud analyst simulator · Data center (DC) · Virtual machines (VM) · User bases (UB) · Load balancing algorithms · Service broker policy · Data center service request time

P. Choubey · B. Mohapatra (✉)
Department of Electrical, Electronics and Communication Engineering, Galgotias University, U.P, Greater Noida, India
e-mail: bmohapatra9@gmail.com

A. K. Dubey et al. (eds.), *Advanced IoT Sensors, Networks and Systems*,
Lecture Notes in Electrical Engineering 1027,
https://doi.org/10.1007/978-981-99-1312-1_15

1 Introduction

In the field of internet-based technology, the cloud computing technology has shown a phenomenal growth due to the advancement of communication technology. It is a novel technology which is based on internet, and it has emerged as substitute for traditional computing technologies for providing services to clients at any time and any location on a pay-per-use basis. Cloud computing technology provides both hardware and software application along with software development platforms and testing tools as resources [1, 2]. There are some giant techies in cloud computing are Amazon, Microsoft, Google, SAP, Oracle, VMware, Sales force, IBM, and others. Many of these cloud service providers are high-tech IT organizations. The cloud computing model is viewed under two different headings. The first one is the service delivery model, which defines the type of the service offered by a typical cloud provider. Based on this aspect, there are popularly following three important service models SaaS, PaaS, and IaaS [3, 4]. The other aspect of cloud computing model is viewed on its scale of use, affiliation, ownership, size, and access. The official "National Institute of Standards and Technology" (NIST) definition for cloud computing outlines four cloud deployment models, namely, private, public, community, and hybrid clouds [5]. A cloud computing model provides better results if its resources are used in the best possible way and if such effective use can be achieved through employing and maintaining proper management of cloud resources. Resource management is ensured through the use of robust resource planning, allocation, and scalability techniques. These resources are delivered to customers as virtual machines (VMs) through a process known as virtualization which uses an entity (software, hardware, or both) known as hypervisor [6]. The greatest benefit of cloud computing is that a single-user physical machine turns into a multi-user virtual machine. [7, 8] (Fig. 1).

The Cloud Service Provider (CSP) plays a vital role in the provision of services to users and is a complex task with available virtual resources. While serving user demands, some virtual machines will get a large amount of user task traffic and some will get less traffic. This leads to load unbalancing problems. The problem of load unbalance is an undesirable event on the CSP side which degrades the performance and efficiency of IT resources along with guaranteed Quality of Service (QoS) on agreed Service Level Agreement (SLA) between customer and service provider. Under such circumstances, load balancing (LB) is necessary and is a specific area of research interest among researchers. Load Balancing is the process of redistributing workload into a distributed system such that no computing machine is over loaded, under-loaded, or idle. In this paper, we have use service broker policy and load balancing algorithms and have compared result of service response time of different service broker policy with different algorithm.

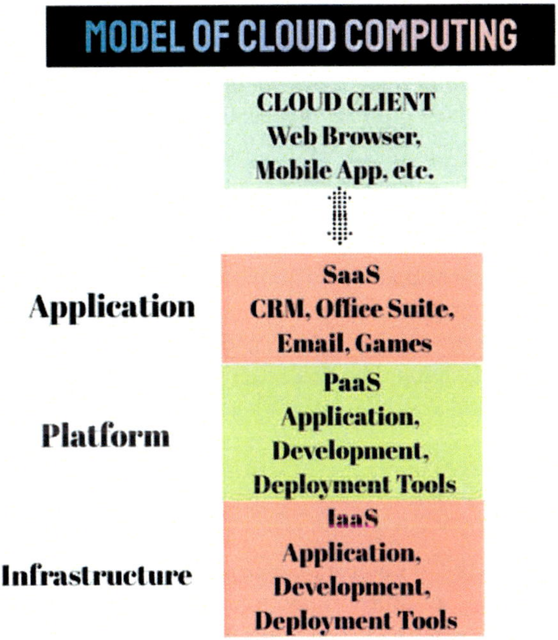

Fig. 1 Generalized cloud computing framework

2 Related Work

In general, a great deal of work has been done in the field of cloud computing, in particular in programming (tasks, VMs, and Compute), resource sourcing, resource management, energy management, and load management.

In Shah et al. [9], the author presents an overview of the load balancing algorithms. The various types of load balancing methods were classified as static and dynamic as a function of system status, homogeneous and heterogeneous as a function of the uniformity of the VM type. The load balancing methods are classified using performance matrix. Further, the advantages and disadvantages of each algorithm were discussed. The paper does not necessarily deal with literature.

In Domanal and Reddy [10], the author proposed "Modified throttled algorithm." It is a load balancing technique which is used for distribution of upcoming service request consistently between the virtual machines. Here Cloud Analyst simulator tool is used for comparison between existing algorithm Round Robin and throttled algorithm. In [11], Kalra and Singh proposed a systemic study of different scheduling algorithms for cloud as well as for grid computing considering five fundamental metaheuristic methods, namely, Ant Colony Optimization (ACO), Genetic Algorithm (GA), Particle Swarm Optimization (PSO), League Championship Algorithm (LCA), and BAT algorithm.

In [12], Neghabi presented a well defined, systematic, and potential review of load balancing techniques in software-defined networks and broadly classified them into deterministic and non-deterministic approaches along with associated metrics being investigated in depth. The study poses some important questions and tries to answer them along dimensions of their significance, metric analysis, role, and challenges being faced in load balancing of software-defined networks. The study carried out by the authors presents the detailed advantages and limitations of existing literature in communication networks. Further the paper holds a strong foundation and solid correlation among load balancing metrics, despite the fact that it does not go specific to cloud computing domain. Also, the study is based on single-level classification rather than hierarchical classification.

In [13], Moradi et al., proposed algorithm called New Time Optimizing Probabilistic Load Balancing. An algorithm is proposed for the management of load and response time should be less and main purpose is to choose the resources based on least completion time and better past status.

3 Proposed Work

In this work, a comparative analysis among three traditional load balancing algorithms (i.e., throttled, round robin, equally spread current execution load) has been analyzed in using three different service broker policies, such as Reconfigure Dynamically, Closet Data Center, and Optimize Response time using cloud analyst simulator. In this paper, we have compared the result (service request time and total cost) of these three algorithms among each other by fixing a service broker policy. We have fixed service broker policy (i.e., closest data center) and changed the algorithm one by one, for example, closest data center with round robin algorithm, closest data center with throttled algorithm, closest data center with equally spread current execution load algorithm, and so on.

The three existing algorithms to distribute the workload across multiple nodes over the network links to achieve optimal resource utilization, minimum data processing time, minimum average response time, and to avoid overload are as follows:

Throttled load balancing algorithm—Throttled algorithm is completely based on virtual machine. Here the client first requests the load balancer to check the right virtual machine which access that load easily and performs the operations which are given by the client [10].

Round Robin load balancing algorithm—Round robin load balancing is a simple way to distribute client requests across a group of servers. A client request is forwarded to each server in turn. The algorithm instructs the load balancer to go back to the top of the list and repeats again.

Equally spread current execution load balancing algorithm—Equally spread current execution algorithm performs on a continuous basis on the queue and randomly passes it to another virtual machine and distributes it by measuring the size of the coming load, then distributes it to a virtual machine that has light work [14].

About cloud analyst simulator—The cloud analyst simulator is used to analyze the different load balancing algorithms. Cloud analyst simulator runs on cloudsim. It is developed by university of Melbourne.

3.1 Main Components of Cloud Analyst Simulator

I. **User base**—it is group of customer or user at a particular place
II. **Region**—it refers to the cloud region in which the "cloud region" describes the actual, real-life geographic location where our public cloud resources are located. ... When we choose a cloud provider, it means we are also choosing a "region," which is where our physical data centers really exist.
III. **Internet characteristics**—the configuration of internet depends upon two matrices, i.e., delay matrix and bandwidth matrix. Delay matrix is about the transmission delay between different regions (units in miliseconds) and bandwidth matrix defines the available bandwidth between regions for stimulated application (units in Mbps).
IV. **Service broker policy**—A service broker policy is basically a data center selection policy. It decides which data center should provide the service to the requests coming from each user. And thus, service broker controls the traffic routing between user and data centers [15].
From Fig. 2, it has been observed that in which kind of scenario service broker policy works in cloud analyst simulator. Here some data centers and users are shown in the figure. When request comes from the user then service broker policy helps to decide which data center will provide service for upcoming request.
The existing three service broker policy of cloud analyst simulator are following:

a. **Closest data center policy**—The data center which is having least proximity from the user is selected. Proximity in terms of least network latency. If more than one Data center has the same proximity, then it will select data center randomly to balance the load.
b. **Optimize response time policy**—optimize response time policy works on response time of each data center and selection of data center depends upon the least response time, i.e., optimum response time. The data center which has least response time will be faster, and it will provide service to the request coming from user.
c. **Dynamically reconfigurable routing with load balancing**—its application is based on currently processing request which is dynamic in nature. This service broker policy considers the current processing times and best processing time whichever is optimum.

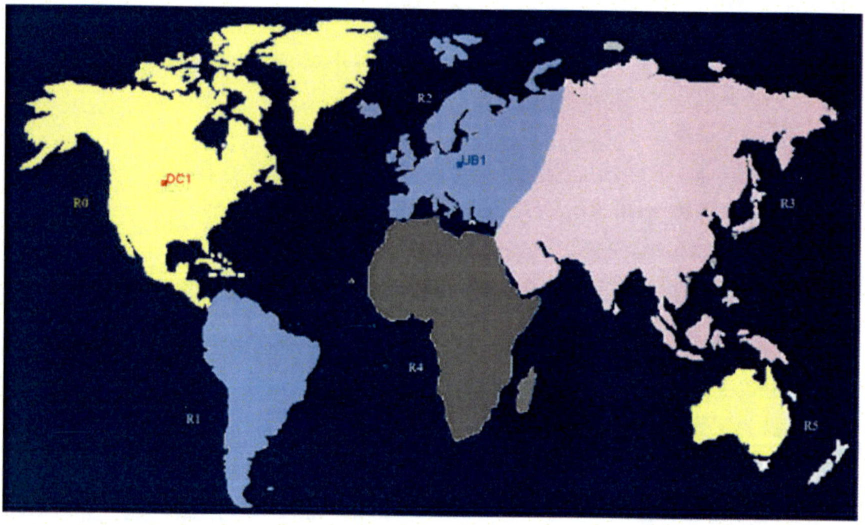

Fig. 2 Geographically distributed data centers

V. **Data center configuration**—it includes architecture, operating system, virtual machine cost, memory cost, no. of physical hardware units, storage cost, data transfer cost of particular data center.
VI. **Graphical user interface (GUI)**—This is a graphical user interface which acts as frontend controller.

3.2 Simulation Scenario

The cloud analyst simulator is used to analyze the existing load balancing algorithms which are provided by cloud sim. To implement the existing three algorithms, we have simulated the environment by taking 10 user base, 5 data center having 5 VM in each data center. Each simulation is performed for 60 min which is default simulation time of cloud analyst simulator. In the context of India, we have considered average peak user 100,000, average off peak user 10,000 in each user base, 100 request per user per hour, and data size per request of 1000 byte. Normally in India peak hour of internet use lies between 7:00 and 11:00 pm therefore we have considered that peak hour starts at 7:00 pm and end at 11:00 pm. Above all the consideration of data has been used in main configuration of cloud analyst simulator which is shown below.

Data center configuration includes architecture, operating system, virtual machine cost, memory cost, no. of physical hardware units, storage cost, data transfer cost of particular data center. We have feeded default data in data center configuration provided by cloud analyst which has shown in Figs. 3 and 4.

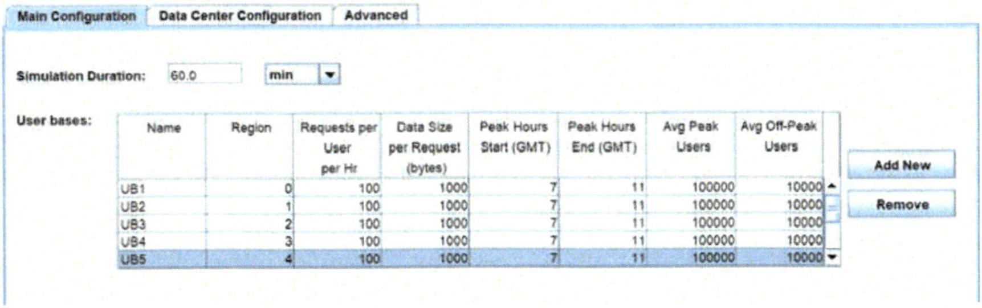

Fig. 3 Main configuration of cloud analyst simulator

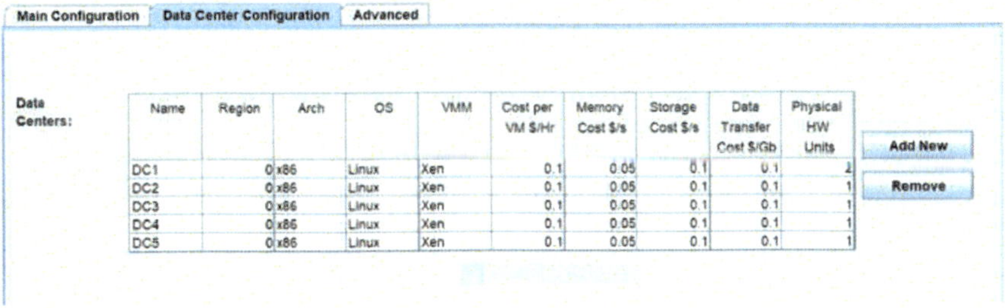

Fig. 4 Data center configuration of cloud analyst simulator

The advance section of cloud analyst simulator mainly deals with the various type of algorithms used by cloud analyst which are throttled algorithm, round robin algorithm, equally spread current execution load algorithm.

We will consider the existing algorithm (used in advance section of cloud analyst) one by one by fixing one service broker policy at a time. We will do the comparative analysis of data center request servicing time and total cost (VM cost + data transfer cost) of a data center obtained by comparing various algorithms fixing service broker policy one at a time.

Data center request servicing time:—it is the time taken by a data center to service a request coming from each user base. Usually, it is in milliseconds in cloud analyst simulator.

Total cost (Virtual Machine Cost and Data Transfer Cost):—Basically it is a service charge of cloud service provider. The total cost depends not only on VM cost, data transfer cost but also depends on various factor like network pricing, VM instance pricing, disk & image pricing, Gpu pricing, VM manager pricing, confidential VM pricing.

The various algorithms are compared by taking one service broker policy at a time. The different scenarios are as follows:

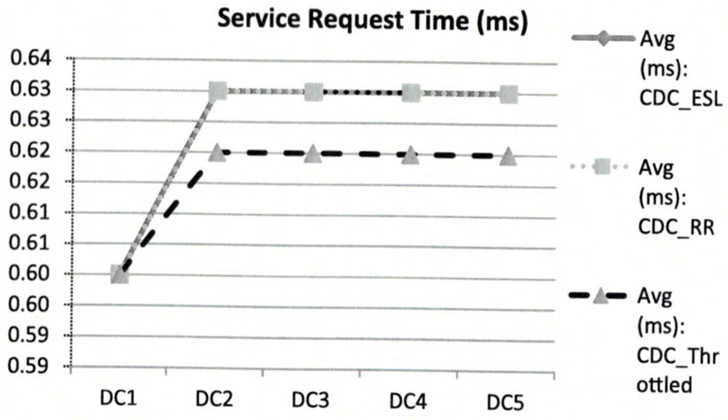

Fig. 5 Data center request service time at different data center

Scenario (I)—when Round robin, Throttled, and Equally spread current execution load algorithms are applied with closest data center service broker policy.

In this scenario round robin, throttled and equally spread current execution load algorithms with the closest data center service broker policy are applied for finding the data center request service time and total cost (VM cost + data transfer cost) for every data center that we have available in cloud analyst simulator, the result is depicted in Figs. 4 and 5.

In Fig. 6, we can observe that the data center service request time for CDC_ESL (closest data center with equally spread current execution load algorithm) and CDC_RR (closest data center with round robin algorithm) are same while the data center service request time for CDC_Throttled (closest data center with throttled algorithm) is less compared to former two algorithms. It clearly indicates that throttled algorithm with the closest data center service broker policy gives better result compared to other two algorithms, i.e., equally spread current execution load algorithm and round robin algorithm.

In Fig. 5, we can observe that the total cost including VM cost and data transfer cost data for CDC_ESL (closest data center with equally spread current execution load algorithm) is less compared to CDC_RR (closest data center with round robin algorithm) and CDC_Throttled (closest data center with throttled algorithm). It clearly indicates that equally spread current execution load algorithm with closest data center service broker policy is less costly compared to other two algorithms, i.e., throttled algorithm and round robin algorithm.

Scenario (II)—when Round robin, Throttled, and Equally spread current execution load algorithms are applied with Optimize response time service broker policy.

In this scenario round robin, throttled and equally spread current execution load algorithms with optimize response time service broker policy are applied for finding

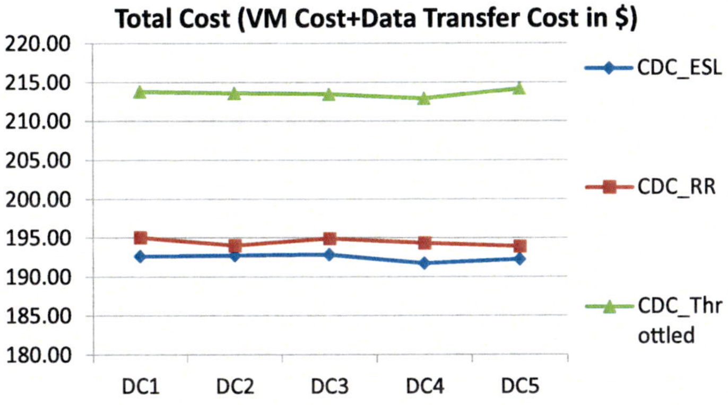

Fig. 6 Total cost at different data center

the data center request service time and total cost (VM cost + data transfer cost) for every data center that we have available in cloud analyst simulator, the result is depicted in Figs. 7 and 8. In Fig. 7, we can observe that the data center service request time for ORT_ESL (optimize response time with equally spread current execution load algorithm), ORT_RR (optimize response time with round robin algorithm), and ORT_Throttled (optimize response time with throttled algorithm) are almost same for every data center because of their overlapping line graph.

In Fig. 8, we can observe that the total cost including VM cost and data transfer cost data for ORT_ESL (Optimize response time with equally spread current execution load algorithm) is less compared to ORT_RR (Optimize response time with round robin algorithm) and ORT_Throttled (Optimize response time with throttled algorithm). It clearly indicates that equally spread current execution load algorithm

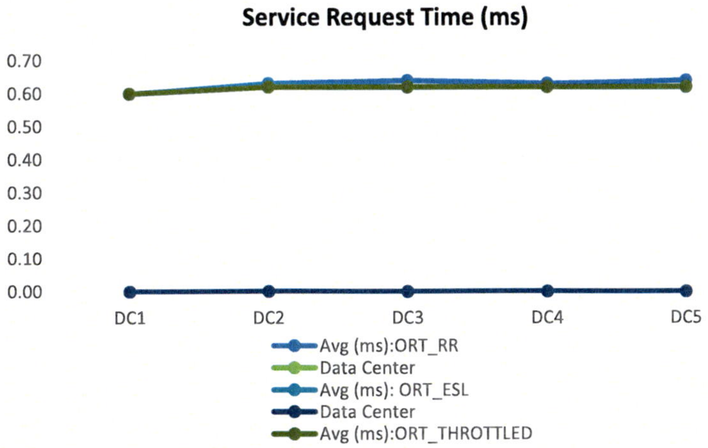

Fig. 7 Data center request service time at different data center

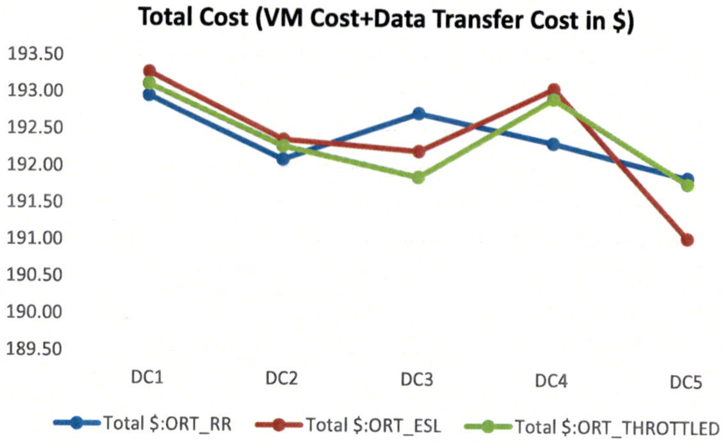

Fig. 8 Total cost at different data center

with optimize response time service broker policy is less costly compared to other two algorithms, i.e., throttled algorithm and round robin algorithm.

Scenario (III)—When Round robin, Throttled, and Equally spread current execution load algorithms are applied with Dynamically reconfigure service broker policy.

In this scenario round robin, throttled and equally spread current execution load algorithms with dynamically reconfigure service broker policy are applied for finding the data center request service time and total cost (VM cost + data transfer cost) for every data center that we have available in cloud analyst simulator, the result is depicted in Figs. 8 and 9. In Fig. 9, we can observe that the data center service request time for RD_Throttled (Reconfigure dynamically with throttled algorithm) is less compared to other two algorithm, i.e., RD_ESL (Reconfigure dynamically with equally spread current execution load algorithm) and RD_RR (Reconfigure dynamically with round robin algorithm. It clearly indicates that throttled algorithm with reconfigured dynamic service broker policy has faster request service time to the user compared to round robin and equally spread current execution load algorithm.

In Fig. 10, we can observe that the total cost including VM cost and data transfer cost data for RD_ESL (Reconfigure dynamically with equally spread current execution load algorithm) and RD_RR (Reconfigure dynamically with round robin algorithm) are almost the same and less costly than RD_Throttled (Reconfigure dynamically with throttled algorithm). It clearly indicates that equally spread current execution load algorithm and round robin algorithm with reconfigure dynamically service broker policy is less costly compared to other algorithm round robin.

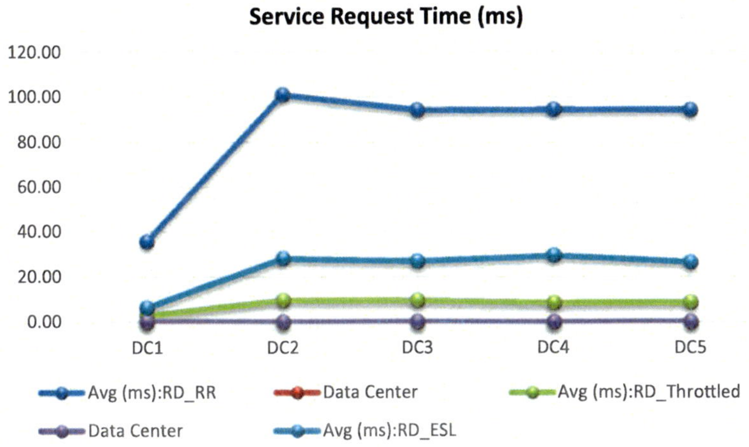

Fig. 9 Data center request service time at different data center

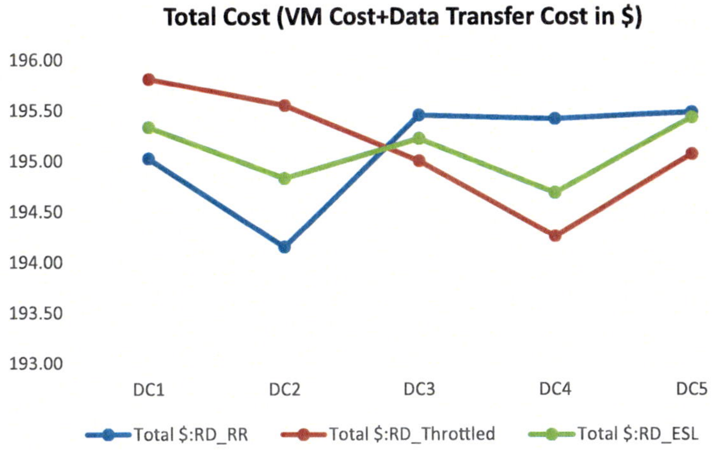

Fig. 10 Total cost at different data center

4 Result Analysis

Comparing the scenario (I), scenario (II), and scenario (III) having existing algorithms and different service broker policy, the Overall data center service request time and the Total cost graphs obtained have shown in Figs. 11 and 12. In Fig. 11, the round robin algorithm has greater data center service request time in comparison to other two algorithms which is not desirable because nowadays cloud service users prefer fastest service request time. In Fig. 12, the throttled algorithm costs more in comparison to other two algorithms.

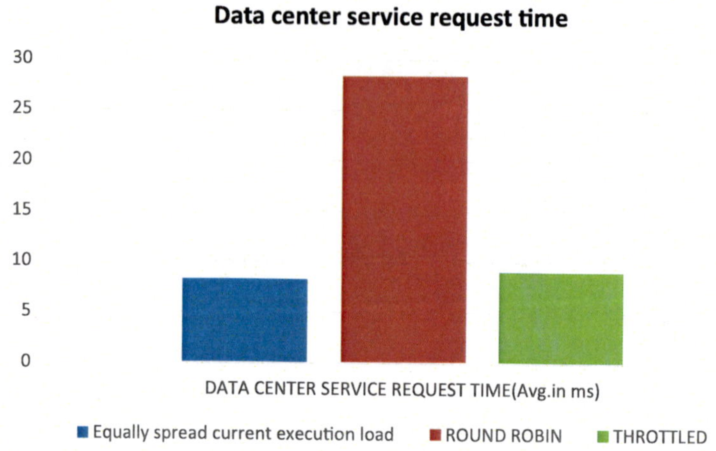

Fig. 11 Overall data center service request time of load balancing algorithms

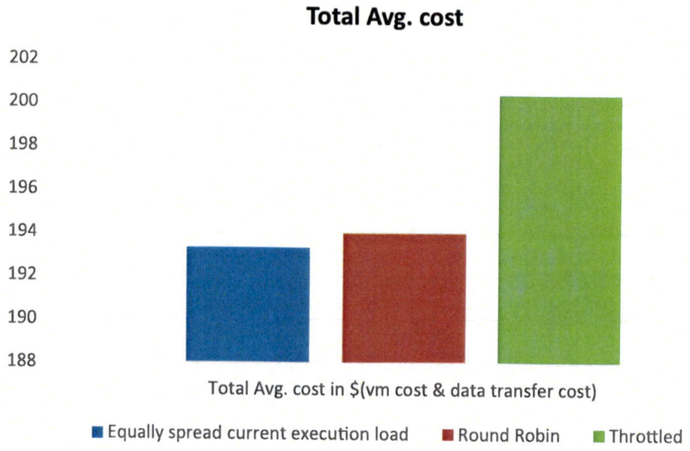

Fig. 12 Overall total average cost of load balancing algorithms

5 Conclusion

From result analysis, we can conclude that among the different load balancing algorithms, the data center service request time of Equally spread current execution load algorithm is good for five data centers and ten userbases in comparison to throttled and round robin algorithms. Also, the total cost (VM cost and data transfer cost) taken by the data center for Equally spread current execution load algorithm is less compared to other two algorithms mentioned above while the throttled algorithm costs more in comparison to other two algorithms.

From the above results analysis, we can say Equally spread current execution load balancing algorithm proves to be better than the other two algorithms. The comparison is performed among the three algorithms that are taken into considerations and taking one service broker policy at a time.

References

1. Kumar P, Kumar R (2019) Issues and challenges of load balancing techniques in cloud computing: a survey 2019. ACM Comput Surv 51, 6, Article 120, 35. https://doi.org/10.1145/3281010
2. Afzal S, Kavitha G (2019) J Cloud Comput Adv Syst Appl 8:22 https://doi.org/10.1186/s13677-019-0146-7
3. Bokhari MU, Shallal QM, Tamandani YK (2016) Cloud computing service models: a comparative study. In: 3rd international conference on computing for sustainable global development (INDIACom). pp 890–895
4. Mahmood Z (2011) Cloud computing: characteristics and deployment approaches. In: 2011 IEEE 11th international conference on computer and information technology (CIT). pp 121–126
5. Buyya R, Vecchiola C, Selvi ST (2013) Mastering cloud computing: foundations and applications programming. Morgan Kaufmann, USA
6. Jain N, Choudhary S (2016) Overview of virtualization in cloud computing. In: Symposium on colossal data analysis and networking (CDAN). pp 1–4
7. Alouane M, El Bakkali H (2016) Virtualization in cloud computing: no hype vs HyperWall new approach. In: 2016 International conference on electrical and information technologies (ICEIT). pp 49–54
8. Rimal BP, Choi E, Lumb I (2009) A taxonomy and survey of cloud computing systems. In: Fifth international joint conference on INC, IMS and IDC, NCM'09. pp 44–51
9. Shah JM, Kotecha K, Pandya S, Choksi DB, Joshi N (2017) Load balancing in cloud computing: methodological survey on different types of algorithm. In: 2017 International conference on trends in electronics and informatics (ICEI). pp 100–107
10. Domanal SG, Reddy GRM (2013) Load balancing in cloud computing modified throttled algorithm IEEE, international conference. CCEM
11. Kalra M, Singh S (2015) A review of metaheuristic scheduling techniques in cloud computing. Egypt Inform J 16(3):275–295
12. Neghabi AA, Navimipour NJ, Hosseinzadeh M (2018) Rezaee A: load balancing mechanisms in the software defined networks: a systematic and comprehensive review of the literature. IEEE Access 6:14159–14178
13. Moradi M, Dezfuli MA, Safavi MH Department of Computer and IT, Engineering, Amirkabir University of Technology, Tehran, Iran. A new time optimizing probabilistic load balancing algorithm in grid computing. IEEE 232–237,20 I O
14. Sahu Y, Pateriya RK (2013) Cloud computing overview with load balancing techniques. Int J Comput Appl 65(24)
15. Patel HV, Patel R (2015) Cloud analyst: an insight of service broker policy. Int J Adv Res Comput Commun Eng 4(1)

Robotics and Embedded Systems

Securely Importing of Back-Up File into Cryptowallet

P. Gnanendra Prasad, P. P. Amritha, and M. Sethumadhavan

Abstract Blockchain is one of the rapidly growing technologies in the current world, based on which various applications are being implemented. Similarly, various cryptowallets are being developed which are used for storing cryptocurrencies and other assets. These cryptowallets will play a major role in every cryptosystem, so a safe and secure wallet creation is the primary objective. Current cryptowallets are made to perform operations like transactions, backing-up cryptocurrencies, storing NFTs, etc. When it comes to backing up and importing the wallet, there are not many sufficient security measures that make sure that the stolen wallet's back-up file is not being imported by the attacker into the attacker's wallet. This paper proposes a safety measure that can be implemented for importing the back-up file of the wallets in a secure way through the secret sharing algorithm.

Keywords Blockchain · Cryptowallet · Cryptocurrency · Secret sharing algorithm · NFT

1 Introduction

Every day various technologies are being introduced into the world, one among them is the Blockchain. Blockchain is a decentralized and distributed ledger which performs immutable transactions through consensus, a rational decision-making process which is performed among multiple untrusted parties. In the blockchain, the transactions are stored in blocks and all those blocks are linked through their

P. G. Prasad · P. P. Amritha (✉) · M. Sethumadhavan
TIFAC-CORE in Cyber Security, Amrita School of Engineering, Amrita Vishwa Vidyapeetham, Coimbatore, India
e-mail: pp_amritha@cb.amrita.edu; cb.en.p2cys200016@cb.students.amrita.edu

P. G. Prasad
e-mail: gnanz06@gmail.com

M. Sethumadhavan
e-mail: m_sethu@cb.amrita.edu

A. K. Dubey et al. (eds.), *Advanced IoT Sensors, Networks and Systems*,
Lecture Notes in Electrical Engineering 1027,
https://doi.org/10.1007/978-981-99-1312-1_16

hash values, which led to the name "Blockchain". It is often managed by a peer-to-peer network, where the nodes in the network often participate to validate the transactions that are being performed in the blockchain. Once a block of transactions is added to the blockchain, it cannot be modified by any means, which provides the property of immutability. The very first application of blockchain that came to the spotlight among the public is Bitcoin, which was developed by an unknown person/group known as Nakamoto [1]. Secret sharing-based handling of social media content using blockchain and distributed storage through IPFS was discussed in [2]. Assorted applications of blockchain are covered in [3–5].

2 Cryptowallet

A cryptowallet can be an application or a device which generates and stores the public and private key pairs that are required for cryptocurrency transactions. Sometimes, a cryptowallet can also perform operations like encrypting and signing an information using the keys. Initially, when a user wants to create a new wallet in the application, a public and private key pair is generated by the application. Among those pairs, the public key will be used as the unique address/identifier of the wallet for receiving cryptocurrencies, whereas the private keys are used for accessing and sending the cryptocurrencies and also for signing. The generated private keys can also be backed up when the user wants to import them into another wallet of her/his own. The public key will be known to everyone, since it will be used as the address of the wallet, whereas the private key should be secured by the user [6]. If the private key's possession is lost, then all the assets that the user holds will be lost forever or can be used by an unauthorized person.

2.1 Types of Cryptowallets

The cryptowallets are categorized by considering the internet connectivity and location of storage [7].

2.1.1 Based on Internet Connectivity

Hot Wallet

The Hot Wallets are used in the presence of internet connectivity. These wallets are often in the form of a software application and are used for regular transactions or short-term holdings. Due to more interaction through internet, these kinds of wallets possess high risk, where the attackers have a way to steal [8]. Examples: Metamask, Bitcoin Core, Phantom, etc.

Cold Wallet

These wallets have rare interaction with the internet, and these are often in the form of a hardware or device and paper. These wallets are used for holding assets for a long term. Unlike Hot Wallet, due to less internet interaction, it has very low risk of losing the assets in it [8]. Examples: Trezor, Ledger Nano, Paper Wallet, etc.

2.1.2 Based on Location of Storage

Paper Wallet

A Paper Wallet is a physically printed or written wallet. The user can either write down all the private keys or else applications like Bitcoin Core generate a QR code which can be printed on a paper and used in later situations. Although they are not prone to attacks, re-usability of such wallets is not possible.

Desktop Wallet

These wallets are installed in computer devices as an ordinary application, like any other wallet, it also generates public and private key pair for transaction purpose.

Some blockchain wallets like Exodus, Ledger, etc., enable the user to just store the assets, whereas other wallets like Bitcoin core, Dogecoin, etc., have an extra feature like adding the device to the blockchain network for the purpose of mining.

Mobile Wallet

These are similar to Desktop Wallets, but are created to support the mobile devices. These are often developed for operating systems like Android, IOS, etc. The mobile wallets are often used for daily transactions. But, it is also vulnerable to various attacks [9]. Examples: Coinbase, Mycelium, etc.

Web Wallet

The Web Wallets are developed to perform transactions through various web browsers like Chrome, Firefox, Brave, etc. These are directly accessed by web browsers and often come in the form of web extensions. Examples: Metamask, Phantom, Coinbase, etc.

3 Problem Statement

In the current cryptowallets, after the user backs up the private keys, those private keys can be imported back into any other wallet, besides the one it was exported from initially. It provides an advantage to an attacker when the backed-up file has been stolen and imported into the attacker's wallet. This paper proposes a method, where during the process of importing, the wallet will check whether the private keys that are being imported belong to only that particular wallet from which it was exported.

4 Literature Survey

This section deals with the various approaches proposed to secure the private keys in a wallet, by researchers around the world. The common grounds that raised most of them to find a solution are the real-world incidents where many assets stored in a cryptowallet are being stolen by attackers, either through attacking the system in a network, or by injecting a malicious software which steals the cryptocurrencies from the victim.

Liu et al. proposed an efficient method in order to enhance the security of the bitcoin wallet [10]. In Fig. 1, we can notice that they have combined random seeds and certain passphrases which will be easy for the user to keep in mind, for generating private keys whenever a new wallet is generated. So, the only other task is to reserve the randomly generated seeds in the local device, rather than storing the private keys itself. It is not possible to construct the whole private keys just by using random seeds and avoiding the passphrase. Thus, this method is convenient to be used by a common user without any professional knowledge. The key management scheme can also be used in operations like signing and authorizing the transactions. And also, a function to recover the passphrase from respective personal questions has also been proposed so that if there is a situation where the user forgets the passphrase, then this function will be useful.

In the paper proposed by Hossein et al. [11], each party will have a public and private key pair and, along with that, a shared secret will be generated on both sides using their respective private key and other party's public key. Using Elliptical Curve Cryptography, the private key will be a random scalar, whereas the public key will be calculated by using the private key and a fixed-point G. Later, during the process of signing, the respective party will share their public key to the other, and the shared secret message is further calculated. Since this method is prone to Man-in-the-Middle Attack, further solution was provided, a working prototype was designed where the user has to provide visual confirmation that both the secret messages are same. In the prototype, the secret messages will be generated by both the back-up wallet and the main wallet. The generated secret message will be displayed on a screen, where the user has to visually confirm that they both are the same. If the user confirms, the back-up wallet enables the user to back-up her/his private keys.

5 Proposed Work

5.1 Keys Generation

The flowchart of key generation is shown in Fig. 2. In this project, the keys are generated by using Shamir Secret Sharing Algorithm [12]. The keys are generated by obtaining the device's characteristics like IP Address, MAC Address, Processor ID, Time and Date, Hashed value of an image, and a random string from the user.

Fig. 1 Enhancement of security aspect in the bitcoin wallet

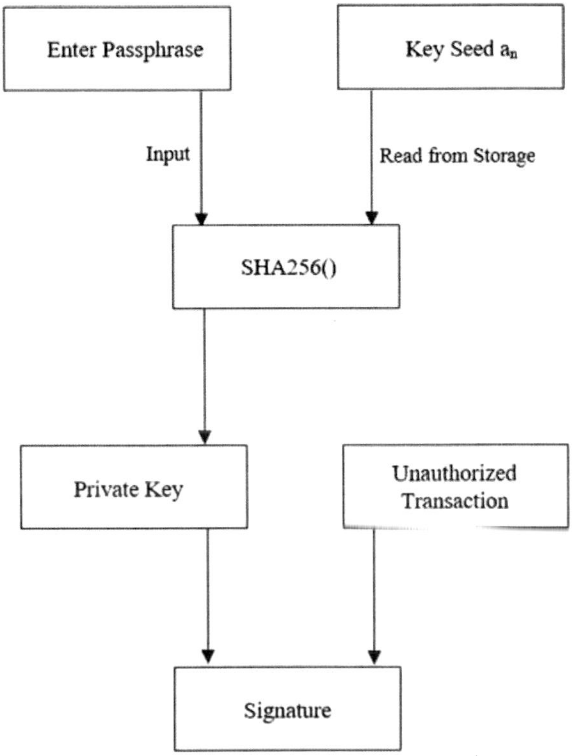

Using the above-mentioned information, respective values are concatenated in a random order and are undergone the SHA3-256 hashing algorithm. The final hash value is used as the secret message in Shamir Secret Sharing Algorithm and the keys are generated.

By using Shamir Secret Sharing Algorithm, we will be generating the keys by following (2, 3), where minimum two keys are required in order to reconstruct the secret message which was initially used for generating the keys. Among those three keys, two keys are used as public and private key pair for the wallet, whereas the third key is used as a verification key, which will be used during the process of importing a wallet. For a single wallet, only one pair of public and private keys and a verification key should be generated. These keys should be stored in the respective way as it is deemed to be.

5.2 *Wallet Importing*

Fig. 3 describes the flow that occurs during the process of importing a wallet that has been backed up. Initially, the respective keys are fetched from the files where they

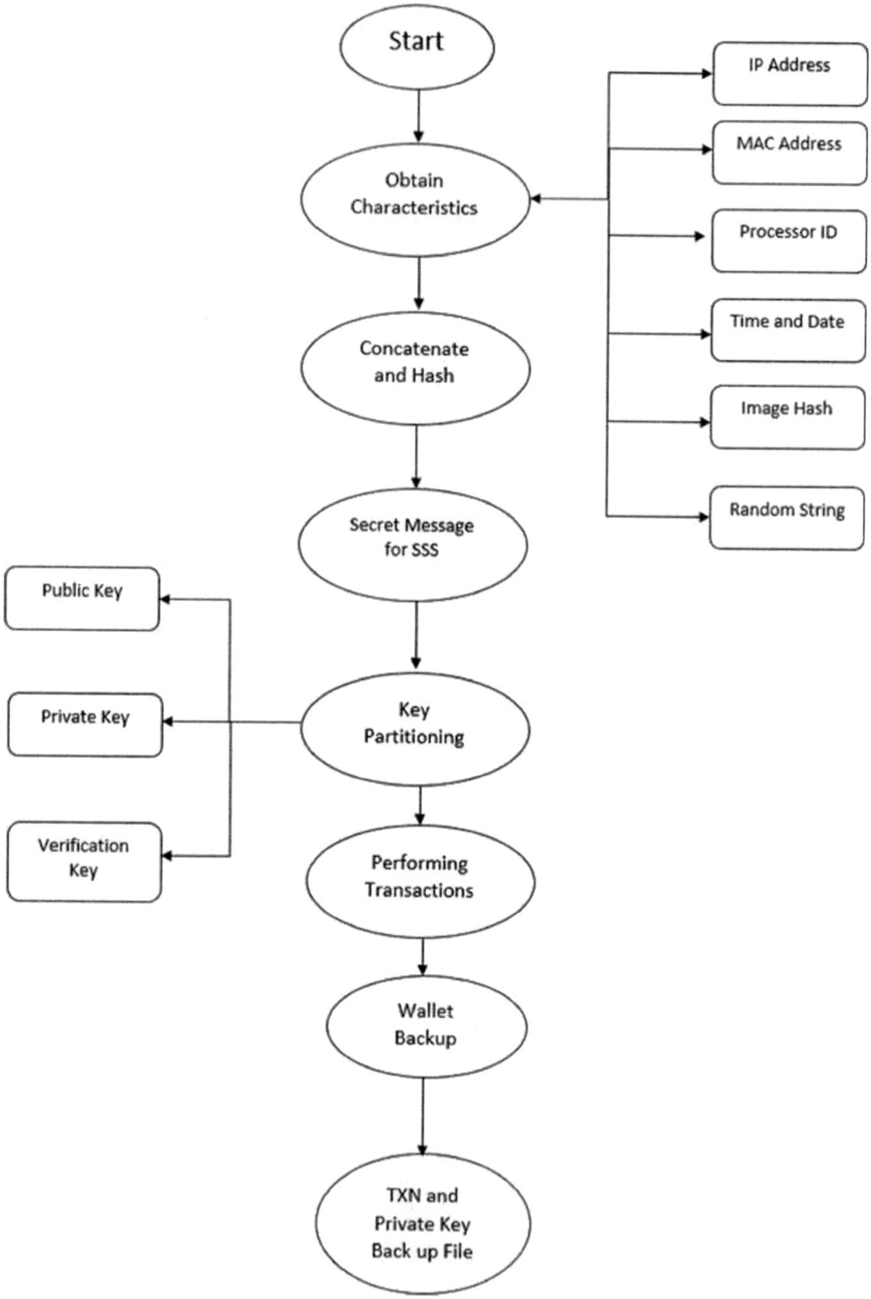

Fig. 2 Key generation flowchart

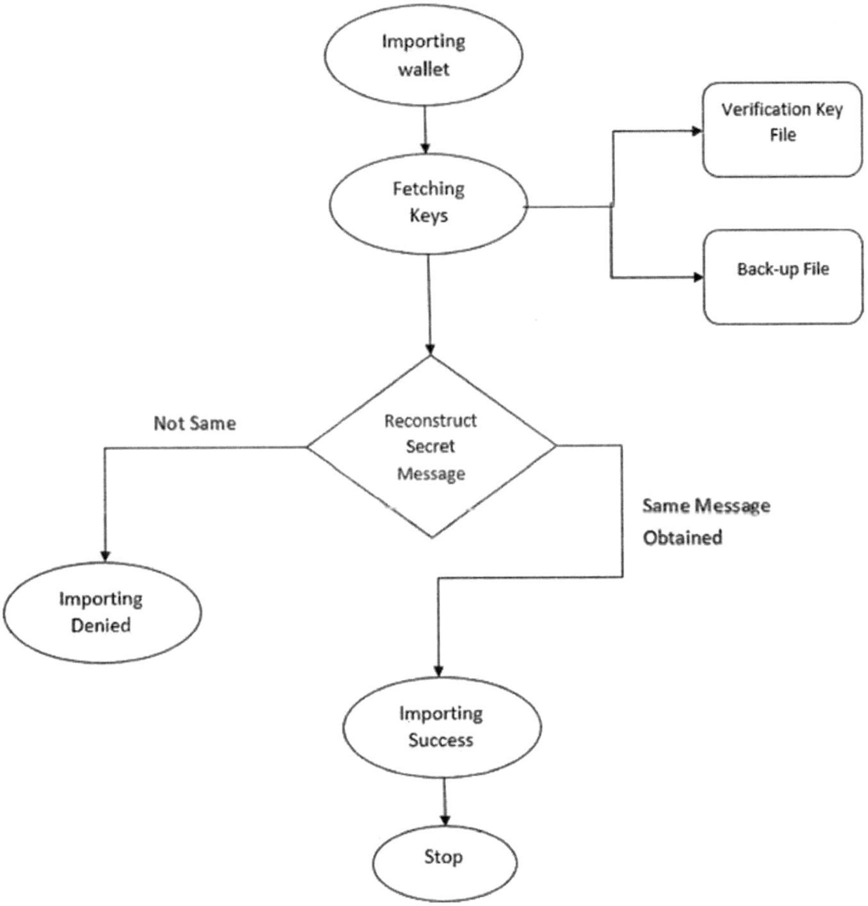

Fig. 3 Wallet importing flowchart

were stored. After fetching it, the wallet tries to reconstruct the secret message and a comparison is done which will be explained in the below section. After reconstructing the message if the secret message has been obtained, then it states that a legit import is being performed. If the secret message isn't reconstructed successfully, then the wallet recognizes it as an invalid import and rejects the process of importing.

5.3 *Verification*

During the process of importing, after the respective public, private and verification keys are fetched from their respective files, reconstruction of the secret message is done twice using public and private pair and verification key and private key pair. If

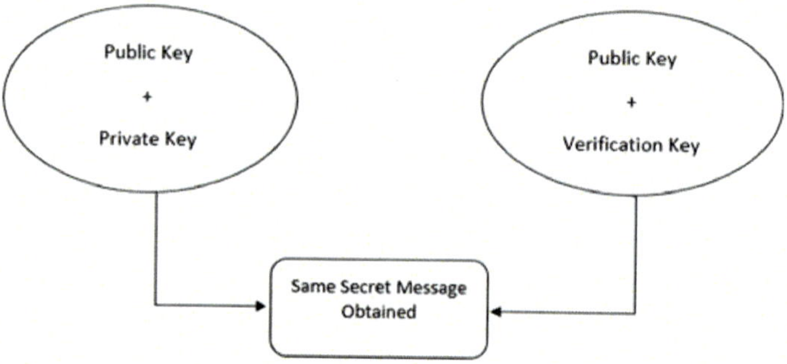

Fig. 4 Verification process flowchart

the reconstructed message from the above-mentioned two pairs is same, then it is a valid importing since the user is holding valid verification keys and the importing is legit, whereas the attacker will be holding wrong verification keys. So, if the attacker tries to import the stolen wallet, then while importing, two different secret messages are reconstructed which shows that an invalid import is being done and thus it will be stopped. The flow of the mentioned verification process can be depicted in Fig. 4.

6 Results and Discussion

A basic implementation using Java programming language has been done and executed. It can be seen in Fig. 5 the program is successfully getting the respective computer's characteristic information, image's hash, and a random string from the user.

And later those values are successfully hashed using the SHA3-256 algorithm which gives the result in hexadecimal. The hexadecimal value has been converted to decimal value so that it can be used for the mathematical calculations in Shamir Secret Sharing Algorithm. After obtaining the secret message, using the SSS algorithm, three keys with their respective indexes have been created as shown in Fig. 6 and those keys are written to their respective files so that they can be fetched during the process of importing.

In Fig. 7, we can notice that the keys are being fetched from their respective files and are being used for reconstructing the secret message for further verification.

The basic implementation gave a brief idea about what will happen during the process of importing and the same algorithm can be used while designing a whole cryptowallet. The size of all the generated keys was observed and all of them were of same size. Various obstacles like storing the keys in a file and accessing it were also faced, but they were eventually solved. The keys were directly dumped into their respective file after getting appended with the timestamp using which they were created. Later, those keys were fetched by using the same timestamp. Timestamp was included because it has randomness and collision doesn't happen.

```
-------------------< com.mycompany:3rd_Sem_Project >--------------------
Building 3rd_Sem_Project 1.0-SNAPSHOT
--------------------------------[ jar ]---------------------------------

--- exec-maven-plugin:3.0.0:exec (default-cli) @ 3rd_Sem_Project ---
Enter Wallet Name:
a1
SHA Value of Image: 58e62540dc83eac36e9bb780a94055dc4af5b78bd4dc40d7135faa285ed5f5e3

IP Address: 192.168.56.1

MAC Address: 0A-00-27-00-00-0F

Processor ID: BFEBFBFF000706E5

Nonce Value (Time and Date): 08103819122021

Enter a random string:
gnanendra

Concatenated String: 192.168.56.10A-00-27-00-00-0FBFEBFBFF000706E5  gnanendra08103819122021

SHA value of concatenated string: b3d43cbd915c4d5e290ba1423a54fb83fd7f51550e8d1cee38cbb2d1786ed12b

Hex Value: 30768134b469ecc46935b70e59c43156a3463b71bcbcb4ae6faff53a480cab22

Decimal Value: 2192039643183638613261233511166995146993469883029184447532578156884260897872
```

Fig. 5 Getting characteristics and generating secret message

```
Generated Key Pairs:
Pair 1 Value: (1,35028303354406743528782317758258024380410880767802936710186188273923490360942)

Generated Key Pairs:
Pair 2 Value: (2,48136210276977100924952300404846097290887062705314028945046594979004371743162)

Generated Key Pairs:
Pair 3 Value: (3,61244117199547458321122283051434170201363244642825121179907001684085253125382)
Data Written to Files!!!
```

Fig. 6 Generating three keys

```
---------RECONSTRUCTED public+private----------

VALUE: 2192039643183638613261233511166995146993469883029184447532578156884260897872

---------RECONSTRUCTED priv+sec----------

VALUE: 2192039643183638613261233511166995146993469883029184447532578156884260897872
Import Success!!!
------------------------------------------------------------------------
BUILD SUCCESS
------------------------------------------------------------------------
```

Fig. 7 Fetching keys and reconstructing successfully

In the current wallets like Bitcoin Core, the private keys are directly stored in the back-up file, and the same back-up file is used for importing into another wallet. In a scenario where, if the attacker steals that particular back-up file from the victim's device, then the attacker can also easily import the victim's assets into his (attacker's) own wallet. In our proposed method, even if the attacker steals the private keys, the attacker should possess the third set of keys for importing the victim's assets. Thus, stealing private keys alone won't make the user to lose his/her own assets.

7 Conclusion

Although various security factors are being implemented in cryptowallets [13], securing the private keys are being one of the major problems when it comes to the safety of the cryptowallet [14]. This method preserves the integrity of the imported backed-up data from the wallet. This work has been done by considering the backing up process in a cryptowallet which can be done only by the authorized personnel. The ingenuity of the solution presented in this paper is achieved through the usage of algorithms based on secret sharing methods. Further work can be done by concentrating on the ways to store the "verification key", i.e., in an IPFS, or anywhere else. Furthermore, by using the concept of the third key, a cryptowallet governance can also be implemented, which will enable an authorized governing body to govern the cryptowallets which are being used for illegal activities.

References

1. Nakamoto S (2008) Bitcoin: a peer-to-peer electronic cash system. Decentralized Business Review. 21260
2. Kripa M, Nidhin Mahesh A, Ramaguru R, Amritha PP (2020) Blockchain framework for social media DRM based on secret sharing. In: International conference on information and communication technology for intelligent systems. Springer, Singapore, pp. 451–458
3. Ramaguru R, Sindhu M, Sethumadhavan M (2019) Blockchain for the internet of vehicles. In: International conference on advances in computing and data sciences. Springer, Singapore, pp 412–423
4. Gopalakrishnan P, Ramaguru R (2019) Blockchain based waste management. Int J Eng Adv Technol 8(5):2632–2635
5. Anjum S, Ramaguru R, Sethumadhavan M (2021) Medical records management using distributed ledger and storage. In: International conference on advances in computing and data sciences, vol 1441. Springer, Cham, pp 52–62
6. Suratkar S, Shirole M, Bhirud S (2020) Cryptocurrency wallet: a review. In: 2020 4th International conference on computer, communication and signal processing (ICCCSP). pp 1–7. https://doi.org/10.1109/ICCCSP49186.2020.9315193
7. Taylor SK, Ariffin A, ZainolAriffin KA, Sheikh Abdullah SNH (2021) Cryptocurrencies investigation: a methodology for the preservation of cryptowallets. In: 3rd international cyber resilience conference (CRC). pp. 1–5

8. Khan AG, Zahid AH, Hussain M, Riaz U (2019) Security of cryptocurrency using hardware wallet and QR code. Int Conf Innov Comput (ICIC) 2019:1–10. https://doi.org/10.1109/ICIC48496.2019.8966739
9. Li C, He D, Li S, Zhu S, Chan S, Cheng Y (2020) Android-based cryptocurrency wallets: attacks and countermeasures. Int Conf Comput, Commun Signal Process (ICCCSP) 2020:9–16. https://doi.org/10.1109/Blockchain50366.2020.00010
10. Liu Y, et al (2017) An efficient method to enhance Bitcoin wallet security. In: 11th IEEE international conference on anti-counterfeiting, security, and identification (ASID). pp 26–29
11. Rezaeighaleh H, Zou CC (2019) New secure approach to backup cryptocurrency Wallets. In: IEEE global communications conference (GLOBECOM). pp. 1–6
12. Bissoli A, d'Amore F (2021) Authentication as a service based on Shamir secret sharing. In: International symposium on computer science and intelligent controls (ISCSIC). pp 368–373
13. Urien P (2021) Innovative countermeasures to defeat cyber attacks against blockchain wallets. In: 2021 5th cyber security in networking conference (CSNet). pp 49–54. https://doi.org/10.1109/CSNet52717.2021.9614649
14. Sapna, Prashar D (2021) Analysis on blockchain vulnerabilities and attacks on wallet. 2021 3rd international conference on advances in computing, communication control and networking (ICAC3N). pp 1515–1521. https://doi.org/10.1109/ICAC3N53548.2021.9725403

Trajectory Optimization of a Satellite

Mahika Sachdeva, M. S. Prasad, and Shivani Verma

Abstract In the past three decades, for economical space transportation, minimum energy transfers in orbit or trajectories is sought. This is done via in-depth research to understand intricate N-body problem dynamical structure. To deduce preferable and promising economical transfer trajectories, dynamical attributes of CRTBP is taken into consideration which perfectly works for multiple binary systems within the solar system. However, due to chaotic dynamics, capture/trap problem around smaller body becomes a challenging and complex problem due to large gravitational impact of larger body in CRTBP. This research can be used to design preliminary lunar missions from Earth parking orbit to B-plane encounter at the moon. The software assumes trans-lunar injection (TLI) occurs impulsively from a circular Earth parking orbit. The B-plane coordinates used for targeting the final conditions at the moon are expressed in a moon-centered (selenocentric) mean equator and IAU node of epoch coordinate system. The results from this scientific simulation can be used as an initial guess for a finite-burn trajectory optimization program.

Keywords CRTBP · TLI · IAU

1 Introduction

Spacecraft is a scientific and practical means of providing useful information to the humans and making life easier. For example, in our everyday lives, we can easily communicate with the people on the other side of the world through communication

M. Sachdeva (✉) · M. S. Prasad · S. Verma
Amity Institute of Space Science and Technology, Amity University Uttar Pradesh, Noida, India
e-mail: mahikasachdeva98@gmail.com

M. S. Prasad
e-mail: msprasad@amity.edu

S. Verma
e-mail: sverma2@amity.edu

A. K. Dubey et al. (eds.), *Advanced IoT Sensors, Networks and Systems*,
Lecture Notes in Electrical Engineering 1027,
https://doi.org/10.1007/978-981-99-1312-1_17

satellites. Or we can anticipate and prevent weather events through meteorological satellites. We can also examine the earth surface more thoroughly through geographical observational satellites. Space telescopes and reconnaissance satellites provide us with a lot of information about space and deep space and form the basis for future space missions. Thus, spacecrafts are very helpful to us in our everyday lives and scientific studies, and therefore it is imperative that these tools be stable in a proper orbit for correct operation.

The above-mentioned spacecrafts are vehicles that work in orbits in Earth's vicinity. But space research is not limited to closeness to the earth. Space Science aims to reach all celestial bodies beyond the Earth in an acceptable and safe manner. There is no need to be a priest to say that mankind can continue its existence in extra-terrestrial life zones in the near or far future. In order to be able to reach to this pinnacle, it is necessary to find the way to make the interplanetary journey short, cheap and safe.

This research can be used to plan preliminary lunar missions ranging from Earth parking orbit to a B-plane contact. Trans-lunar injection (TLI) occurs impulsively from a circular Earth parking orbit, according to the software. The final conditions at the moon are targeted using B-plane coordinates in a moon-centered (selenocentric) mean equator and IAU node of epoch coordinate system. This scientific simulation's results can be utilized as a starting point for a finite-burn trajectory optimization algorithm.

The first part of the project solves for the minimum TLI delta-v using a two-body Lambert solution for the transfer trajectory from the Earth parking orbit to the center of the moon. The second part of the project implements a shooting method that attempts to minimize the TLI delta-v while numerically integrating the spacecraft's n-body equations of motion and targeting user-defined constraints. In the shooting algorithm, the spacecraft motion model includes the Earth's non-spherical gravity effect and the point-mass perturbations of the sun and moon. The B-plane targets are enforced via a user-defined periapsis radius and orbital inclination of the arrival hyperbola relative to the moon.

The MATLAB script uses a nonlinear programming method from the Optimization Toolbox for both optimization tasks required for the lunar transfer problem implemented in this script. The solar and lunar coordinates required by this script are computed using the JPL DE430 ephemeris.

2 Literature Survey

In celestial mechanics, one of the most interesting problems is N-Body Dynamics which was first formulated by Newton and since then undergone in-depth mathematical research on the cars in hands of Poincare, Lagrange, Delaunay, Newcomb, Lindstedt, Sundman and many others [1].

However, 2BP provides an elegant solution but TBP provides more accurate solution with its eighteenth order system and ten exact integrals (6 mass center motion integrals, 3 angular momentum integrals, 1 internal energy integral). Langrange's equilateral triangle and collinear solution are special solutions for general 3BP. Later, the general 3BP was formulated to a relatively more practical CRTBP [2].

In CRTBP, two large bodies are circularly around each other, thus they are bound in a rotating frame, and a third body whose mass is very less or almost negligible with respect to other two masses, moves in the rotating frame.

CRTBP model imitates celestial system dynamics as many solar system bodies makes nearly circular motion around the sun [3] but CRTBP has quite chaotic nature of N-Body Dynamics. Complex CRTBP motions are analyzed, same as nonlinear dynamical behavior, via 2 ways:

- CRTBP's equilibrium points dynamics
- Mapping Technique.

These two methods help not only in designing smart and fuel economical space missions, but also in understanding the motion dynamics inclusive of capture/trap problem. CRTBP's equilibrium points helps in obtaining periodic trajectories and orbits, also more insight in CRTBP notion is gained through mapping technique [4].

Due to chaotic behavior of N-Body Dynamics, the motion intricacy is generally examined through a set of periodic orbits connected to the equilibrium or Lagrange points. To achieve in-depth and better understanding of CRTBP, much detailed research has been carried out about the dynamics of Lagrange or equilibrium point. This is done through linearization of Lagrange or equilibrium points and thus identifying families of Lissajous and Halo orbits [5]. The first researchers to work in this domain were Dunham and Farquhar, McGehee and Conley continued work from where it was left and became the first to describe phase space regions between Lagrange points L1 and L2 and gave the accurate classification of the types of trajectories.

Their work was further researched at Barcelona School, Spain by Martinez, Llibre, Simo, Masdemont and Gomez, Lastly, by taking past research into account, Lo, Ross, Koon, Marsden along with F. Topputo, F. Bernelli- Zazerra and M. Massari became first to present the space mission design application using Lagrange or equilibrium points [6].

Looking into the practical aspect, there are various space missions which demonstrate the application of dynamics of Lagrange or equilibrium points. For missions like sun exploration mission where there is a requirement of station keeping at libration point L1 for CRTBP, in-depth understanding of Lagrange points dynamics plays a very vital part especially in application domains like establishing communication link. This is due to the reason that periodic orbits (Lyapunov or Halo orbit) around equilibrium point L1 (within Sun-Earth region) are the best regions for examination point of view of the sun's shape in determining angular deviation from the sun. For space telescope positioning, Sun-Earth system equilibrium points were taken into account for missions like FIRST and PLANK [7].

For mission like Terrestrial Path Finder, quasi-periodic Halo orbits were taken into consideration along with formation flying trajectories about Lagrange point and the control methodology. Also, a higher order expansion was carried out by Masdemont of an invariant manifold of Lagrange point orbits as an analytical solution technique to them, along with a detailed analysis of complete set of periodic (Halo orbits) and quasi-periodic orbits around Lagrange point L1 or L2 [8].

Also, other than the use of equilibrium points of CRTBP in station keeping missions, they are also used low energy transfers and capture/trap transfers which are carried out through the use of trajectories which are asymptotic in nature around equilibrium or Lagrange points of periodic orbits which are also known as Lagrange point's invariant manifold tubes. A space highway is formed in between the interconnection of two manifold tubes which allows spacecraft to jump from one tube to another.

However, there is a major problem associated with invariant manifold tubes which arises when manifold tube approaches the secondary body, chaotic nature of its motion around secondary body is observed, which requires a huge and expensive trajectory corrective burn for safety and stability of trajectory [9]. Practical solution of this problem lies with a suitable map, which is used widely in Non-Linear and Chaotic Dynamics. Classical Poincare Mapping technique provides an optimum practical approach for the analysis and use of these chaotic regions of CRTBP.

3 Working Description

The first part of this MATLAB script solves for the minimum TLI delta-v using a two-body Lambert solution for the transfer trajectory from the Earth's parking orbit to the center of the moon. The second part of the script implements a shooting method that attempts to minimize the TLI delta-v while numerically integrating the spacecraft's n-body equations of motion and targeting to user-defined constraints. In the shooting algorithm, the spacecraft motion model includes the Earth's non-spherical gravity effect and the point-mass perturbations of the sun and moon. The B-plane targets are enforced via a user-defined periapsis radius and orbital inclination of the arrival hyperbola relative to the moon [10].

The code is "data-driven" by a simple text file created by the user. The first four lines of any input file are reserved for user comments. These lines are ignored by the software. However, the input file must begin with four and only four initial text lines.

The software allows the user to specify an initial guess for the TLI calendar date and lower and upper bounds on the actual date found during both the two-body and numerically integrated TLI delta-v optimization processes. For any guess, for the TLI time TLI t and user-defined lower and upper bounds l Δt and u Δt, the actual TLI time t is constrained according to

$$t_{\mathrm{TLI}} - \Delta t_l \leq t \leq t_{\mathrm{TLI}} + \Delta t_u. \tag{1}$$

3.1 Solving the Two Body Lambert Problem

The algorithm used in the MATLAB script is based on the method described in "A Procedure for the Solution of Lambert's Orbital Boundary-Value Problem" by R. H. Gooding, Celestial Mechanics and Dynamical Astronomy 48: 145–165, 1990. This iterative solution is valid for elliptic, parabolic and hyperbolic transfer orbits which may be either posigrade or retrograde, and involve one or more revolutions about the central body [11].

3.2 Modeling the Spacecraft's n-Body Trajectory

The trajectory analysis implements a special perturbation technique which numerically integrates the vector system of second-order, nonlinear differential equations of motion of a spacecraft given by:

$$\vec{a}(\vec{r}, \vec{v}, t) = \ddot{\vec{r}}(\vec{r}, \dot{\vec{r}}, t) = \vec{a}_g(\vec{r}) + \vec{a}_m(\vec{r}, t) + \vec{a}_s(\vec{r}, t) \tag{2}$$

where

$t =$ UTC time
$\vec{r} =$ inertial position vector of the spacecraft
$\vec{v} =$ inertial velocity vector of the spacecraft
$\vec{a}_g =$ accelerationduetotheEarth's gravity
$\vec{a}_m =$ acceleration due to the Moon
$\vec{a}_s =$ acceleration due to the Sun

The spacecraft's orbital motion is modeled with respect to the Earth's true-of-date coordinate system.

This computer program uses a spherical harmonic representation of the Earth's geopotential function given by

$$\begin{aligned}\Phi(r, \phi, \lambda) &= \frac{\mu}{r} + \frac{\mu}{r}\sum_{n=1}^{\infty} C_n^0 \left(\frac{R}{r}\right)^n P_n^0(u) \\ &+ \frac{\mu}{r}\sum_{n=1}^{\infty}\sum_{m=1}^{n}\left(\frac{R}{r}\right)^n P_n^m(u)\left[S_n^m \sin m\lambda + C_n^m \cos m\lambda\right]\end{aligned} \tag{3}$$

where ϕ is the geocentric latitude of the spacecraft, λ is the geocentric east longitude of the spacecraft and

$$r = |\mathbf{r}| = \sqrt{x^2 + y^2 + z^2} \tag{4}$$

is the geocentric distance of the spacecraft. In this expression, the S's and C's are unnormalized harmonic coefficients of the geopotential, and the P's are associated Legendre polynomials of degree n and order m with argument u = $\sin\phi$.

The software calculates the spacecraft's acceleration due to the Earth's gravity field with a vector equation derived from the gradient of the potential function expressed as

$$\mathbf{a}_g(\mathbf{r}, t) = \nabla\Phi(\mathbf{r}, t) \tag{5}$$

This acceleration vector is a combination of pure two-body or point mass gravity acceleration and the gravitational acceleration due to higher order non-spherical terms in the Earth's geopotential. In terms of the Earth's geopotential Φ, the inertial rectangular cartesian components of the spacecraft's acceleration vector are as follows:

$$\begin{aligned}
\ddot{x} &= \left(\frac{1}{r}\frac{\partial\Phi}{\partial r} - \frac{z}{r^2\sqrt{x^2+y^2}}\frac{\partial\Phi}{\partial\phi}\right)x - \left(\frac{1}{x^2+y^2}\frac{\partial\Phi}{\partial\lambda}\right)y \\
\ddot{y} &= \left(\frac{1}{r}\frac{\partial\Phi}{\partial r} - \frac{z}{r^2\sqrt{x^2+y^2}}\frac{\partial\Phi}{\partial\phi}\right)y + \left(\frac{1}{x^2+y^2}\frac{\partial\Phi}{\partial\lambda}\right)x \\
\ddot{z} &= \left(\frac{1}{r}\frac{\partial\Phi}{\partial r}\right)z + \left(\frac{\sqrt{x^2+y^2}}{r^2}\frac{\partial\Phi}{\partial\phi}\right)
\end{aligned} \tag{6}$$

The three partial derivatives of the geopotential with respect to r, ϕ, λ are given by

$$\begin{aligned}
\frac{\partial\Phi}{\partial r} &= -\frac{1}{r}\left(\frac{\mu}{r}\right)\sum_{n=2}^{N}\left(\frac{R}{r}\right)^n (n+1)\sum_{m=0}^{n}\left(C_n^m \cos m\lambda + S_n^m \sin m\lambda\right)P_n^m(\sin\phi) \\
\frac{\partial\Phi}{\partial\phi} &= \left(\frac{\mu}{r}\right)\sum_{n=2}^{N}\left(\frac{R}{r}\right)^n \sum_{m=0}^{n}\left(C_n^m \cos m\lambda + S_n^{mn} \sin m\lambda\right) \\
&\quad \left[P_n^{m+1}(\sin\phi) - m\tan\phi P_n^m(\sin\phi)\right] \\
\frac{\partial\Phi}{\partial\lambda} &= \left(\frac{\mu}{r}\right)\sum_{n=2}^{N}\left(\frac{R}{r}\right)^n \sum_{m=0}^{n} m\left(S_n^m \cos m\lambda - C_n^m \sin m\lambda\right)P_n^m(\sin\phi)
\end{aligned} \tag{7}$$

where

R = radius of the Earth
r = geocentric distance of the spacecraft
S_n^m, C_n^m = harmonic coefficients
ϕ = geocentric latitude of the spacecraft = $\sin^{-1}(z/r)$
λ = longitude of the spacecraft = $\alpha - \alpha_g$

α = right ascension of the spacecraft = $\tan^{-1}(y/x)$
α_g = right ascension of Greenwich

Right ascension is measure positive east of the vernal equinox, longitude is measured positive east of Greenwich, and latitude is positive above the Earth's equator and negative below.

For m = 0, the coefficients are called zonal terms, when m = n the coefficients are sectorial terms, and for n > m ≠ 0 the coefficients are called tesseral terms.

The Legendre polynomials with argument sin ϕ are computed using recursion relationships given by:

$$\begin{aligned} P_n^0(\sin\phi) &= \frac{1}{n}\left[(2n-1)\sin\phi\, P_{n-1}^0(\sin\phi) - (n-1)P_{n-2}^0(\sin\phi)\right] \\ P_n^n(\sin\phi) &= (2n-1)\cos\phi\, P_{n-1}^{n-1}(\sin\phi), \quad m \neq 0, m < n \\ P_n^m(\sin\phi) &= P_{n-2}^m(\sin\phi) + (2n-1)\cos\phi\, P_{n-1}^{m-1}(\sin\phi), \quad m \neq 0, m = n \end{aligned} \tag{8}$$

where the first few associated Legendre functions are given by

$$P_0^0(\sin\phi) = 1, \quad P_1^0(\sin\phi) = \sin\phi, \quad P_1^1(\sin\phi) = \cos\phi \tag{9}$$

and $P_i^j = 0$ for $j > i$.

The trigonometric arguments are determined from expansions given by

$$\begin{aligned} \sin m\lambda &= 2\cos\lambda\sin(m-1)\lambda - \sin(m-2)\lambda \\ \cos m\lambda &= 2\cos\lambda\cos(m-1)\lambda - \cos(m-2)\lambda \\ m\tan\phi &= (m-1)\tan\phi + \tan\phi \end{aligned} \tag{10}$$

In this computer program, the heliocentric coordinates of the sun and moon are based on the JPL Development Ephemeris DE430. These coordinates are evaluated in the Earth's mean equator and equinox of J2000 coordinate system (EME2000). The following figure illustrates the geometry of the EME2000 coordinate system. The origin of this Earth-centered-inertial (ECI) inertial coordinate system is the geocenter and the fundamental plane is the Earth's mean equator. The z-axis of this system is normal to the Earth's mean equator at epoch J2000, the x-axis is parallel to the vernal equinox of the Earth's mean orbit at epoch J2000, and the y-axis completes the right-handed coordinate system (Fig. 1).

4 Results and Conclusion

4.1 Results

The output is organized by the following major sections:

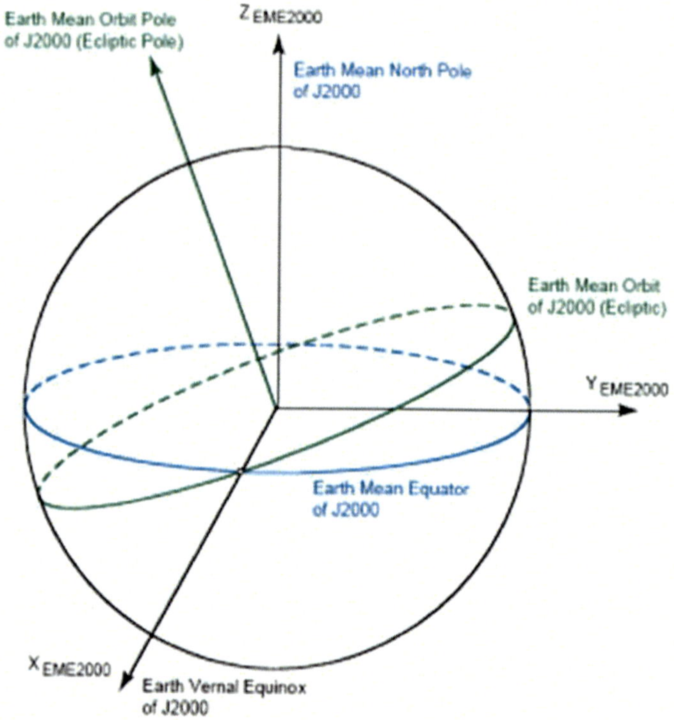

Fig. 1 Earth mean equator and equinox of J2000 coordinate system

- First pass

1. Two body Lambert solution
2. TLI delta-v vector and magnitude

- Targeting pass

3. Pre-TLI and post-TLI flight conditions
4. TLI delta-v vector and magnitude
5. Time and conditions at lunar closest approach
6. Classical orbital elements of the lunar transfer trajectory

The first output section summarizes the optimized two-body Lambert solution. The solution is presented in the Earth's true-of-date coordinate system. The trajectory characteristics are given before and after the impulsive TLI maneuver.

The next section of the program output is created after the B-plane targeting problem has been solved. It includes a summary of the solution, the TLI delta-v vector and magnitude, the final B-plane coordinates and the orbital elements, state vector and flight path angle of the encounter hyperbola.

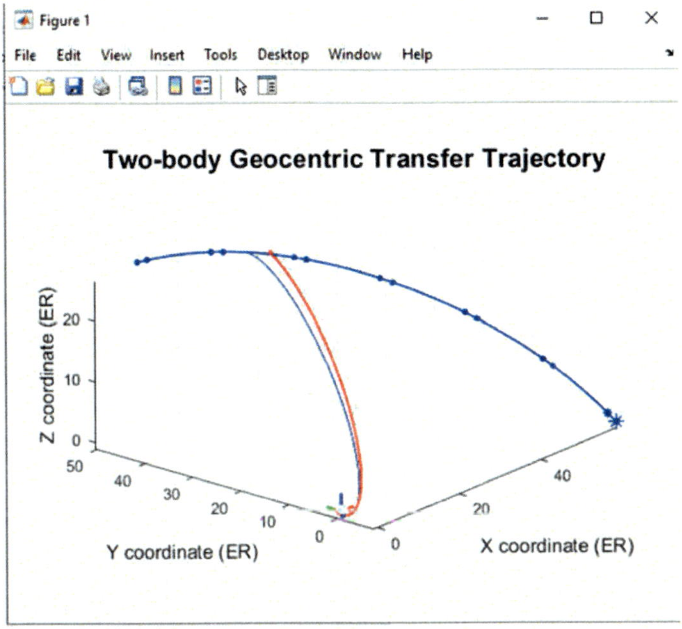

Fig. 2 Two body geocentric transfer trajectory

After the n-body optimization finishes, the software will verify the solution by numerically integrating the equations of motion using a Runge–Kutta-Fehlberg numerical method. The integration starts with the position and velocity vector of the trans-lunar transfer orbit immediately after the TLI impulsive maneuver and propagates to the closest approach to the moon predicted by the software.

The important results from solving this orbital initial value problem (IVP) are the selenocentric periapsis radius and orbital inclination, and the flight path angle. For a valid solution, the periapsis radius and orbital inclination should be very close to the user-defined values and the selenocentric flight path angle at the closest approach should be very near zero.

The code can also create graphic displays of the geocentric transfer orbit and the selenocentric hyperbola within the sphere-of-influence (SOI) of the moon.

The following is a plot of the geocentric transfer trajectory. The coordinates are displayed in the units of Earth radii (ER). The asterisk symbol is the position of the moon at the moment of trans-lunar injection and the moon's orbit is marked with a small blue dot symbol at 24 h intervals. The park orbit trace is red (Fig. 2).

This plot is a "zoomed" display of the first plot closer to the Earth. The initial park orbit is displayed in red, the transfer trajectory is blue, and an inertial, Earth-centered coordinate system is on the plot. The x-axis of this system is red, the y-axis green and the z-axis blue. The location on the park orbit at which TLI occurs is marked with a small dot symbol (Fig. 3).

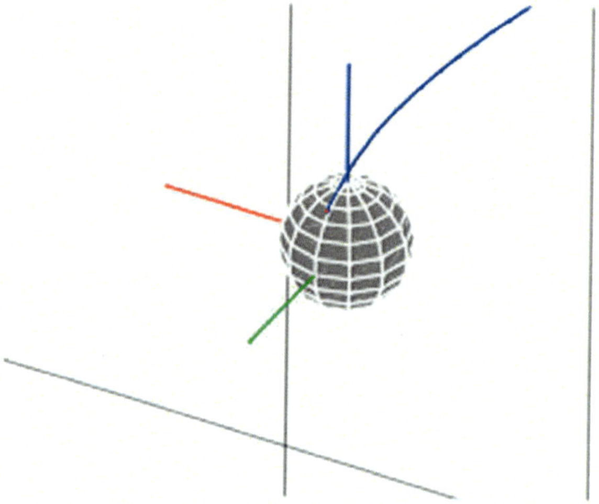

Fig. 3 Zoomed display of Fig. 2

The following is a plot of the selenocentric hyperbola within the moon's sphere-of-influence. The coordinate units are lunar radii (LR). The entry into the SOI is marked with an asterisk. In this MATLAB script, the radius of the moon's SOI is "hardwired" to a value of 64,000 km (Fig. 4).

The final plot is a "zoomed" display of the previous plot. This display is labeled with a selenocentric, inertial coordinate system. The x-axis is red, the y-axis green and the z-axis blue. The small red dot at the end of the blue orbit trace is the periapsis of the approach hyperbola (Fig. 5).

4.2 Conclusion

This targeting option attempts to determine lunar encounter conditions such that the nodal crossing of the arrival hyperbola is also located at the periapsis of the arrival hyperbola. The periapsis radius for this option is specified by the user. This type of mission design is useful for two types of propulsive maneuvers. The first maneuver is performed at periapsis which creates an elliptical orbit about the moon. Since the line of nodes of the subsequent orbit is aligned with periapsis/apoapsis, an orbital plane change maneuver can be performed at apoapsis of the capture orbit.

If we examine the state vector and orbital elements at the closest approach to the moon, we can verify these three conditions have been satisfied by the optimization. In the data display, the closest approach is at periapsis because the true anomaly is nearly zero. Furthermore, the z-component of the selenocentric position vector is nearly zero and the z-component of the velocity vector is positive. These two conditions define an ascending node. The periapsis radius is 1837.999988 km.

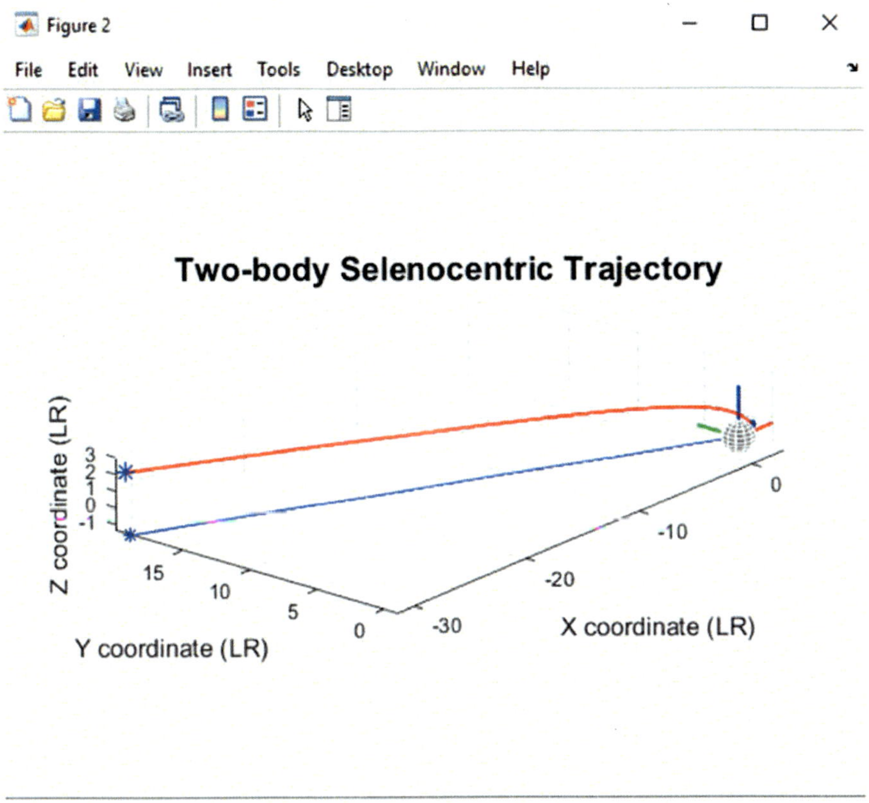

Fig. 4 Two body selenocentric trajectory

Fig. 5 Zoomed display of Fig. 4

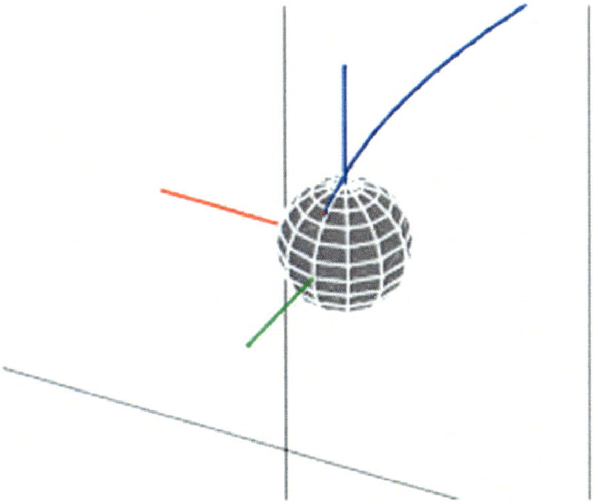

4.3 Future Scope

Trajectory Optimization (TO) is a model-based optimization approach, built upon Optimal Control theory, which are becoming increasingly popular in robotics. They offer an automatic way to synthesize and stabilize highly dynamic and contact-rich motions: a possibility which is appealing to any robotic system but in particular to legged robots, as humanoids or quadrupeds.

While TO is commonly used to plan complex open-loop trajectories over a long prediction horizon, MPC enables fast replanning and feedback stabilization, over a shorter horizon.

However, it remains an open question how to effectively employ TO on real legged systems. This requires smart formulations of the contact-constrained robot dynamics, convenient models of the environment, as well as computationally efficient and real-time optimizations algorithms.

References

1. Hoffman T () GRAIL: Gravity Mapping the Moon. In: Proceedings of the 2009 IEEE aerospace conference. Big Sky, Montana
2. Richon K, Newman L (1995) Flight dynamics support for the clementine deep space program science experiment (DSPSE). Paper No. AAS 95-444, AAS/AIAA Astrodynamics Specialist Conference, Hailifax, Nova Scotia, Canada
3. Folta D (1999) The lunar prospector mission. AAS Paper No. 99-397, AAS/AIAA Astrodynamics Specialist Conference. Girdwood, Alaska
4. SELENE: SELenological ENgineering Explorer KAGUYA. [Online Publication], http://www.jaxa.jp/pr/brochure/pdf/04/sat16.pdf. Accessed 4 April 2009
5. Pieters C, Goswami J, Clark R et al (2009) Character and spatial distribution of OH/H_2O on the surface of the moon. Science 326(5952):568–572
6. Poincar´e H (1993) Les M´ethods de la M´echanique C´eleste. (English Translation). History of modern physics and astronomy, vol 13. American Institute of Physics, Woodbury, NY
7. Diacu F (1996) The solution of the n-body Problem. Math Intell 18(3):66–70
8. Lorenz E (1963) Deterministic nonperiodic flow. J Atmos Sci 20(2):130–141
9. H´enon M, Heiles C (1964) The applicability of the third integral of motion: some numerical experiments. Astron J 69(1):73–79
10. Lunar Reconnaissance Orbiter (LRO) Leading NASA's Way Back To the Moon, Lunar Crater Observation and Sensing Satellite (LCROSS): NASA's Mission to Search for Water on the Moon," [Online Publication], http://www.nasa.gov/pdf/360020main LRO LCROSS presskit2.pdf [retrieved 4 April 2009].
11. "Chandrayaan: Lunar Mission by the Indian Space Research Organisation. [Online Publication], http://www.chandrayaan-i.com/index.php. Accessed 4 April 2009

Artificial Intelligence-Aided Trajectory Planning for UAV

Sanjoy Debnath, Achanta Sai Mani Teja, and Miriyala Naveen

Abstract In this work, we investigate the multi-UAV trajectory planning for optimal resource allocation and energy-efficiency maximization. Considering different terrain models, evaluate effective path for UAV to avoid the obstacle and maximize the energy efficiency of the UAV. With the motivation of reducing the total length of UAV trajectory path while maintaining high QoS and energy efficiency, a stochastic obstacle avoidance game based on q-learning is modeled. Here we reframe the reward function of q-learning algorithm for optimal trajectory path evaluation. Based on the designed reward function, the optimal trajectory path for the UAV deployment is evaluated from the stochastic analysis. It is observed from the analysis that UAV trajectory is improved by zero collision with the obstacle.

Keywords UAV trajectory · Q-learning algorithm · Artificial intelligence

1 Introduction

Learning-based algorithms are very useful tools to train the model and evaluate the optimal solution in a multi-disciplinary environment. Q-learning is one of the re-reinforcement learning algorithms that work on the value of action, state of the agent, and reward value of the agent for a particular action. In this learning technique, the algorithm's agent performs the iteration-wise *action* Q(*cs, ns*), in which the action helps to move the agent from the *current-state* (*cs*) to the *next-state* (*ns*). Based on

S. Debnath (✉) · A. S. M. Teja · M. Naveen
Vel Tech Rangarajan Dr. Sagunthala R&D Institute of Science and Technology, Chennai, India
e-mail: drsanjoydebnath@veltech.edu.in; sanjoydebnath80@gmail.com

A. S. M. Teja
e-mail: vtu14217@veltech.edu.in

M. Naveen
e-mail: vtu11592@veltech.edu.in

A. K. Dubey et al. (eds.), *Advanced IoT Sensors, Networks and Systems*,
Lecture Notes in Electrical Engineering 1027,
https://doi.org/10.1007/978-981-99-1312-1_18

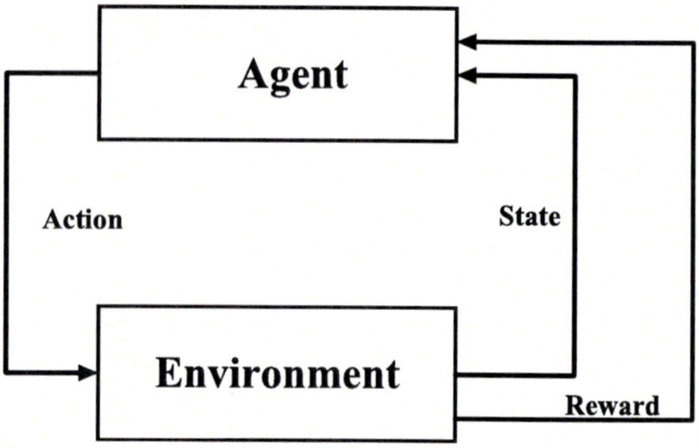

Fig. 1 Agent environment interaction model diagram

the environmental conditions, the agent can move between the *current-state* (cs) and the *next-state* (ns) or may expect any obstacles in between the moves (Fig. 1).

Suppose the agent moves from the *current-state* (cs) to the *next-state* (ns) without any interference. In that case, the agent will receive good feedback in the form of a reward, for example, $r(i, j) = 1$, where i, j are the row and column numbers of reward matrix, respectively. If the agent finds any interference in the *next-state* (ns) movement, then the reward will be $r(i, j) = -\infty$. With the motivation to maximize the reward function, agents continuously evaluate the possible next-state.

In recent multi-disciplinary research, topics include the development and implementation of UAV-based communication networks for the expansion of various wireless services. Nowadays, the idea of aerial or flying base stations to support ground users and mobile networks has received a lot of interest from researchers. UAVs acting as flying base stations (FBS) to support wireless communication networks is a recent research trend, mainly for the application possibility of UAVs in public safety and emergency rescue operation after-disaster situations [1]. The key advantage of using FBSs is laying in parameters like aerial movement and accessibility to not reachable places. Due to these benefits, these communication networks deployed in after-disaster situations are becoming famous for public safety and rescue operations as well as in the offloading operation of cellular traffic load.

It has been observed that during disasters (natural or manmade) such as earthquakes, floods, Olympic games, or war situations, where a large public gathering is occurring for a particular time, then there is a high demand for communication if the existing cellular communication network is not capable to fulfill the demand of the ground user [2]. At that moment the cellular network needs some assistance to provide the required service and coverage [1]. In such instances, using UAVs as base stations to provide better quality-of-service (QoS) to ground users is tremendously beneficial. In trajectory planning, based on UAV's energy consumption, a fairness-based model is developed for UAV's trajectory path planning by deep reinforcement learning [3]. The developed model consists of the rotor's parameter in the energy consumption for optimal trajectory path evaluation. Whereas Multi-agent deep reinforcement learning for multi-UAV trajectory designs is modeled considering the ground user QoS on the Internet of UAVs [4]. It is observed that the sensory data can be transferred to cellular devices over cellular networks via UAV to Device (U2D) connections or directly by the base station (BS). The only limitation of the above model is less computational efficiency, which restricts the application in an emergency situation.

A multi-agent decentralized learning algorithm is used in [5] for the trajectory planning based on the reward function to ensure the maximum secrecy rate of the architecture. Where to minimize the computational time complexity of the algorithm, a multi-agent deep deterministic policy gradient (MADDPG) technique is advocated for the creation of an artificial intelligence system called simultaneous target assignment and path planning (STAPP) [6]. The suggested MADDPG system can effectively cope with dynamic settings because it simply needs the positions of UAVs, targets, and danger regions for execution. An obstacle avoidance method for UAVs in indoor and street environments is proposed in [7], where the author discusses the utilization of a deep reinforcement learning scheme to evaluate the exact interference measurement of collusion. The developed algorithm ensures the adequate navigation of the UAV swarm network.

It is observed that most of the literature works exclude the energy life as well as the storage part of the UAVs. Thus in [8], the authors include these two vital components and proposed a model considering content recharge and energy constraint of UAVs. The proposed strategy enables the cache-assisted UAV network's continuous functioning as well as responsiveness to changes in content popularity in the ground user's demand fulfillment.

In [9], the author proposed a hierarchical UAV swarm model to investigate the 3-Dimensional uneven terrain surface covering problem. Create a 3-Dimensional model of randomly uneven terrain and suggest a geometric method for projecting the surface of the 3-Dimensional terrain into several weighted 2-Dimensional patches. Furthermore, in [10], the author presents a model which deals with the balancing of UAV's limited energy and traffic offloading while performing vehicle-to-UAV communication. The same work also deals with the optimal trajectory planning of the UAV while maximizing the energy efficiency of the UAV network.

A communication system in which a swarm of unmanned aerial vehicles (UAVs) and a base station (BS) are separated by a large number of interferers. A cost-based multi-parameter programming-based modified Q-Learning technique for striking UAVs must strike a balance between mobility and communication performance [11]. The UAVs in the system can leverage degrees of freedom in the frequency, motion, and antenna spatial domain to improve communication quality in the receiving area. An improved reinforcement Q-learning for finding the optimal trajectory in a cellular UAV network is presented in [12]. The proposed algorithm is designed to encounter the issues that arise in the optimal UAV trajectory path evaluation.

However, considering the computational efficiency of the algorithm into account, a Hybrid meta-heuristic algorithm namely differential evolution (DE) with quantum-behaved particle swarm optimization (DEQPSO) is used to solve the optimal trajectory problem of UAVs [13]. The proposed model formulates the optimization problem considering the route of UAVs on the sea. Results depict the most effective safe and flyable path of UAV in different terrain conditions. Downlink wireless networks using non-orthogonal multiple access (NOMA) with quality-of-service (QoS) limits in UAV-based networks are modeled and presented in [14]. To enhance the entire energy efficiency of the UAV communication network, resource allocation and UAV trajectory optimization are applied. The formulated NOMA-based UAV trajectory is solved through the meta-heuristic optimization technique namely sine–cosine algorithm (SCA), to find the optimal trajectory solution at a low computational cost. However, to reduce the total length of the trajectory flight of UAVs while ensuring the QoS demand of users, an energy-efficiency maximization problem incorporating recharge scheduling is proposed in [15].

Toward the collision avoidance of UAVs, the author discusses innovative approaches with which the problem of collusion avoidance problem of UAVs can be solved [16, 17]. One common path planning for UAV consists of an artificial potential field (APF) algorithm [18]. However, due to convergence into local minima, APF is further modified for efficient collision-free path planning method eAPF-CPP for effective path planning of UAVs. After that, Gao et al. suggested a trouble-free method based on a Q-learning algorithm for enabling the real-time distance between the target location and UAV information. So that it reduces the failure and learning costs, and its algorithm use for more investigation [19]. In [20], the author makes decisions based on environmental conditions, which cause obstacles and unnecessary situations for UAVs in path planning. Fast deployment of UAVs to confirm user density, data rate, and coverage area is a significant issue in public safety and emergency communication that must be addressed. Thus, in a scenario where numerous UAVs must follow diverse trajectories to cover different regions in an area in order to maximize network resource utilization, determining the ideal deployment location with an optimal trajectory path is very important but not available in the literature. Hence, in this work, we perform the optimal UAV trajectory planning by maximizing the wireless network resource in the network function virtualization paradigm.

2 Proposed Method

In Q-Learning algorithm, the value of Q is updated based on action (A) taken by the agent to update their state from S_t to S_{t+1}. The optimal combination of state and action will provide maximum reward.

$$Q_t : (S_t,\ A_t) \rightarrow R \tag{1}$$

Initially, the value of Q is randomly fixed and updated throughout the iteration based on the reward value. In Q-learning, the value of Q is updated based on the Bellman equation as follows:

$$\begin{aligned} Q_{t+1}(S_t,\ A_t) = {} & Q_t(S_t,\ A_t) \\ & + \alpha\ \times\ (R_t + \gamma\ \times\ (Q_{Max}(S_{t+1},\ A)\ -\ Q_t(S_t,\ A_t))) \end{aligned} \tag{2}$$

where R_t is the reward received when the agent moves from the current-state (S_t) to the next-state (S_{t+1}), α and γ are, respectively, the learning rate and discount factor lies between 0 to 1. From the Bellman Eq. (2) it can be concluded that the change of Q_t to Q_{t+1} depends on three factors (i) current Q_t weighted by learning rate, (ii) Reward R_t , and (iii) estimated future value of S_{t+1}.

After Δt steps, the agent will decide to move to the next-state (S_{t+1}) based on the evaluated reward.

$$R_t : (S_t, A_t) \rightarrow S_{t+1} \tag{3}$$

By using the above Eq. (3), we can maximize the reward function by adding all the rewards which we get as feedback by all the number of possible actions.

$$R^{\Delta t} = R_0 + R_1 + R_2 + R_3 + \cdots \tag{4}$$

Using the above functions, we get the Q-values for reaching the target in the table. Initially, we start with the values in Q-table as all zeros.

$$\begin{bmatrix} 0\,0\,0 \\ 0\,0\,0 \\ 0\,0\,0 \end{bmatrix} \tag{5}$$

Using the above Q-function, get better and better estimations by updating the values in the Q-table through iterative process. Initially, obstacle is created by “X” and random ones “1” in Q-table which is shown in Fig. 3.

$$\begin{bmatrix} 1\ \mathrm{X}\,\mathrm{X} \\ \mathrm{X}\ 1\ \mathrm{X} \\ \mathrm{X}\,\mathrm{X}\ 1 \end{bmatrix} \tag{6}$$

In the above matrix, “1” is the empty space (no obstacle), the agent can freely move, and “X” is the representation of obstacles.

To initialize the Q-values in Q-table, we need to create an $(n \times n)$ matrix, where n is the number of actions. If there are direct movements in between states, then we update Q-values with 0 or else we update with $-\infty$.

$$\begin{bmatrix} -\infty & 0 & 0 \\ 0 & -\infty & 0 \\ 0 & 0 & -\infty \end{bmatrix} \tag{7}$$

In this way, and by using Eq. (1), we can update and maximize the Q-value to reach the better approximation of the UAV goal.

Initially, UAVs do not know about environmental conditions. We assume there are eight actions to select by a UAV from up, down, left, right, diagonally north-east

(NE), diagonally north-west (NW), diagonally south-east (SE), and diagonally south-west (SW). When the Q-values update, the UAV knows which action has to choose from those eight actions. During this process, UAVs will get accurate assumptions in estimating the Q-values. The system model and diverse terrain model are presented in Sect. 3. In Sect. 4 of the paper, the results and discussion part is presented, whereas the conclusion is drawn in Sect. 5.

3 System Model

The proposed system model is presented in Fig. 2, where it is shown that a black-out area is created due to the occurrence of malfunction in Macro Base Station (MBS). The star shows the optimal goal (destination) location where the UAV has to deploy to cater the network service to black-out users. Here the essential work is to find the optimal trajectory path (dash line) of this goal localization.

In Fig. 2b, "X" shows the obstacle location that one UAV has to avoid while finding the optimal trajectory path. In our observation, a different terrain model is created with a diverse position of obstacle. The reward function is designed in such a way to get $-\infty$ for each undesired movement (steps) of UAVs toward an obstacle.

The proposed flowchart of the UAV trajectory planning is shown in Fig. 2c given.

Initialization of all the parameters like learning rate, discount factor, and the number of iterations are performed at the beginning of the algorithm. To evaluate the optimal path of UAVs, the first footstep of the UAV is to randomly choose steps, and

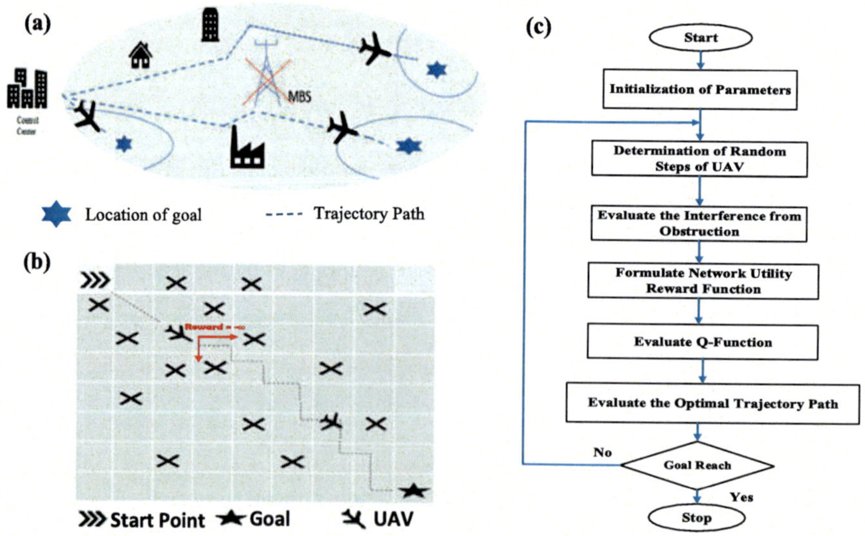

Fig. 2 **a** System model; **b** Model for reward evaluation; **c** Flowchart of the proposed model

based on that optimal selection of steps, UAV receives a reward. In path planning, the best-received reward value by UAV finalizes the optimal path. The second prior step of the UAV is to evaluate the interference from the designed terrain model. That interference factor formulates the reward value. The network utility reward function defines by Eq. (3), where the utility term is used to depict the low interference from the nearby obstacle of the terrain model. Based on the received reward value, Q-function evaluates the Q-values. Those Q-values are listed in the matrix of size (n $\times$ n); where n is the number of actions, and the UAV takes toward the trajectory path evaluation. Finally, the optimal path is evaluated from the matrix defined by the Q-values.

The different terrain condition to check the UAV trajectory is shown in Fig. 3.

As shown in Fig. 3, three different terrain models are considered with the diverse location of obstacles (blue lines). Here we study the UAV path planning considering three terrain models as shown in Fig. 3a–c. We define three irregular terrain models to

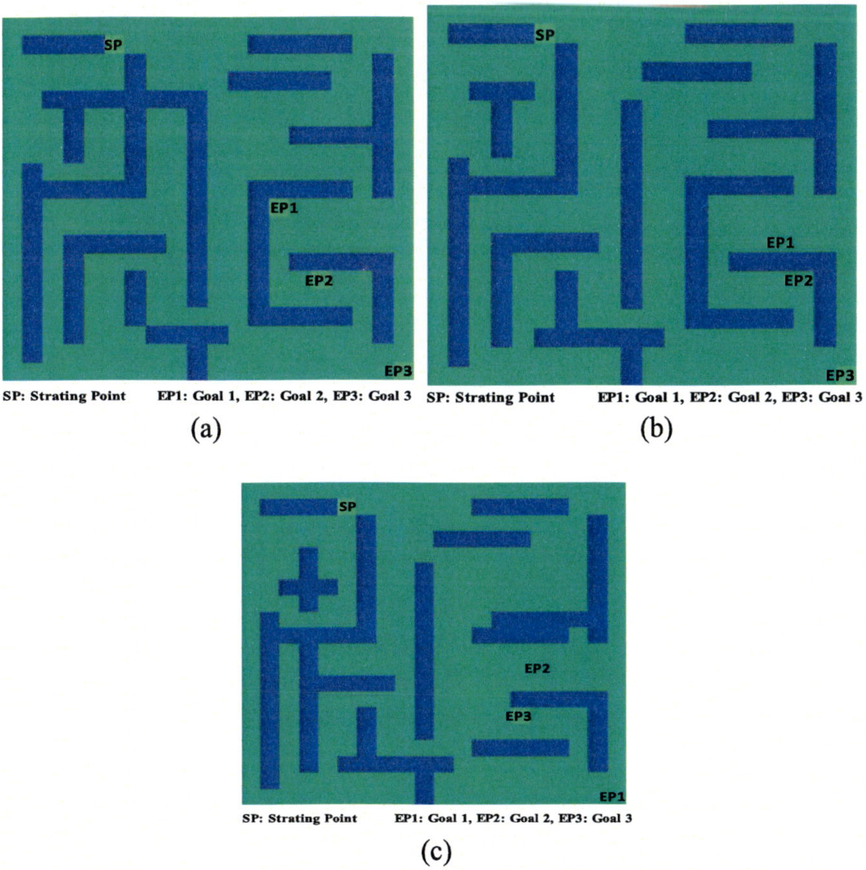

Fig. 3 **a** Terrain Model 1, **b** Terrain Model 2, and and **c** Terrain Model 3

mimic the actual geo-graphical scenario containing mountains, trees, and high-rise buildings. The terrain model is a graphical structure of obstacles that UAVs have to avoid while finding the destination. Here we form three different terrain models with a diverse set of obstacles. Those obstacle locations are depicted by blue lines in Fig. 3, where SP and EP define the Starting Point and Ending Point, respectively.

Then based on the Bellman equation, the Q-values in the different terrain models are initialized. Initially, the state of the UAV is Starting Point (SP); at this state, it does not become aware of environmental conditions (how to move to the next-state from the current-state). To move from the current-state to the next-state, UAVs need to use any of the actions such are up, down, left, right, diagonally NE, diagonally NW, diagonally SE, or diagonally SW, and accordingly, the reward value is assigned.

UAV with an improved learning strategy having the knowledge of goal is able to solve the problem of identifying the best action for reward. The evaluation of the trajectory path consists of the highest value of reward matrix of the particular selected position row. That index value is considered the next action that helps move to the next-state. This iterative process of finding the maximum q-value will provide updated q-value solutions. By using these updated q-values, UAVs will iteratively reach toward the goal.

4 Results and Discussion

Utilizing the proposed algorithm, we analyze the trajectory path for the considered three terrain models. While setting different goals in the considered terrain models 1, 2, and 3, the computational complexity of the algorithm is evaluated for maximum UAV flight time.

4.1 Terrain Model-1

Figures 4, 5, and 6 given below show the trajectory path evaluation of terrain model 1, respectively, for goals 1, 2, and 3.

It is observed from the results shown in Fig. 4 (End Point (EP-1)), Fig. 5 (Ep-2), and Fig. 6 (EP-3) that the UAVs are able to avoid the obstacle shown in blue color with high accuracy and time efficiency. The red line (optimal trajectory path) shown above in Figs. 4, 5, and 6 depicts the UAV's positions in the different stages of the goal-finding process.

Fig. 4 Optimal path for terrain model 1 goal 1

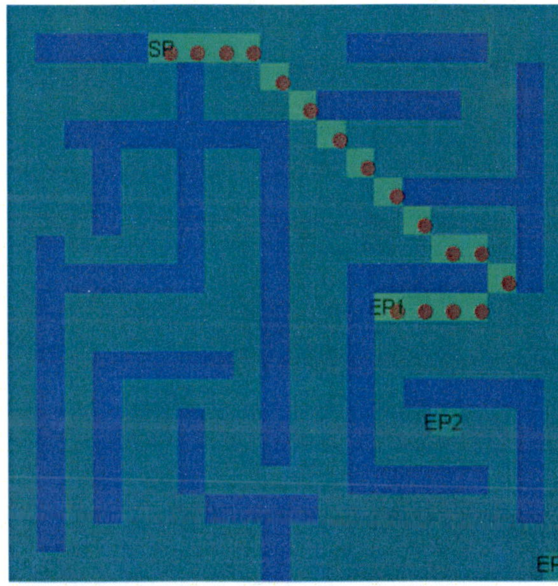

SP: Strating Point EP1: Goal 1, EP2: Goal 2, EP3: Goal 3

Fig. 5 Optimal path for terrain model 1 goal 2

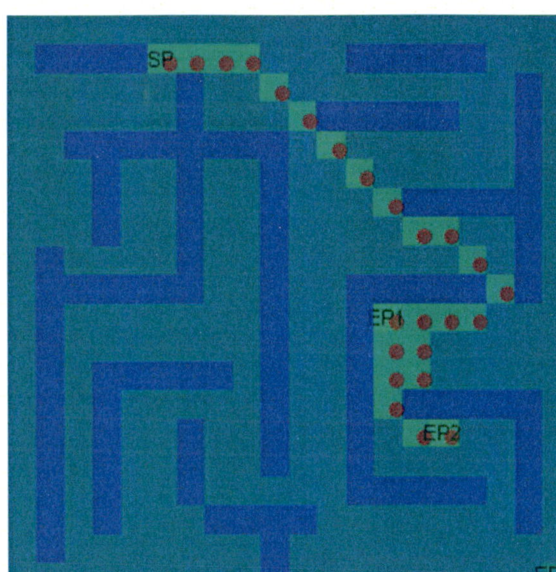

SP: Strating Point EP1: Goal 1, EP2: Goal 2, EP3: Goal 3

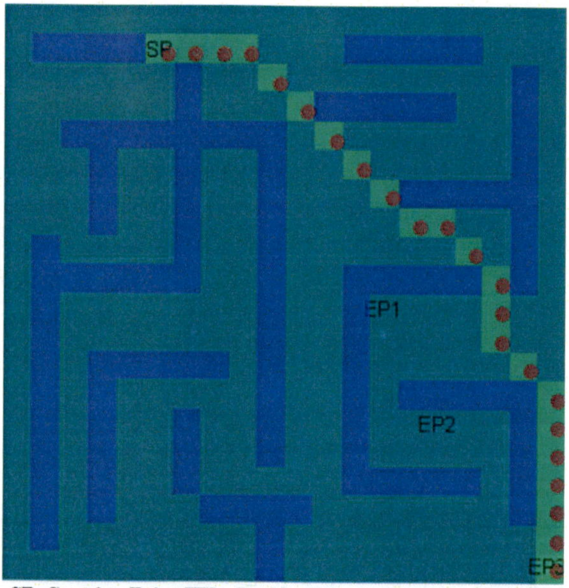

Fig. 6 Optimal path for terrain model 1 goal 3

SP: Strating Point EP1: Goal 1, EP2: Goal 2, EP3: Goal 3

4.2 Terrain Model-2

Here considering terrain model 2, evaluate the optimal trajectory path while setting three different goal positions (EP). It is observed from Figs. 7, 8, and 9 that UAVs are able to find the optimal trajectory path with high accuracy and no interference to the obstacle.

From Figs. 7, 8, and 9, it is noticed that the UAV can find the best trajectory path while avoiding the blue barrier. It has also been observed that the UAV can find the best trajectory path in an average time period of 26.83 s.

4.3 Terrain Model-3

Here considering terrain model 3, the optimal trajectory path is evaluated while setting three different goal positions (EP) as above. It is observed from Figs. 10, 11, and 12 that the UAVs are able to find the optimal trajectory path (shown in red color) with high accuracy and no collusion with the blue obstacle.

From Figs. 11 and 12, it is observed that the UAVs are able to find the optimal trajectory path with avoiding obstacles. It is also observed that the UAVs are able to find optimal trajectory path in the stipulated time period of 24.66 s.

Lastly, we analyze the computational time acquired by UAV in different models to reach the respective goal. The experiment is performed in MATLAB® 2015 on a computer having an intel i3 processor (2.4 GHz clock frequency) with 6 GB ram. As

Fig. 7 Optimal path for terrain model 2 with goal 1

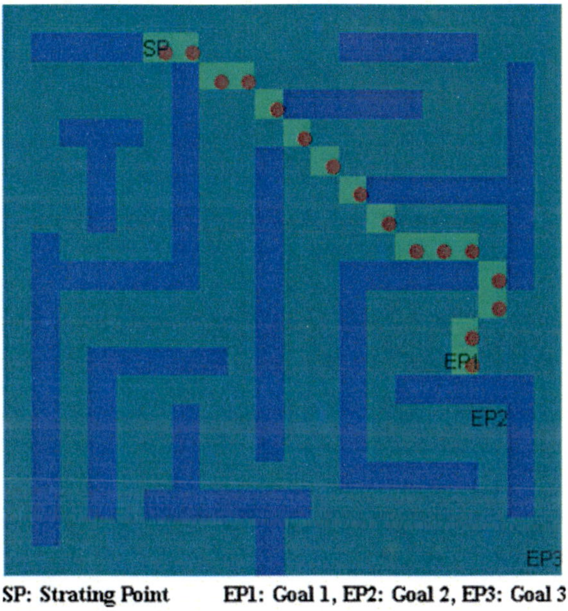

Fig. 8 Optimal path for terrain model 2 with goal 2

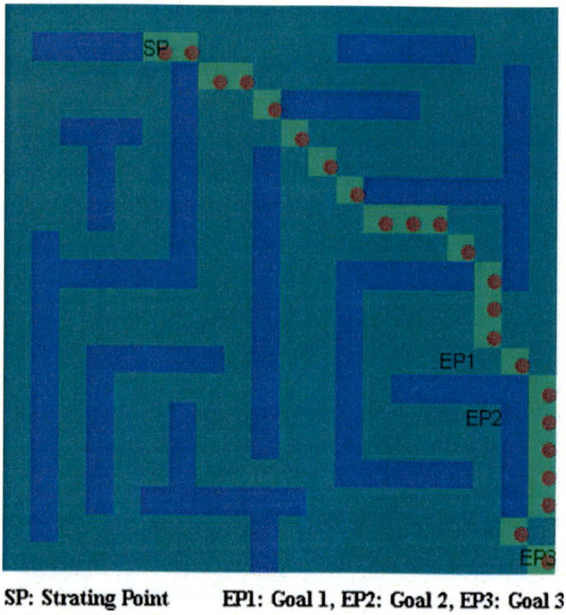

Fig. 9 Optimal path for terrain model 2 with goal 3

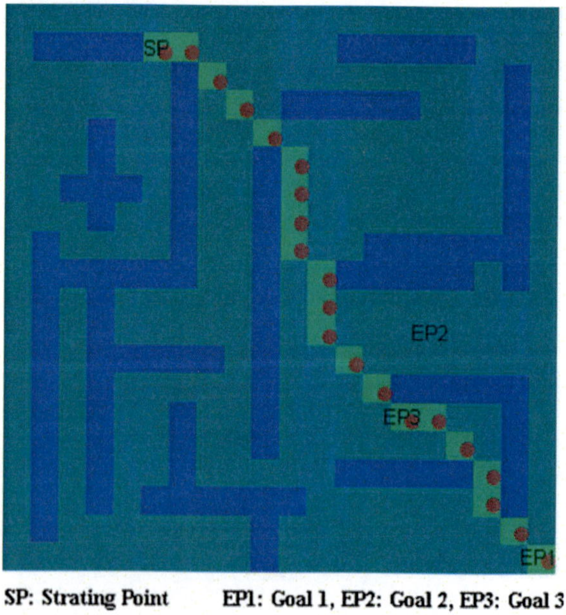

Fig. 10 Optimal Path for Terrain Model 3 with Goal 1

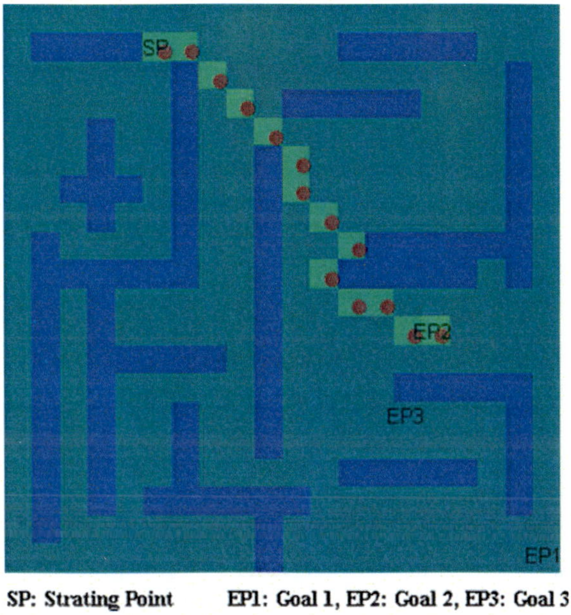

Fig. 11 Optimal Path for Terrain Model 3 with Goal 2

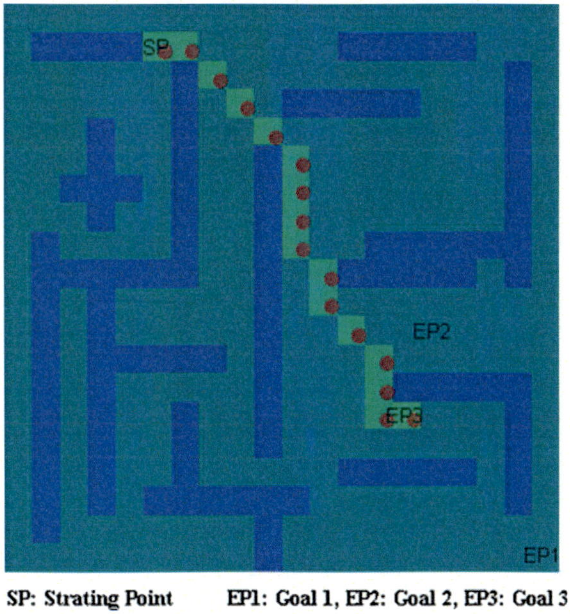

Fig. 12 Optimal Path for Terrain Model 3 with Goal 3

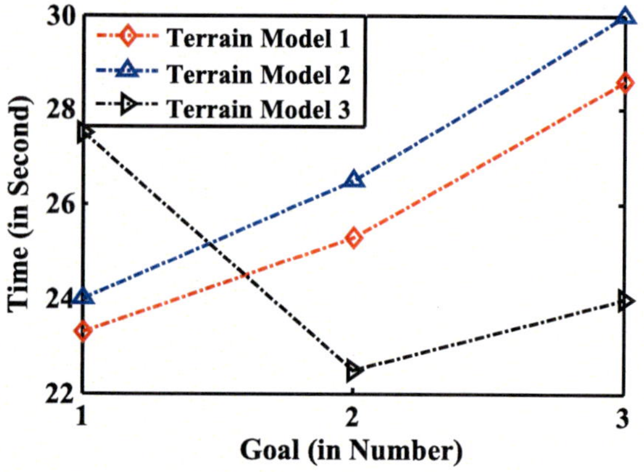

Fig. 13 Computational time in different terrain models

the structural complexity is different in each model, we analyze the computational time required by a UAV to reach the goal. Time complexity depicts the effectiveness of the algorithm in evaluating the optimal trajectory path (Fig. 13).

It is observed that the time required by a UAV to reach the goal position depends on the terrain model complexity and the distance of the goal position. Observed that a UAV acquires 23.3 s, 25.3 s, and 28.6 s to reach goals 1, 2, and 3 in terrain model 1. In contrast, a UAV requires 24 s, 26.5 s, and 30 s to reach goals 1, 2, and 3, respectively, in terrain model 2. The time required to reach the goal in model 2 is more as compared to model 1 because of the complexity of the model. Terrain model 3 is a little different than the other two models in respect of goal position; thus, in terrain model 3, 27.5 s, 22.5 s, and 24 s are required by a UAV to reach goals 1, 2, and 3. The evaluated results are intuitively justified as in terrain model 3, the position of goal 1 is at a higher distance than goal 2 and goal 3.

5 Conclusion

In this work, the optimal trajectory planning algorithm based on Q-learning was proposed for UAVs. It is observed that the proposed algorithm is able to evaluate the optimal UAV's trajectory while avoiding impediments on the way to the goal in all three considered terrain models. It is also observed that the average time required by a UAV to reach the goal in terrain model 1 is 25.73 s, terrain model 2 is 26.83, whereas around 24.66 s in the terrain model 3. The time complexity of terrain model 3 is comparatively less due to aerial path availability and high in terrain model 2. The achieved results are intuitively justified as the complexity of the terrain model 2 is higher than models 1 and 3. The results depict that the proposed approach can evaluate the UAV's optimal trajectory path with minimal time complexity in all the considered terrain models.

References

1. Bucaille I, Hethuin S, Munari A, Hermenier R, Rasheed R, Allsopp S (2013) Rapidly deployable network for tactical applications: aerial base station with opportunistic links for unattended and temporary events absolute example. In: MILCOM 2013-2013 IEEE military communications conference. San Diego, CA, USA, pp 1116–1120
2. Debnath S, Arif W, Roy S, Baishya S, Sen D (2021) A comprehensive survey of emergency communication network and management. Wireless Pers Commun Springer 124:1375–1421
3. Ding R, Gao F, Shen XS (2020) 3D UAV trajectory design and frequency band allocation for energy-efficient and fair communication: a deep reinforcement learning approach. IEEE Trans Wireless Commun 19(12):7796–7809
4. Wu F, Zhang H, Wu J, Song L (2020) Cellular UAV-to-device communications: trajectory design and mode selection by multi-agent deep reinforcement learning. IEEE Trans Commun 68(7):4175–4189
5. Zhang Y, Mou Z, Gao F, Jiang J, Ding R, Han Z (2020) UAV-enabled secure communications by multi-agent deep reinforcement learning. IEEE Trans Veh Technol 69(10):11599–11611
6. Qie H, Shi D, Shen T, Xu X, Li Y, Wang L (2019) Joint optimization of multi-UAV target assignment and path planning based on multi-agent reinforcement learning. IEEE Access 7:146264–146272
7. Singla A, Padakandla S, Bhatnagar S (2021) Memory-based deep reinforcement learning for obstacle avoidance in UAV with limited environment knowledge. IEEE Trans Intell Transp Syst 22(1):107–118
8. Chai S, Lau VKN (2021) Multi-UAV trajectory and power optimization for cached UAV wireless networks with energy and content recharging-demand driven deep learning approach. IEEE J Sel Areas Commun 39(10):3208–3224
9. Mou Z, Zhang Y, Gao F, Wang H, Zhang T, Han Z (2021) Deep reinforcement learning based three-dimensional area coverage with UAV swarm. IEEE J Sel Areas Commun 39(10):3160–3176
10. Al-Hilo A, Samir M, Assi C, Sharafeddine S, Ebrahimi D (2021) UAV-assisted content delivery in intelligent transportation systems-joint trajectory planning and cache management. IEEE Trans Intell Transp Syst 22(8):5155–5167
11. Peng J, Zhang Z, Wu Q, Zhang B (2019) Anti-jamming communications in UAV swarms: a reinforcement learning approach. IEEE Access 7:180532–180543
12. Hu J, Zhang H, Song L (2019) Reinforcement learning for decentralized trajectory design in cellular UAV networks with sense-and-send protocol. IEEE Internet Things J 6(4):6177–6189
13. Fu Y, Ding M, Zhou C, Hu H (2013) Route planning for unmanned aerial vehicle (UAV) on the sea using hybrid differential evolution and quantum-behaved particle swarm optimization. IEEE Trans Syst Man Cybern Syst 43(6):1451–1465
14. Li Y, Zhang H, Long K, Jiang C, Guizani M (2021) Joint resource allocation and trajectory optimization with QoS in UAV-based NOMA wireless networks. IEEE Trans Wireless Commun 20(10):6343–6355
15. Zhao C, Liu J, Sheng M, Teng W, Zheng Y, Li J (2021) Multi-UAV trajectory planning for energy-efficient content coverage: a decentralized learning-based approach. IEEE J Sel Areas Commun 39(10):3193–3207
16. Yasin JN, Mohamed SAS, Haghbayan M, Heikkonen J, Tenhunen H, Plosila J (2020) Unmanned aerial vehicles (UAVs): collision avoidance systems and approaches. IEEE Access 8:105139–105155
17. Liu X, Chen M, Liu Y, Chen Y, Cui S, Hanzo L (2021) Artificial intelligence aided next-generation networks relying on UAVs. IEEE Wireless Commun 28(1):120–127
18. Xie R, Meng Z, Wang L, Li H, Wang K, Wu Z (2021) Unmanned aerial vehicle path planning algorithm based on deep reinforcement learning in large-scale and dynamic environments. IEEE Access 9:24884–24900

19. Gao Y, Li Y, Guo Z (2021) A Q-learning based UAV path planning method with awareness of risk avoidance. In: 2021 China automation congress (CAC). IEEE, Beijing, China, pp 669–673
20. Tu GT, Juang JG (2021) Path planning and obstacle avoidance based on reinforcement learning for UAV application. In: 2021 international conference on system science and engineering (ICSSE). IEEE, Vietnam, pp 352–355

Sensors and Sensor Technology

9T1R nvSRAM Cell with Improved Read Delay and Margin

Damyanti Singh, Neeta Pandey, and Kirti Gupta

Abstract This paper presents a 9T1R cell which gives the advantage of high read speed with better read stability. To improve the read delay, it uses dynamic MOS (DTMOS) technique. It uses two row shared transistors to boost the read and write performance of the cell. The non-volatile operation is performed through a transistor connected to a memristor. The proposed 9T1R cell shows an increment of 48.31% and 17.24% in read speed and read margin of the cell, respectively. The write margin is boosted by 10.04% and write power consumption is degraded by 17.39%. The proposed cell reduces the store delay by 14.02% and store energy consumption by 10.78%, while the restore delay is reduced by 17.36% and restore energy consumption is reduced by 10.27%. The supply voltage variation study is also carried out for completeness. The proposed cell is simulated using 32 nm PTM CMOS model and TiO_2 based linear model of memristor is used for the simulation.

Keywords DTMOS · HRS · LRS · Memory · Memristor · nvSRAM

1 Introduction

The advancement in semiconductor industry has led to the development of battery oriented mobile devices. For their long operation, it should be ensured that mobile devices consume minimum power. It can be achieved by dynamic voltage scaling

D. Singh (✉) · N. Pandey
Electronics and Communication Engineering, Delhi Technological University, Delhi 110042, India
e-mail: damyanti_singh@rediffmail.com

N. Pandey
e-mail: neetapandey@dce.ac.in

K. Gupta
Electronics and Communication Engineering, Bharati Vidyapeeth's College of Engineering, Delhi 110063, India
e-mail: kirti.gupta@bharatividyapeeth.edu

A. K. Dubey et al. (eds.), *Advanced IoT Sensors, Networks and Systems*,
Lecture Notes in Electrical Engineering 1027,
https://doi.org/10.1007/978-981-99-1312-1_19

method (DVS) [1, 2]. However, at lower voltages, the leakage is increased and stability of the cell is degraded. The power in non-volatile memory (NVM) can be saved by switching OFF device during idle state, however, NVM's require more time to access the data [3, 4]. The SRAM cells provide on-chip storage and requires more power consumption. Most of the power is consumed during idle state. If power supply is turned OFF, SRAM cell lose the internal node data due to their volatile nature. To introduce non-volatile feature in SRAM cell, NVM is connected to the internal nodes of the cell and it results in development of non-volatile SRAM (nvSRAM) cell [5, 6]. In literature, there are various nvSRAM cells which uses either one or two NVM's to perform non-volatile operation [7–23].

In existing nvSRAM cells, most of the cell uses conventional 6T SRAM cell as a core to perform read/write operations [7–18]. These cells suffer from read/write conflict issue which affects the stability of nvSRAM cell. To maintain the stability of the cell and avoid read/write conflict issue, the isolated port is used to execute read operation [19–23]. However, there are issues of slow read, store and restore time in these cells. In this work, a 9T1R nvSRAM cell is presented which employs dynamic threshold MOS (DTMOS) technique to improve read delay. The differential read operation and isolated read access transistors boosts the read margin of the cell. The differential write operation and isolation of NVM from internal nodes increases write stability of the proposed cell. The non-volatile operation is performed with the help of a transistor and a NVM which needs two additional control signals. In the proposed work, memristor is used as NVM due to its small size and fast operation among the existing NVMs [24].

The remaining paper is summarized as follows: Sect. 2 gives brief information on memristor and existing nvSRAM cells. Section 3 gives detailed structural explanation and operational description of the proposed 9T1R nvSRAM cell. In Sect. 4, the performance comparison of existing 8T1R [8], MS 8T1M [22], 9T2R [19] nvSRAM cells, as shown in Fig. 1, and the proposed 9T1R cell is carried out. The supply voltage variation study is also performed for the sake of completeness. Section 5 gives conclusion of the paper.

2 Memristor and Existing nvSRAM Cells

This section provides a brief introduction of memristor. It is followed by literature review on existing nvSRAM cells.

2.1 *Memristor*

Memristor has emerged as a promising candidate for next generation of semiconductor technology. It behaves as two terminal resistor which can retain its previous data in form of resistance. L.O. Chua introduced memristor in 1971, however, it was

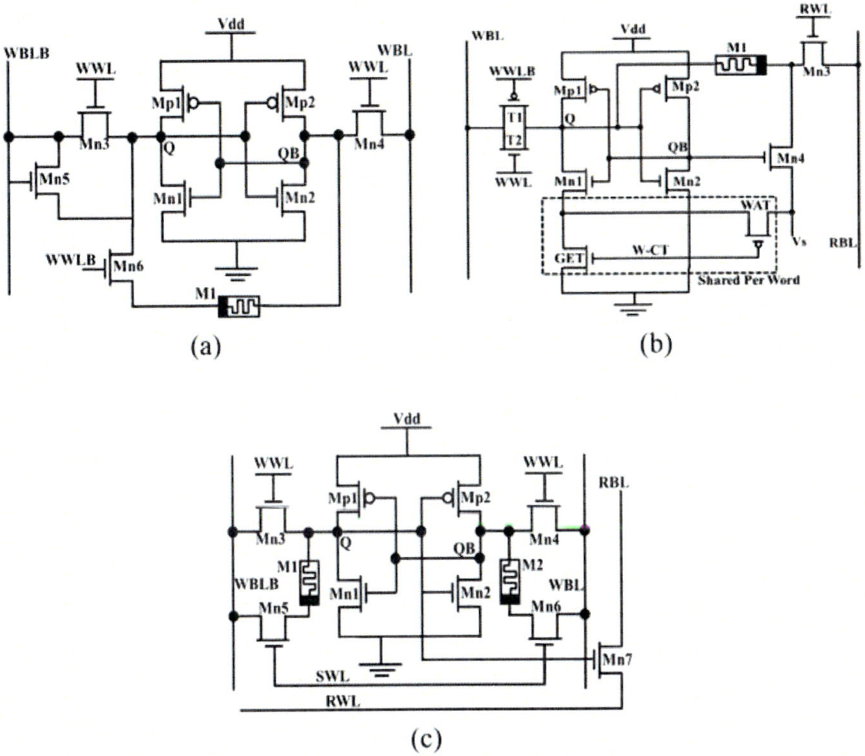

Fig. 1 Existing **a** 8T1R [8] **b** MS 8T1M [22] and **c** 9T2R [19] nvSRAM cells

not physically available at that time [25]. William along with his co-worker from HP lab first introduced physical memristor in 2008 [26]. It's physical and symbolic representations are presented in Fig. 2a, b, respectively. It is a simple metal-oxide-metal (MOM) structure where metal is platinum sandwiching doped and undoped titanium di-oxide (TiO_2) layer. Due to the positive supply across memristor, the oxygen deficiencies of doped $TiO_{2\text{-}X}$ layer are moved toward undoped TiO_2 layer making the whole device more conductive and less resistive. It results in Low Resistance State (LRS) of memristor and denoted as R_{ON}. With the application of negative supply, the oxygen deficiencies of $TiO_{2\text{-}X}$ layer are attracted toward negative terminal and results in increasing width of pure TiO_2 layer. It makes the whole device less conductive and more resistive resulting in High Resistance State (HRS) of memristor denoted by R_{OFF}. The oxygen deficiencies maintain their previous position if no supply is given to memristor and in this way, the previous data is retained in the device.

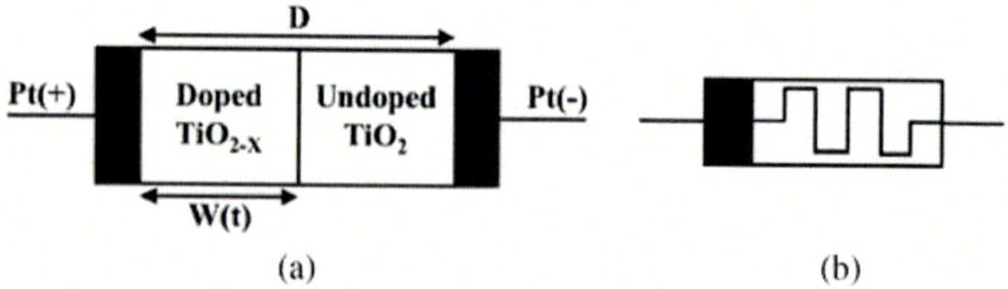

Fig. 2 **a** Physical and **b** symbolic representation of memristor

2.2 *Existing nvSRAM Cells*

In literature, there are different nvSRAM cells in which the conventional 6T SRAM cell is used as the core. The memristor is connected either directly or through the transistor to add non-volatile feature. In 6T2R cell [7], memristor is connected at each internal node to introduce non-volatility in the cell and it is controlled by a control signal. However, the direct connection of memristor increases leakage and reduces the cell stability. Another 6T2R cell is introduced in Ref. [8] which replaces write pass transistors with transmission gates (TG) and load PMOS transistors with memristors. The TG helps in fast write operation; however, it degrades stability during read operation [9]. To improve the cell stability and reduce leakage, a control transistor is added in 6T2R cell [7] and it leads to the development of 7T2R cell [10]. The controlling transistor control memristors operation to maintain the cell stability, but the leakage issue is still present. To reduce leakage, 1T1M (one transistor with memristor) structure is linked at each internal node [11]. It gives advantage in terms of margin and leakage at the cost of increased energy consumption and degraded restore yield. Another 8T2R cell [12] is introduced in literature in which memristor is connected between internal node and bitline through control transistor. It improves restore yield of the cell, but energy consumption is still very high. To reduce energy consumption and for "instant-on" operation, 7T1R cell is introduced [13]. Memristor is attached to internal node Q via control transistor. It helps in reduction of energy consumption and gives faster ON/OFF speed, but restore yield and write stability of the cell are deteriorated. In 8T1R cell [14], the control transistors are connected at each internal node and single memristor is connected to both transistors. It improves stability, and store and restore performances of the cell. An additional circuitry is needed during restore operation which increases power consumption of the cell. To reduce power consumption, improve stability and store/restore performances, 7T1R cell is introduced [15]. It uses dual power supply to improve stability and store/restore performances. To further enhance the performance of the cell, avg.7T1R cell is introduced [16]. It uses column shared transistors which increases write-ability of the cell. It further gives advantage in terms of store/restore energy consumption and restore yield. In 7T2M cell [17], one memristor is connected to the source of pull-down transistors which increases read/write stability of the cell. The 1T1M structure connected to internal node Q to perform non-volatile operation. The 8T1R cell introduced in [8] uses complementary wordline signal to perform the non-volatile operation, however, its complex operation increases the power and energy consumption. In 8T1M cell [18], the write power consumption is reduced to an extent by breaking the back-to-back connection of feedback transistors. In the discussed nvSRAM cells, the read and write operations are performed through same port which requires proper sizing of transistors otherwise the read/write conflict issue can cause the read/write failure in the cell [9]. To avoid this, the read decoupled (RD) nvSRAM cells are introduced in literature [19–23].

The 9T2R cell [19] uses isolated read port which is controlled by internal node Q. Although, the internal nodes remain unaffected during read operation to improve the cell stability, the opposite data is sensed by sense amplifiers and needs additional inverter to read correct data. It leads to increased read power consumption and read access time. In RD 8T1R cell [20], the isolated port is controlled by internal node QB. It uses shared transistors to refine the write and restore performances of the cell. The MS 7T1R cell [21] performs single ended write operation to reduce write power consumption. In MS 8T1M cell [22], the write performance is further improved by the use of transmission gate instead of write pass transistor. In RD 8T1M cell [23], the pull-down transistor is grounded so that the cell can operate at the near threshold voltage. The read assist circuit is used to perform the read operation.

In the above discussed nvSRAM cells, the stability, power consumption and store/restore performances are improved. The read access time is still an issue in these cells. To reduce read delay and improve read margin, differential read operation with DTMOS technique is performed in this paper. The operation and performance analysis of the proposed cell is carried out in further sections.

3 Proposed 9T1R nvSRAM Cell

The schematic of the proposed 9T1R nvSRAM cell is shown in Fig. 3. It comprises of two CMOS inverters (Mn1–Mp1 and Mn2–Mp2) connected back-to-back, two write access transistors (Mn3 and Mn4) and two read pass transistors (Mn5 and Mn6). The write wordline signal WWL control access transistors (Mn3–Mn4) and connects bitlines WBLB/WBL to internal nodes Q/QB. The read operation is performed through transistor ST2 controlled by read wordline RWL. The transistor ST2 follows DTMOS technique, i.e., the gate and body terminals are biased together. The pass transistor ST1 is shared in a row and connected to the source of pull-down transistors Mn1 and Mn2. It is controlled by control signal CTL which is computed as exclusive-OR (XOR) of WWL and RWL signals. The pass transistor Mn7, controlled by signal CTL1 and memristor M1, controlled by signal CTL2 perform non-volatile operation. The different operations of the proposed cell are discussed in the following subsections. The status of control signals during different operations are summarized in Table 1.

Fig. 3 Schematic of proposed 9T1R nvSRAM cell

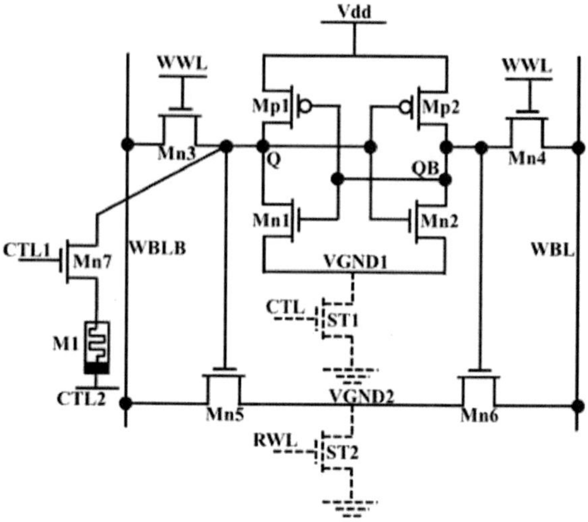

Table 1 Status of control signals during different operations

Control signal / Operation	WWL	RWL	CTL	CTL1	CTL2
Read	'0'	'1'	'1'	'0'	'0'
Write	'1'	'0'	'1'	'0'	'0'
Store	'0'	'0'	'0'	'1'	'0'
Restore	'0'	'0'	'0'	'1'	'1'

3.1 *Read Operation*

The proposed cell performs differential read operation and read path is isolated from internal nodes Q and QB. The control signals are asserted as described in Table 1. The precharged bitline WBL/WBLB discharges through read pass transistor Mn5/Mn6 to perform read operation. The row shared transistor ST2 follows DTMOS technique which reduces the threshold voltage of the transistor and increases read current of the cell. It causes faster discharging of one of the bitlines and supports fast read operation.

3.2 *Write Operation*

During write operation, one of the precharged bitlines WBLB/WBL is discharged to GND. The control signals are asserted as mentioned in Table 1. With the assertion of WWL signal, the write pass transistors Mn3/Mn4 are turned ON and the internal nodes Q/QB flip their stored data. The transistors ST1 remain ON during write operation and the stack of transistors (Mn1-ST1 or Mn2-ST1) is formed. It reduces pull-down strength of transistor Mn1/Mn2 which results in reduced write speed, however, write margin of the cell is increased.

3.3 Store Operation

The status of control signals during store operation are given in Table 1. The assertion of CTL1 signal turns ON transistor Mn7, while CTL2 signal is LOW. The logic '1' at internal node Q causes the potential difference across memristor M1 due to which state is changed from HRS to LRS. While, memristor M1 remain in HRS state when internal node Q is at logic '0' as there is no potential difference across memristor M1.

3.4 Power Down Operation

During power down, the power supply is turned OFF and control signals are asserted LOW. It results in loss of internal node data and stored data is changed to logic '0', however, memristor M1 maintain its state.

3.5 Restore Operation

The restore operation is followed by power down operation. The power supply is turned ON again while the control signals remain LOW except for CTL1 and CTL2. The LRS state of memristor M1 allows a large flow of current through it. This current is sufficient to raise the internal node Q to logic '1'. Whereas, a small current flows through memristor M1 when it is in HRS state. This current is not enough to raise the potential of internal node Q and it remains at logic '0'. With the completion of restore operation, the memristor M1 is reset to HRS state i.e., its initial state.

4 Performance Analysis

The proposed 9T1R nvSRAM cell is implemented and SPICE simulations are performed using 32 nm CMOS PTM model at Vdd = 1.0 V. The TiO_2 based linear model of memristor is used [27] which uses the following parameters: $R_{ON} = 100\ \Omega$, $R_{OFF} = 16\ K\Omega$, mobility $\mu_v = 10^{-14} m^2 V^{-1}\ s^{-1}$ and thickness D = 10 nm. The timing waveform for different operations of the proposed cell is shown in Fig. 4.

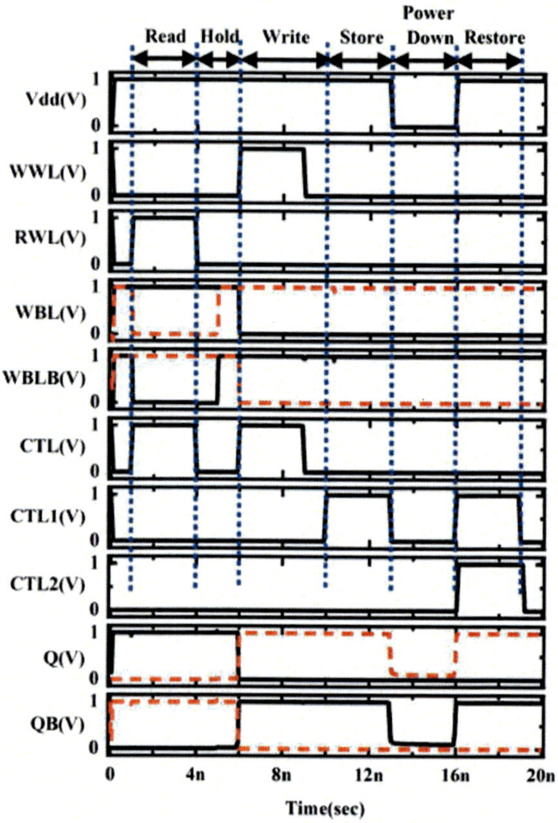

Fig. 4 Timing waveform for different operations of proposed 9T1R cell

The performance of the proposed cell is analyzed in terms of delay, stability margin, power and energy consumption. The results are compared with the existing 9T2R cell [19] at different supply voltages. Table 2 shows the comparison of different parameters of various nvSRAM cells at Vdd = 1.0 V. For a fair comparison, the dimension of all transistors of proposed and existing cells are kept same i.e., 72 nm/36 nm. The layout of the proposed 9T1R cell is shown in Fig. 5. It is observed that the area occupied by proposed cell is 0.943488 μm^2.

4.1 *Read Performance*

In proposed 9T1R cell, the current flows through transistors Mn5/Mn6-ST2. The transistor ST2 has reduced threshold voltage transistor as it follows DTMOS technique. Due to this, the read current of proposed cell is increased and read delay is reduced. The 8T1R cell [8] act as conventional 6T SRAM cell during read operation, while isolated port is used in MS 8T1M cell [22]. In existing 9T2R cell [19], the read

Table 2 Comparison of performance parameters for different nvSRAM cells at Vdd = 1.0 V

Performance parameter	8T1R cell [8]	MS 8T1M cell [22]	9T2R cell [19]	Proposed 9T1R cell
Read delay (ps)	48.56	31.65	39.84	20.59
Write delay (ps)	41.25	25.65	28	31.63
Read margin (mV)	265	325	320	355
Write margin (mV)	370	615	565	595
Read power consumption (μW)	14.56	30.52	23.22	42.27
Write power consumption (μW)	102.63	133.65	121.08	112.13
Store delay (ps)	43.23	43.85	42.56	36.59
Restore delay (ps)	42.36	38.65	39.74	31.84
Store energy consumption (fJ)	870	865	865	815
Restore energy consumption (fJ)	805	805	790	725

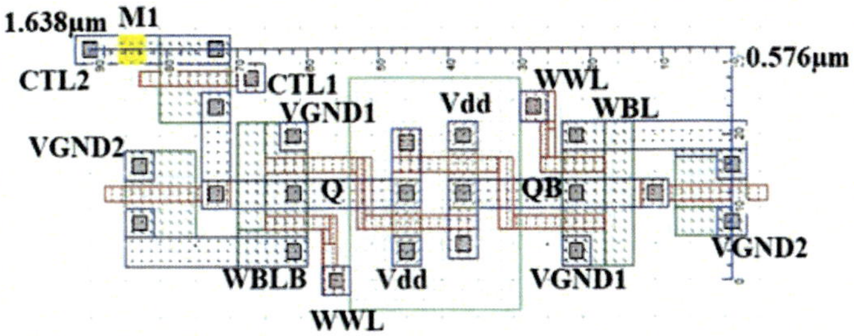

Fig. 5 Layout of proposed 9T1R nvSRAM cell

pass transistor is controlled by internal node Q and opposite data is sensed by the sense amplifier. Hence, an additional inverter is required for reading the original data that enhances read delay of existing 9T2R cell. The comparison in read delay, read power consumption and read margin for different cells at different supply is shown in Fig. 6a–c, respectively. The proposed cell reduces read delay by (57.59%, 34.94% and 48.31%), while the read power consumption and read margin are enhanced by (65.55%, 27.79% and 45.06%) and (25.35%, 17.05% and 17.24%), respectively, in comparison to existing (8T1R [8], MS 8T1M [22] and 9T2R [19]) cells at Vdd = 1.0 V. The increase in read current enhances the read power consumption, while, the differential read operation increases read margin of the proposed cell in comparison

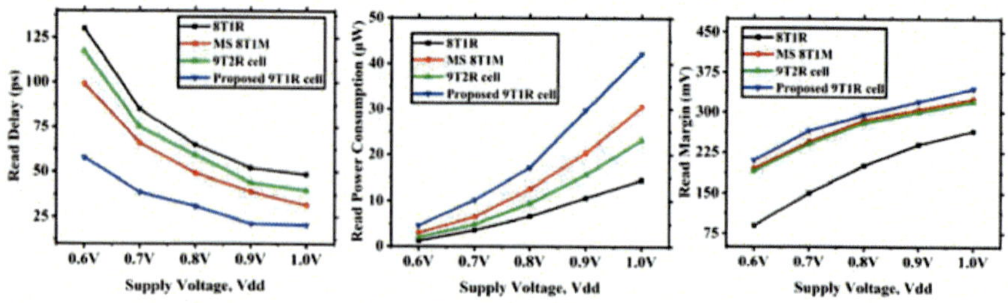

(a) (b) (c)

Fig. 6 **a** Read delay **b** read power consumption and **c** read margin of different nvSRAM cells

to existing nvSRAM cells. It is also analyzed that increase in supply voltage reduces read delay, while enhances read power consumption and read margin.

4.2 Write Performance

The 8T1R cell [8] acts as conventional 6T SRAM cell during write operation. In MS 8T1M cell [22], additional circuitry is used to support write performance. In existing 9T2R cell [19], the additional write path is formed through transistor Mn5-memristor M1 and transistor Mn6-memristor M2 which enhances the write speed of existing 9T2R cell. Hence, the reduction in write delay of MS 8T1M [22] and 9T2R [19] cell is observed by 18.91% and 11.47%, respectively, while write delay of 8T1R cell [8] is increased by 23.32% at Vdd = 1.0 V in comparison to proposed 9T1R cell and it is shown in Fig. 7a. The two write paths and additional circuitry in existing 9T2R [19] and MS 8T1M [22] cell enhances the write current and hence, the write power consumption. The proposed 9T1R cell consumes 17.39% and 21.35% less write power at Vdd = 1.0 V than existing 9T2R cell [19] and MS 8T1M [22] cell, respectively, while 8T1R cell [8] consumes 10.47% less power which is shown in Fig. 7b. In addition to this, the proposed 9T1R cell shows 10.04% and 37.82% better write margin than 9T2R [19] and 8T1R [8] cell, while MS 8T1M cell [22] shows 3.25% improvement at Vdd = 1.0 V and it is shown in Fig. 7c. It is also noticed from Fig. 7a–c that increases in supply voltage decreases write delay, while increases write power consumption and write margin.

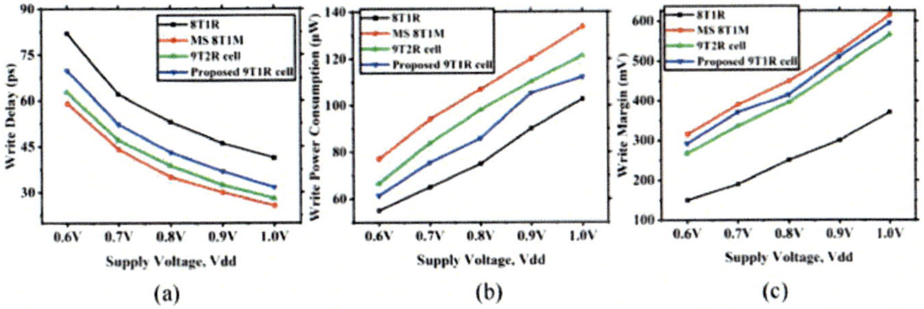

Fig. 7 **a** Write delay **b** write power consumption and **c** write margin of different nvSRAM cells

4.3 *Store Performance*

In existing 9T2R cell [19], the store operation is performed through two transistors (Mn5 and Mn6) and two memristors (M1 and M2). In 8T1R cell [8], the store operation requires two transistor and one memristor. While, MS 8T1M cell [22] requires one transistor and one memristor to perform store operation. The existing cells also require complex SET/RESET operation to perform store operation. The complex programming of memristors increases the store delay and store energy also. In the proposed 9T1R cell, single transistors Mn7 and single memristor M1 is used to perform the store operation. It reduces the store delay and energy consumption by 14.02% and 10.78%, respectively, at Vdd = 1.0 V than the existing nvSRAM cells. Figure 8a and b also shows that the increment in supply voltage decreases the store delay, while increases the store energy consumption.

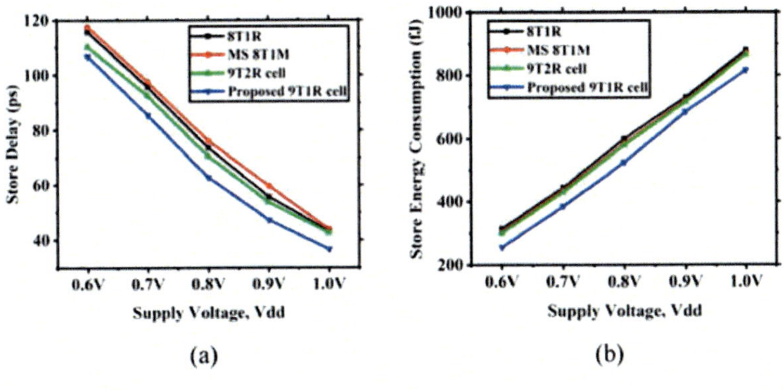

Fig. 8 **a** Store delay and **b** store energy consumption of different nvSRAM cells

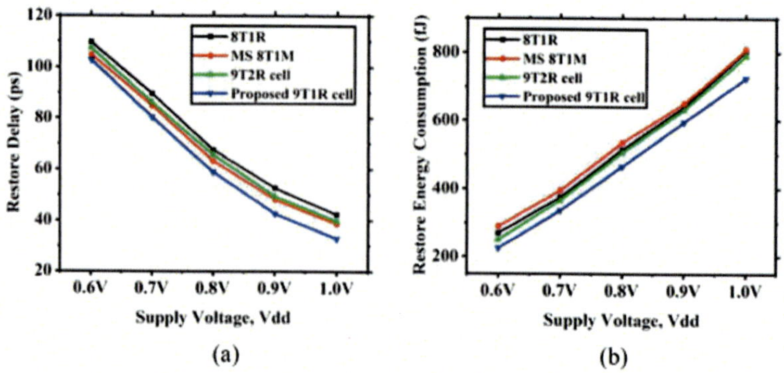

Fig. 9 **a** Restore delay and **b** restore energy consumption of different nvSRAM cells

4.4 Restore Performance

The restore delay and restore energy consumption of the proposed 9T1R and existing nvSRAM cells at different supply voltages are shown in Fig. 9a and b, respectively. In the existing nvSRAM cells [8, 19, 22], the complex restore operation increases the restore delay by 17.36% and restore energy consumption by 10.27% at Vdd = 1.0 V in comparison to the proposed 9T1R cell. In addition to this, the decreasing trend in restore delay and increasing trend in restore energy consumption is observed with supply voltage.

5 Conclusion

This paper introduces 9T1R cell which performs differential read and write operations. It uses DTMOS technique to reduce the read delay by 57.59% and differential read operation enhances the read margin by 25.35% than the exiting nvSRAM cells. However, the read power consumption of the proposed cell is increased by 65.55%. The write margin is improved by 37.82% while, write power consumption is reduced by 21.35% than the existing 9T2R [19] and 8T1R [8] cells. The proposed cell reduces the store and restore delay by 14.02% and 17.36%, respectively. The store energy consumption is reduced by 10.78% and restore energy consumption is degraded by 10.27% in comparison to existing nvSRAM cells.

References

1. Burd TD, Pering TA, Stratakos AJ, Brodersen RW (2000) A dynamic voltage scaled microprocessor system. IEEE J Solid-State Circuits 35(11):1571–1580
2. Chang MF, Yang SM, Chen KT (2008) Wide Vdd embedded asynchronous SRAM with dual-mode self-timed technique for dynamic voltage systems. IEEE Trans Circuits Syst I Regul Pap 56(8):1657–1667
3. Lue HT, Hsiao YH, Hsieh KY, Wang SY, Yang T, Chen KC, Lu CY (2011) Scaling feasibility study of planar thin floating gate (FG) NAND Flash devices and size effect challenges beyond 20 nm. In: 2011 International electron devices meeting. IEEE, pp 9–2
4. Ielmini D (2016) Resistive switching memories based on metal oxides: mechanisms, reliability and scaling. Semicond Sci Technol 31(6):063002
5. Kursun V, Tawfik SA, Liu Z (2007) Leakage-aware design of nanometer SoC. In: 2007 IEEE international symposium on circuits and systems. IEEE, pp 3231–3234
6. Yamamoto S, Shuto Y, Sugahara S (2009) Nonvolatile SRAM (NV-SRAM) using functional MOSFET merged with resistive switching devices. In: 2009 IEEE custom integrated circuits conference, pp 531–534
7. Wang W, Gibby A, Wang Z, Chen TW, Fujita S, Griffin P, Nishi Y, Wong S (2006) Nonvolatile SRAM Cell. In: 2006 International electron devices meeting, pp 1–4
8. Bazzi H, Harb A, Aziza H, Moreau M, Kassem A (2021) RRAM-based non-volatile SRAM cell architectures for ultra-low-power applications. Analog Integr Circ Sig Process 106(2):351–361
9. Singh J, Mohanty SP, Pradhan DK (2012) Robust SRAM designs and analysis. Springer
10. Sheu SS, Kuo CC, Chang MF, Tseng PL, Chih-Sheng L, Wang MC, Lin CH, Lin WP, Chien TK, Lee SH et al (2013) A ReRAM integrated 7T2R non-volatile SRAM for normally-off computing application. In: 2013 IEEE Asian solid-state circuits conference (A-SSCC). IEEE, pp 245–248
11. Turkyilmaz O, Onkaraiah S, Reyboz M, Clermidy F, Anghel C, Portal JM, Bocquet M (2012) RRAM-based FPGA for "normally off, instantly on" applications. In: Proceedings of the 2012 IEEE/ACM international symposium on nanoscale architectures, pp 101–108
12. Chiu P-F, Chang M-F, Wu C-W, Chuang ChingHao, Sheu S-S, Chen Y-S, Tsai M-J (2012) Low store energy, low VDDmin, 8T2R nonvolatile latch and SRAM with vertical stacked resistive memory (memristor) devices for low power mobile applications. IEEE J Solid-State Circuits 47(6):1483–1496
13. Wei W, Namba K, Han J, Lombardi F (2014) Design of a nonvolatile 7T1R SRAM cell for instant-on operation. IEEE Trans Nanotechnol 13(5):905–916
14. Tosson AM, Neale A, Anis M, Wei L (2016) 8T1R: a novel low-power high-speed RRAM-based non-volatile SRAM design. In: 2016 International great lakes symposium on VLSI (GLSVLSI). IEEE, pp 239–244
15. Lee A, Chang MF, Lin CC, Chen CF, Ho MS, Kuo CC, Tseng PL, Sheu SS, Ku TK (2015) RRAM-based 7T1R nonvolatile SRAM with 2x reduction in store energy and 94x reduction in restore energy for frequent-off instant-on applications. In: 2015 Symposium on VLSI circuits (VLSI circuits). IEEE, pp C76–C77
16. Peng C, Xiao S, Lu W, Zhang J, Wu X, Chen J, Lin Z (2017) Average 7T1R nonvolatile SRAM with R/W margin enhanced for low-power application. IEEE Trans Very Large Scale Integr (VLSI) Syst 26(3):584–588
17. Singh J, Raj B (2019) Design and investigation of 7T2M-NVSRAM with enhanced stability and temperature impact on store/restore energy. IEEE Trans Very Large Scale Integr (VLSI) Syst 27(6):1322–1328
18. Singh D, Gupta K, Pandey N (2022) A novel low-power nonvolatile 8T1M SRAM cell. Arab J Sci Eng 47(3):3163–3179
19. Chang MF, Chuang CH, Chen MP, Chen LF, Yamauchi H, Chiu PF, Sheu SS (2012) Endurance-aware circuit designs of nonvolatile logic and nonvolatile SRAM using resistive memory (memristor) device. In: 17th Asia and south pacific design automation conference, pp 329–334

20. Lin Z, Wang Y, Peng C, Lu W, Li X, Wu X, Chen J (2019) Read-decoupled 8T1R non-volatile SRAM with dual-mode option and high restore yield. Electron Lett 55(9):519–521
21. Lin Z, Wang Y, Peng C, Wu X, Li X, Chen J (2019) Multiple sharing 7T1R nonvolatile SRAM with an improved read/write margin and reliable restore yield. IEEE Trans Very Large Scale Integr (VLSI) Syst 28(3):607–619
22. Singh D, Pandey N, Gupta K (2022) MS 8T1M nvSRAM cell with improved write performance. In: 2022 International mobile and embedded technology conference (MECON), pp 425–429
23. Singh D, Gupta K, Pandey N (2022) A novel read decoupled 8T1M nvSRAM cell for near threshold operation. Microelectron J 126:105496
24. Burr GW, Kurdi BN, Scott JC, Lam CH, Gopalakrishnan K, Shenoy RS (2008) Overview of candidate device technologies for storage-class memory. IBM J Res Develop 52(4.5):449–464
25. Chua L (1971) Memristor-the missing circuit element. IEEE Trans Circuit Theory 18(5):507–519
26. Strukov DB, Snider GS, Stewart DR, Williams RS (2008) The missing memristor found. Nature 453(7191):80–83
27. Biolek D, Di Ventra M, Pershin YV (2013) Reliable SPICE simulations of memristors, memcapacitors and meminductors. Radioengineering 22(4):945–968

Fractional-Order Capacitor Realization Based upon Active Inductor

Garima Varshney, Neeta Pandey, and Rajeshwari Pandey

Abstract Fractional-Order elements (FOEs) are the basic building blocks of Fractional-Order (FO) systems. A new active inductor-based realization to approximate the behaviour of FO capacitor (FOC) is proposed in this work. Proposed structure enables higher order approximation of FOC either by parallel connection or by impedance multiplication. The behaviour of the proposed FOC has been verified through SPICE with 0.18 μm CMOS technology model parameters. Further, application of the proposed FOC to realize a fractional-order oscillator (FOO) is also shown to verify the functionality of the proposed FOC.

Keywords Fractional-Order capacitor · Fractional-Order emulator · Active inductor · Fractional-Order oscillator

1 Introduction

The inherent capability of fractional-order calculus to describe more complex system dynamics drew researcher's attention since last few decades. Fractional-order element (FOE) also known as constant phase element (CPE) is a fundamental component for designing the fractional-order analog circuits. Attempts have been made to develop off-shelf fractional-order capacitor (FOC) and make it available in market just like normal capacitor with reliable performance [1–3]. However, these implementations are bulky and non-reconfigurable. Consequently, approximated FOEs are available in literature using suitable passive/active circuitry network, aiming to circuit design and testing [4–9]. Due to the approximation methods, the response of FOE

G. Varshney (✉) · N. Pandey · R. Pandey
Delhi Technological University, New Delhi, India
e-mail: gvarshn1@gmail.com

N. Pandey
e-mail: neetapandey@dce.ac.in

R. Pandey
e-mail: rpandey@dce.ac.in

A. K. Dubey et al. (eds.), *Advanced IoT Sensors, Networks and Systems*,
Lecture Notes in Electrical Engineering 1027,
https://doi.org/10.1007/978-981-99-1312-1_20

Table 1 Comparison table

Refs.	Topology used	ABB used	Component count (Active+Passive)	Application explored
Fouda et al. [5]	Active inductor	PMOS	8 (2+6)	Yes (FO multivibrator)
Bertsias et al. [6]	Intuitive	CMOS	22 (20+2)	Yes (RC series circuit)
Tsirimokou et al. [7]	IFLF	OTA	10 (8+2)	Yes (Parallel resonator)
Kapoulea et al. [8]	MFS	OTA	9 (7+2)	Yes (RC series circuit)
Kapoulea et al. [9]	Foster network	OTA	5 (3+2)	No
Proposed	Active inductor	NMOS	8 (2+6)	Yes (FO oscillator)

tends to show ripples around theoretical response, over a certain range of operational frequency and this frequency range widens as the order of approximation increases. Choosing suitable approximation (roots/coefficients of the impedance/transfer function) [10] is the key factor which has direct impact on precision of the FOE.

A summary of approximated FOE available in open literature [5–9] based upon topology used, component count and application explored is given in Table 1 and following are the observations:

- PMOS [5], CMOS [6] and Operational Transconductance Amplifier (OTA) [7–9] base FOE emulators are reported in respective references.
- Active inductor topology is used in [5], inverse follow the leader (IFLF) topology is used in [7] while multiple feedback structure (MSF) topology is used in [8]. Kapoulea et al. [9] uses Foster network and [6] realizes FOE using intuitive approach.

In present work, an active inductor-based realization to approximate the behaviour of FOC with minimal component count is proposed, which enables higher order approximation of FOC either by parallel connection or by impedance multiplication. This circuit is a modified version of the circuit proposed in [5]. The circuit in [5] consists of PMOS transistor, the proposed circuit is superior to [5] in terms of mobility and fabrication process as proposed circuit uses NMOS transistor, further, proposed circuit employs grounded capacitor which has easier fabrication steps than the floating capacitor.

The behaviour of the proposed FOC has been verified through SPICE with 0.18 μm CMOS technology model parameters. Further, application of the proposed FOC to realize a FOO is also shown to verify the functionality of the proposed FOC.

Rest of the manuscript is arranged as follows: Sect. 2 talks about FOEs followed by proposed active inductor-based FOC in Sect. 3. Section 4 contains simulation results and FOO as an application of the proposed FOC. The paper is concluded in Sect. 5.

2 Fractional-Order Elements

The vital element in analog fractional-order circuit design is called fractional-order element (FOE). The impedance of a FOE is the broad form of conventional impedances and is important for precise modelling of various circuits used in variety of applications. The generalized impedance of FOE is expressed as [11]:

$$Z(s) = ks^{\alpha} = k(\omega)^{\alpha} e^{j\frac{\alpha\pi}{2}} = k(\omega)^{\alpha}\left(cos\left(\frac{\alpha\pi}{2}\right) + jsin\left(\frac{\alpha\pi}{2}\right)\right) = |Z|\angle\phi \quad (1)$$

s^{α} is called as fractional Laplacian operator and k is a constant with unit $\Omega.sec^{\alpha}$. The α is called order of FOE and its value lies between -1 and 1. For $-1 < \alpha < 0$ FOE behaves as a fractional-order capacitor (FOC) and $0 < \alpha < 1$ FOE behaves as a fractional-order inductor (FOI). ϕ denotes the phase of the FOE and equals $\alpha\pi/2$. Phase of FOE does not depend on frequency and constant for a particular value of α; hence FOE is also known as constant phase element. The symbols for FOC and FOI are shown in Fig. 1.

3 Proposed Active Inductor-Based FOC Emulator

In this section, the proposed structure of active inductor and two methods to achieve higher order transfer functions to approximate FOC are discussed.

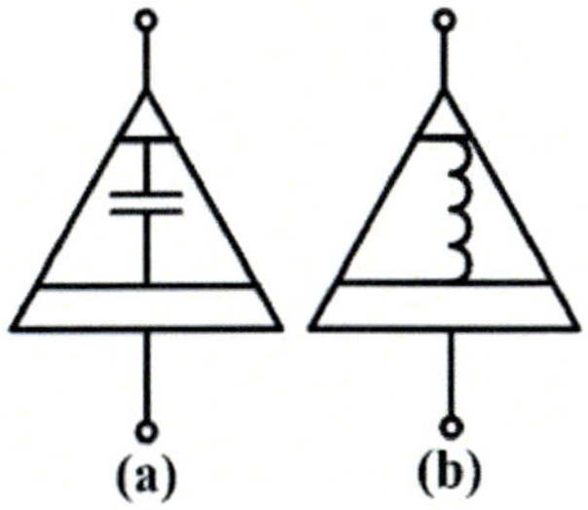

Fig. 1 Symbol of **a** FOC **b** FOI

Fig. 2 Proposed active inductor circuit

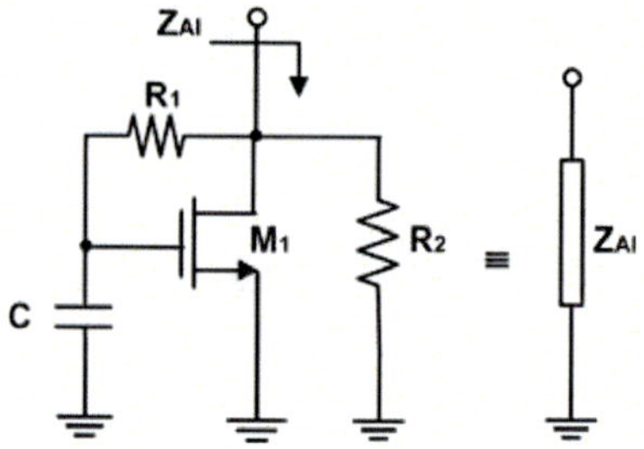

3.1 Proposed Active Inductor

Figure 2 shows the active inductor circuit which uses one NMOS transistor (M_1), two resistors (R_1 and R_2) and one grounded capacitor (C_1). This circuit is a modified version of the circuit proposed in [5]. The circuit in [5] consists of PMOS transistor, the proposed circuit is superior to [5] in terms of mobility and fabrication process as proposed circuit uses NMOS transistor. Further, proposed circuit employs grounded capacitor which has easier fabrication steps than the floating capacitor. By routine network analysis, the input impedance of the proposed active inductor is found out and given in Eq. (2):

$$Z_{AI} = \frac{R_2(1 + sC_1R_1)}{1 + g_{m1}R_2 + sC_1(R_1 + R_2)} \tag{2}$$

Here g_{m1} is the small signal transconductance of transistor M_1. Equation (2) can be rewritten as

$$Z_{AI} = \frac{R_1R_2}{R_1 + R_2} \cdot \left[\frac{s + \frac{1}{C_1R_1}}{s + \frac{(1+g_{m1}R_2)}{C_1(R_1+R_2)}} \right] = K \cdot \left[\frac{s + \omega_z}{s + \omega_p} \right] \tag{3}$$

The zero and pole frequency of the proposed active inductor are defined as $\omega_z = \frac{1}{C_1R_1}$ and $\omega_p = \frac{(1+g_{m1}R_2)}{C_1(R_1+R_2)}$ respectively.

3.2 Higher Order Approximations to Realize FOC

It is eminent that the proposed active inductor can be utilized to approximate FOC. A minimum of second-order transfer function can be optimized to realize an approximated FOC, i.e. minimum two poles and two zeros are required to approximate a FOC [10]. The location of pole and zero of the proposed active inductor can be adjusted by choosing proper values of passive components and transconductance albeit retaining the pole zero alternate interlacing, i.e. $\omega_{z1} < \omega_{p1} < \omega_{z2} < \omega_{p2}$ [10]. The second-order approximation of FOC is adequate to realize narrow-band applications, like FOOs.

Among various methods, two general methods to realize higher order approximation of FOC, are discussed below.

3.2.1 Parallel Connection of Impedances

In this method, n stages of active inductor proposed in Sect. 3.1 are connected in parallel as shown in Fig. 3a. Every stage contributes a pair of zero and pole and to approximate a FOC at least two poles and two zeros are required. These poles and zeros must be placed in alternate manner. The total input admittance of this scheme (Fig. 3a) is given as

$$Y_{in} = \Sigma_{i=1}^{n} \frac{1}{Z_{AIi}} \quad (4)$$

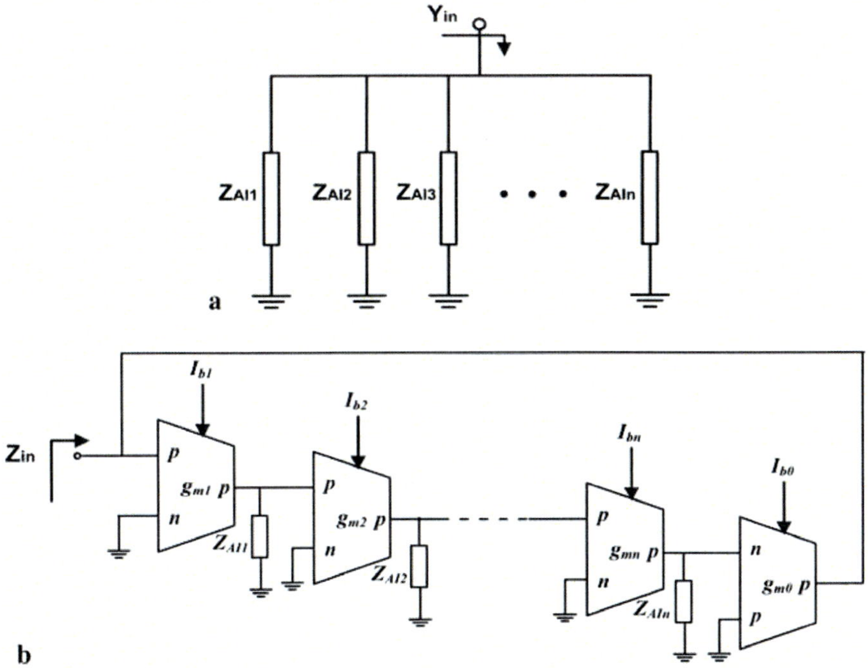

Fig. 3 Higher order approximation of FOC **a** Parallel connection of impedances **b** Impedance multiplication

3.2.2 Impedance Multiplication

In this method, the n stages of active inductor, proposed in Sect. 3.1, are multiplied using a generalized circuit of impedance multiplier shown in Fig. 3b, it is operational transconductance amplifier (OTA) based and inspired by [12]. This impedance multiplier circuit has many advantages over operational amplifier (Opamp)-based GICs: such as, all impedances are grounded and it can multiply n numbers of impedances. The total input impedance of this scheme (Fig. 3b) is given as

$$Z_{in} = \frac{1}{g_{m0} \Pi_{i=1}^{n} g_{mi} Z_{AIi}} \tag{5}$$

Z_{AI} is given in Eq. 3.

The order of approximation to realize FOC depends upon the number of stages (n) in above two methods of higher order approximation of FOC. To procure approximated FOC, we used Valsa's algorithm [10] and recalculated the location of poles and zeros of the proposed structure.

4 Simulation Results and Application

To prove the functionality of the proposed active inductor and FOC, SPICE simulations have been performed with 0.18 μm CMOS technology parameters. The active inductor proposed in Fig. 2 is simulated and magnitude and phase responses are shown in Fig. 4 for two different combinations of passive components $C = 17.5\,\text{nF}$, $R_1 = 40\,\text{k}\Omega$ and $R_2 = 140\,\text{k}\Omega$ (stage-1) while, $C = 5\,\text{nF}$, $R_1 = 13\,\text{k}\Omega$ and $R_2 = 70\,\text{k}\Omega$ (stage-2).

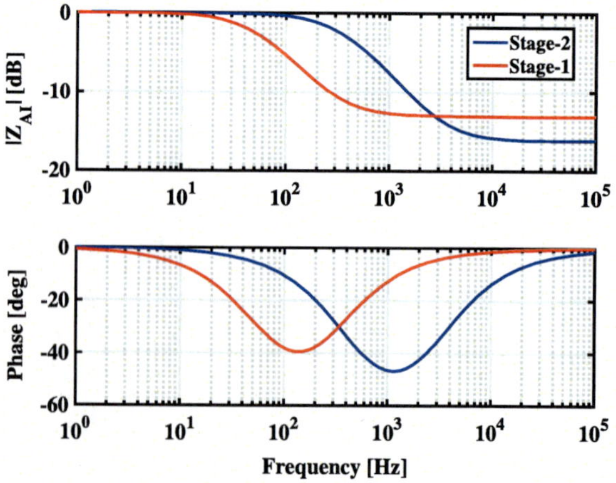

Fig. 4 Magnitude and phase response of active inductor

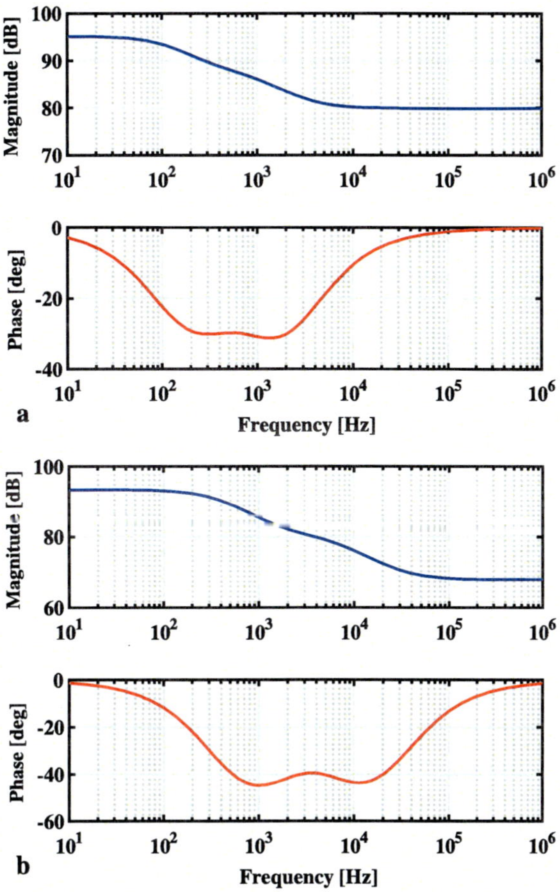

Fig. 5 Simulation results for second-order approximation of FOC based upon two stage parallel connection of active inductor for **a** $\phi = -30°$ **b** $\phi = -45°$

The magnitude and phase responses for second-order approximation of FOC using first method, i.e. parallel connection of two active inductors (*n* is taken as 2 in Fig. 3a) for $\phi = -30°$ and $-45°$ have been shown in Fig. 5a and b respectively. The values used for various components are given in Table 2. The simulated and theoretical slope of magnitude is -7.1 dB/decade and -6.66 dB/decade for $\phi = -30°$ respectively, while for $\phi = -45°$, it is -9.6dB/decade and 10dB/decade respectively. The error in the slope is within 6% while the error in the phase is $\pm 1.5°$.

Further, second-order approximation of FOC using the second method to realize higher order approximation (i.e. impedance multiplier) has been implemented, again the value of *n* is taken as 2 in Fig. 3b. The magnitude and phase responses are shown in Fig. 7 for $\phi = -45°$. The OTA circuit given in [13] has been taken to realize Fig. 3b, the CMOS structure of OTA is shown in Fig. 6. The transconductances of all the three OTAs are taken as same and equal to $61\mu A/V$. The component settings for Z_{AI1} are $C = 33$ nF, $R_1 = 2.4\,\text{k}\Omega$ and $R_2 = 10\,\text{k}\Omega$ while the same for Z_{AI2} are $C = 480$ nF, $R_1 = 4.8\,\text{k}\Omega$ and $R_2 = 2\,\text{k}\Omega$. The simulated and theoretical slope of

Table 2 Simulation settings for two stage parallel connection of active inductor

# Stage	Component	$\phi = -30°$ ($\alpha = 1/3$)	$\phi = -45°$ ($\alpha = 1/2$)
–	$(W/L)_{1,2}$	10	10
Z_{AI1}	C	$17.5nF$	$7nF$
	R_1	$40k\Omega$	$15k\Omega$
	R_2	$140k\Omega$	$130k\Omega$
Z_{AI2}	C	$5nF$	$1.9nF$
	R_1	$13k\Omega$	$3k\Omega$
	R_2	$70k\Omega$	$65k\Omega$

Fig. 6 CMOS structure of OTA [13]

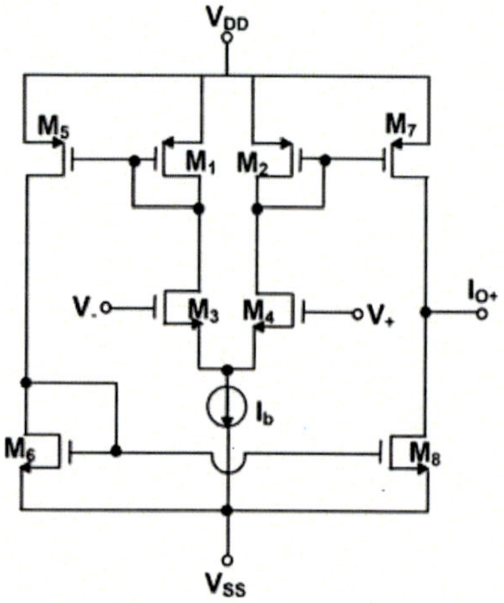

Fig. 7 Simulation results for second-order approximation of FOC based upon impedance multiplication of two active inductor for $\phi = -45°$

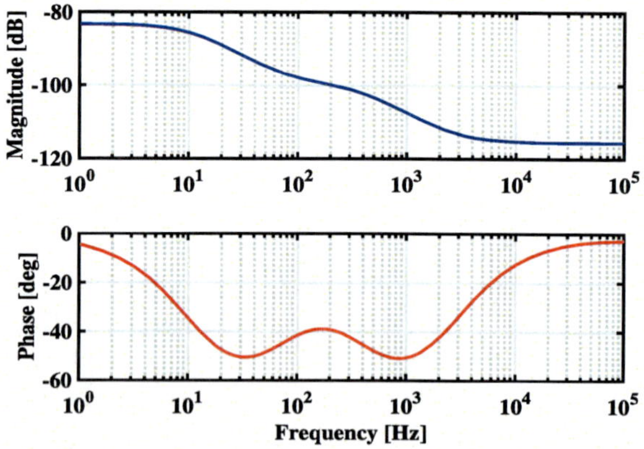

magnitude is -10.6 dB/decade and -10 dB/decade for $\phi = -45°$ respectively, the error is 6%, while it is for phase is $\pm 4°$.

Additionally, an application of proposed FOC as OTA-based fractional-order oscillator (FOO) [14] (see Fig. 8) has also been examined to verify FOC's func-

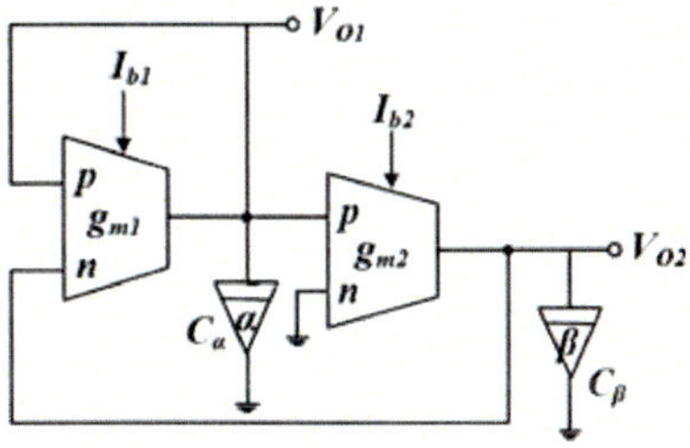

Fig. 8 Circuit diagram of fractional-order oscillator [14] (FOC used is two stage parallel connection of active inductor with $\alpha = 1/2$)

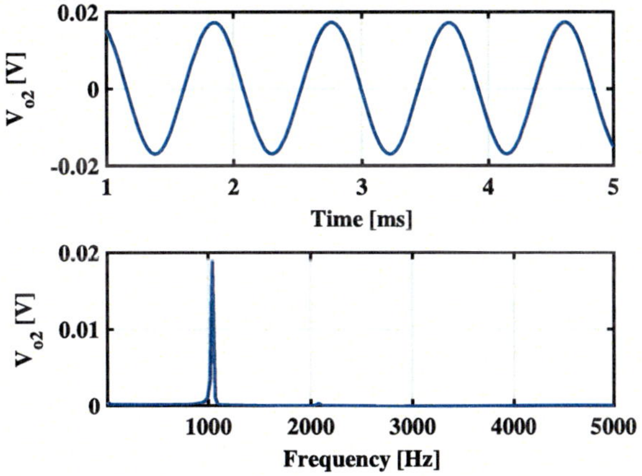

Fig. 9 FOO's transient response and frequency spectrum for $\alpha = \beta = 0.5$

tionality. The frequency of oscillation (ω_0) and condition of oscillation for the given FOO are given below:

$$\textbf{FO:} \quad \omega_0 = \left[\frac{g_{m1}\, sin \frac{\beta\pi}{2}}{C_\alpha\, sin \frac{(\alpha+\beta)\pi}{2}} \right]^{\frac{1}{\alpha}} \tag{6}$$

$$\textbf{CO:} \quad g_{m1} = \left[\frac{C_\alpha \sin \frac{(\alpha+\beta)\pi}{2}}{\sin \frac{\beta\pi}{2}} \right] \left[\frac{g_{m2}}{C_\beta \left(\cos \frac{\beta\pi}{2} - \frac{\sin \frac{\beta\pi}{2}}{\tan \frac{\alpha+\beta\pi}{2}} \right)} \right]^{\frac{\alpha}{\beta}} \tag{7}$$

The phase difference between V_{O1} and V_{O2} is equal to $\frac{\beta\pi}{2}$. The FOO is designed for $f_0 = \frac{\omega_0}{2\pi} = 1$kHz and phase difference between V_{O1} and V_{O2} as 45°. C_α and C_β are taken same and equals $1.65\mu\mho sec^{\alpha}$ and $g_{m1} = 185\mu$A/V, the value of g_{m2} to satisfy condition of oscillation is calculated as 92.5μA/V for $\alpha = \beta = 0.5$. The transient response and frequency spectrum of FOO is shown in Fig. 9. The simulated and theoretical frequency of oscillation are 1069 Hz and 1000 Hz respectively for $\alpha = \beta = 0.5$, the error between simulated and theoretical values of frequency of oscillation lies within 7%. The transient response of V_{O1} and V_{O2} is shown in Fig. 10,

Fig. 10 Transient response of V_{O1} and V_{O2} for $\alpha = \beta = 0.5$

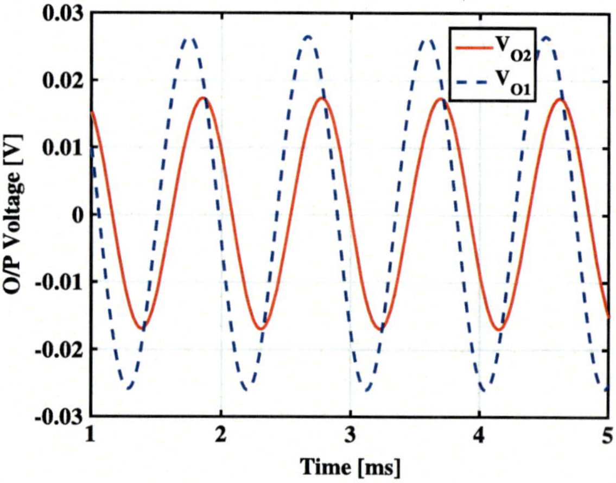

which verifies the phase difference between the two voltages is 44° for $\alpha = \beta = 0.5$ with an error of 2.22% from the theoretical value.

5 Conclusion

A new active inductor-based realization to approximate the behaviour of FOC has been proposed in this work. Proposed structure enables higher order approximation of FOC either by parallel connection or by impedance multiplication. The behaviour of the proposed FOC has been verified through SPICE with $0.18\mu m$ CMOS technology model parameters and the errors are found within 6% for slope of the magnitude. Moreover, a fractional-order oscillator has been verified as an application of the proposed FOC.

References

1. Tu S, Jiang Q, Zhang X, Alshareef HN (2019) Solid state mxene based electrostatic fractional capacitors. Appl Phys Lett 114(23):232903
2. Agambayev A (2019) Design fabrication and application of fractional order capacitors. PhD Thesis, King Abdullah University of Science and Technology, Thuwal, Kingdom of Saudi Arabia
3. Zhang L, Kartci A, Elwakil A, Bagci H, Salama KN (2021) Fractional-order inductor: design, simulation, and implementation. IEEE Access 9:73695–73702
4. Kapoulea S, Bertsias P, Psychalinos C, Elwakil AS (2022) Chapter One—MOS realizations of fractional-order elements. In: Emerging methodologies and applications in modelling, fractional-order design. Academic Press, vol 3, pp 1–33

5. Fouda ME, AboBakr A, Elwakil AS, Radwan AG, Eltawil AM (2019) Simple MOS transistor-based realization of fractional-order capacitors. In: IEEE international symposium on circuits and systems-2019 (ISCAS), pp 1–4
6. Bertsias P, Psychalinos C, Elwakil A, Safari L, Minaei S (2019) Design and application examples of CMOS fractional-order differentiators and integrators. Microelectron J 83:155–167. https://doi.org/10.1016/j.mejo.2018.11.013
7. Tsirimokou G, Psychalinos C, Elwakil AS, Salama KN (2018) Electronically tunable fully integrated fractional-order resonator. IEEE Trans Circuits Syst II: Express Briefs 652:166–170
8. Kapoulea S, Tsirimokou G, Psychalinos C, Elwakil AS (2020) Generalized fully adjustable structure for emulating fractional-order capacitors and inductors of orders less than two. Circuits, Syst, Signal Process 39:1797–1814. https://doi.org/10.1007/s00034-019-01252-5
9. Kapoulea S, Psychalinos C, Elwakil AS (2019) Realizations of simple fractional-order capacitor emulators with electronically-tunable capacitance. Integration 69:225–233
10. Valsa J, Dvorak P, Friedl M (2011) Network model of the CPE. Radioengineering 20:619–626
11. Nakagava M, Sorimachi K (1992) Basic characteristics of a fractance device. IEICE Trans Fundamentals E75-A(12):1814–1818
12. Geiger RL, Sanchez-Sinencio E (1985) Active filter design using operational transconductance amplifiers: a tutorial. IEEE Circ Dev Mag 12:20–32
13. Tsukutani T, Sumi Y, Fukui Y (2006) Electronically controlled current-mode oscillators using MO-OTAs and grounded capacitors. Frequenz 60(11–12):220–223
14. Varshney G, Pandey N, Pandey R (2022) OTA based fractional-order oscillator with controlled phase difference. In: IEEE international symposium on circuits and systems (ISCAS)

Modeling of Martian Ascent Launch Vehicle

Mahima Soota and Sudhir Nadda

Abstract Interplanetary research has advanced far beyond expectations in the last few decades. At this stage, a sustainable long-term system for profitable space travel needs to be planned to clarify the points of research gaps and feasibility in the coming years. The system model of a Martian Ascent Vehicle that carries 3 humans to an altitude of 250 km LMO (Low Mars Orbit) where they will be transported to the Mars Return Vehicle has been presented. The body of the launch vehicle would then land back on the red planet like a reusable frame that can be refueled and launch the next batch of the crew. The entire mission is optimized to have minimum infrastructural and cost requirements with constrained fuel, mass, power, and life support requirements.

Keywords Mars launch vehicle · Space travel · Reusable rocket

1 Introduction

1.1 Space Travel

Humankind has always been fascinated about the exploration of stars and planets in the night sky. This fascination has evolved with humans into present-day Lunar and Martian exploration missions, in addition to fly-by missions to other planets around the solar system. As nations equip themselves with advanced technologies for mass destructions, changes in Earth's environment, and ever-growing human population, planetary exploration is not merely driven by fascination, but also by interests like human colonization of planets and space travel. The future where space

M. Soota (✉) · S. Nadda
Amity University, Noida, Uttar Pradesh, India
e-mail: mhmsoota@gmail.com

S. Nadda
e-mail: snadda@amity.edu

A. K. Dubey et al. (eds.), *Advanced IoT Sensors, Networks and Systems*,
Lecture Notes in Electrical Engineering 1027,
https://doi.org/10.1007/978-981-99-1312-1_21

travel expands beyond the objective of just research and exploration may not be that far today. It has already been over a few decades since the discussions for human colonies and potential habitats have been researched and tested on software [1, 2]. Space sectors like NASA, ISRO, and ESA have been partnering with other space corporations and organizations [3] to develop system designs seeking expansions of human presence into the solar system in a sustainable way. To reduce complications with designs and manufacturing, several attempts to develop the same or similar habitats have also evolved. For instance, [4] examines the advantages and difficulties of a shared environment for these kinds of transit and ground operations.

Earlier ideologies and philosophies of space travel pointed toward the impossibility of exploration missions that have already become a reality today [5]. Fast forward to just a few decades back, [6] summarizes the space exploration taken by different space agencies across the world, proving those hypotheses wrong. These accomplishments are a motivation to investigate and calculate the feasibility of further space exploration missions.

1.2 Mars Exploration Missions

Despite the challenges, the world's leading space agencies have been making remarkable progress in paving the way for Mars exploration missions [7]. From the first successful Mars mission by the Soviet Union's spacecraft, Mars 3, to the United Arab Emirates' successful landing of the Hope Probe in February 2021, the world is getting closer and closer to understanding the mysteries of the Red Planet. In this blog post, we'll take a look at the pioneers of Mars exploration missions, the role of ISRO, NASA, and UAE in these missions, the different rovers and satellites used, the challenges faced, and the plans for colonization of Mars.

Pioneers of Mars Exploration Missions: The first successful Mars mission was launched by the Soviet Union in 1971. Their spacecraft, Mars 3, was the first spacecraft to land on Mars and transmit data back to Earth. The mission ended soon after due to a technical malfunction, but it was a major milestone for space exploration.

In 1976, the United States launched the Viking 1 and Viking 2 orbiters and landers, which were the first successful missions to both orbit and land on Mars. The Viking mission yielded valuable data on the Martian climate, atmosphere, and soil composition. In 1996, NASA launched the Mars Pathfinder mission, which landed the first rover on Mars in 1997. The rover was called Sojourner and it was the first time a rover had been used to explore another planet. The mission also sent back the first-ever close-up pictures of the Red Planet.

Rovers and landers are used to explore the Martian surface and collect data, while orbiters are used to map the planet and take pictures from space. The rovers and satellites used in Mars exploration missions are equipped with high-tech instruments and cameras that allow them to analyze the Martian surface and atmosphere.

One of the biggest challenges is communication. Due to the distance between Earth and Mars, it takes up to 20 min for a signal to travel between the two planets

[8]. This means that any data sent back by spacecrafts has to be stored on board and then transmitted back to Earth when the signal strength is strong enough. Another challenge is the extreme temperatures on the Red Planet. The temperatures can range from a scorching 86 °F to a frigid −225 °F. As a result, the spacecrafts used in Mars exploration missions have to be designed to withstand such extreme temperatures [9].

NASA is currently working on the Space Launch System, which is a powerful rocket that will be used to transport humans to Mars. The rocket will be powered by a combination of liquid oxygen and liquid hydrogen and will be able to transport a large payload to the Red Planet. NASA is also working on the Deep Space Habitat, which is a prototype habitat that will be used to house astronauts on Mars. The habitat will be designed to simulate the environment of Earth and will be equipped with life support systems such as air circulation and water recycling.

To colonize Mars, a more feasible and long-term strategy is needed, one that is simple to maintain and potentially goes on for eternity [9]. An example of this would be SpaceX, which has a plan for a single spacecraft to go to Mars and come back by refueling in the middle of the journey [10].

1.3 Problem Statement

As space exploration shifts to more sustainable methods for Earth-based missions, it is essential to safeguard Mars before arriving there. This might be expensive and take a long time to set up, but it will pay off in the end. The plan for the future encompasses cost-effective and modular space tourism for research and traveling. Several methods of gravity assists will be used, such as the Moon and Mars's moons, as well as a dedicated Mars Return Vehicle. The Mars Ascent Vehicle (MAV) system proposed in [11] is a Single-Stage to Launch (SSTL) vehicle, with a reusable booster powered by fuel and a payload module that can hold 3 people in a parking orbit 250 km away from Mars's surface. The booster stage is reusable, meaning it can land on Mars to be refueled and transport the next team of travelers. This system is intended to be employed multiple times each sol and is modeled in further sections of the paper.

2 Martian Atmosphere and Geographic Model

The Red Planet is indeed different when compared to our home planet. Not only do the temperature and gravitational forces vary, but also air pressure, air density, and specific impulse of fuel employed also vary. Missions illustrated in Sect. 1.2 have delivered sufficient data to record the atmospheric and geographic model for Mars. Based on [12], temperature, air pressure, and density of the Martian atmosphere can be expressed as in Eqs. (1)–(6) where T represents temperature, p represents Air

Pressure, ρ represents Air Density, and v represents the Speed of Sound at h altitude above surface.

For altitudes (h) less than 7000 m,

$$T(°C) = -31 - 0.000998h \tag{1}$$

$$p(KPa) = 0.699 * e^{-0.00009h} \tag{2}$$

and for altitudes (h) equal and over the value of 7000 m.

$$T(°C) = -23.4 - 0.00222h \tag{3}$$

$$p(KPa) = 0.699 * e^{-0.00009h} \tag{4}$$

In both cases, air density can be calculated as:

$$\rho\left({}^{kg}/_{m^3}\right) = \frac{p}{0.1921 \times (T + 273.1)} \tag{5}$$

Another critical factor being the speed of sound has been calculated in imperial units at [13], which complicates the calculations being SI units across the project. Using the Universal Gas Equations, the speed of sound can be calculated as

$$\rho\left({}^{kg}/_{m^3}\right) = \frac{p}{0.1921 \times (T + 273.1)} \tag{6}$$

$$v\left({}^{m}/_{s}\right) = \sqrt{\frac{\gamma RT}{M}} \tag{7}$$

Here γ is the Heat Capacity Ratio = *1.2941* and R_{gas} *(J/gm/K)* is the Gas constant = *44.01* on Mars [14], and M is the Mean Molecular Weight = *43.4 g/mol* [15]. All the equations can be illustrated in a graphical representation in Fig. 3.

Besides the aerodynamic factors that influence launch velocity, the geographic aspect is important as well. The launch velocity is impacted by the rotational velocity of the launch site (Fig. 1). Eberswalde crater is chosen as the launch site for this mission, instead of the popular Jezero crater as they are both near to the equator. The main reason for choosing the Eberswalde crater instead is the importance of the Jezero crater highlighted across research that portrays it to be an ideal human settlement location. Eberswalde crater is however less explored, keeping it at a safe distance from potential base camps. The latitude for this crater is 24 °S, and the rotational velocity imparted can be seen as illustrated in Fig. 2.

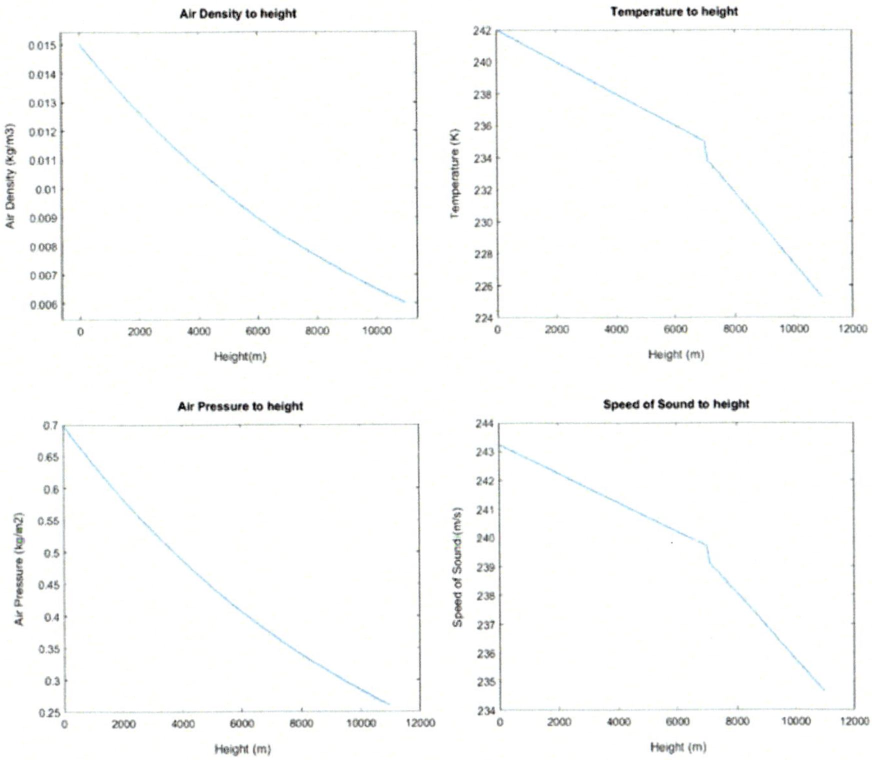

Fig. 1 Martian atmosphere model

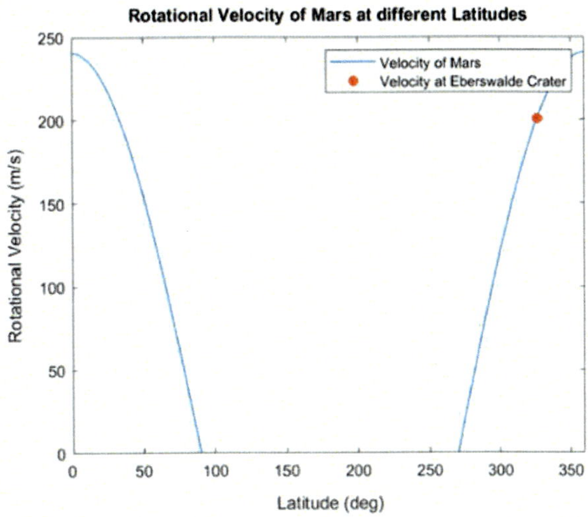

Fig. 2 Rotational velocity of mars added to launch velocity depending on the latitude of the launch site

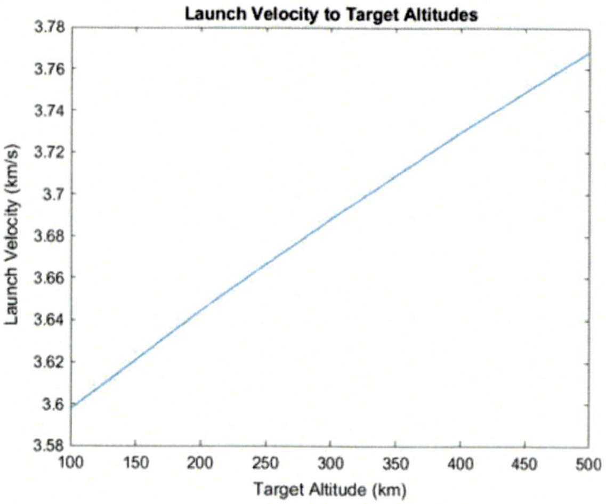

Fig. 3 Launch velocity to target altitudes

3 Trajectory Parameters

In order to calculate the trajectory of the launch vehicle, it is required that factors be considered for optimization and constraints to the mission objectives. Therefore, to formulate the mission statement, the target altitude for the mission needs to be calculated. As represented in Fig. 3, this relation is almost linear indicating the increase in required launch velocity with respect to the target altitude. Achieving a launch velocity above 3.5 km/s is a difficult number on the surface of Mars given the scarce availability of fuel and specifically oxygen. Such numbers and constraints highlight the need to consider a second stage for the Launch Vehicle, which might be too complex for such a low velocity and simple mission profile. It can be therefore convenient and feasible to illustrate the Reusable Module as the second stage such that it travels a sufficient distance beyond the gravity of Mars.

The calculations for adding Mars velocity to launch delta v must take place in the South-East-Zenith (SEZ) reference frame, where the rotational velocity of Mars must always be directed toward East, thanks to the rotational motion of Mars. It can therefore be appropriate to conclude that the launch should be tilted eastward to reap the benefits of launching from the specified launch site. Inertial Velocity can be represented in SEZ reference frame using the following equations:

$$V_S = -\Delta V_{1,i} cos(\phi_{bo}) \cos(Az_i) \tag{8}$$

$$V_E = \Delta V_{1,i} cos(\phi_{bo}) \sin(Az_i) - V_{Mars} \tag{9}$$

$$V_z = \Delta V_{1,i} sin(\phi_{bo}) \tag{10}$$

It is important to note how the rotational velocity of Mars has been subtracted from the *East* component and thus contributes toward the reduction of overall launch velocity. Here, Az_i denotes the Launch Azimuth that is required to achieve the desired orbital inclination, *i*, in the inertial frame. This can be calculated as:

$$Az_i = sin^{-1}[\cos(i)\cos(lat)] \tag{11}$$

Impulsive, idealistic value of launch velocity can be calculated as the magnitude of velocity in SEZ reference frame and can be observed graphically prior to accounting for gravity, drag, and steering (non-significant) losses.

Second burn velocity to circularize the MAV once it has achieved its apex, which would be executed by payload module after abandoning the reusable module (which executes another burn to reach the ground in a stable manner), can be calculated as

$$\Delta V_2 = V_c - \Delta V_1 \frac{r_m}{r_m + h} \cos \phi_{bo} \tag{12}$$

These velocities can be represented graphically as shown in Fig. 4 to analyze the appropriate Launch Angle for the Launch Vehicle and get an approximation of initial and second burn velocity for the Launch Vehicle, while also understanding the impact of Martian Rotational Velocity on Launch Velocity.

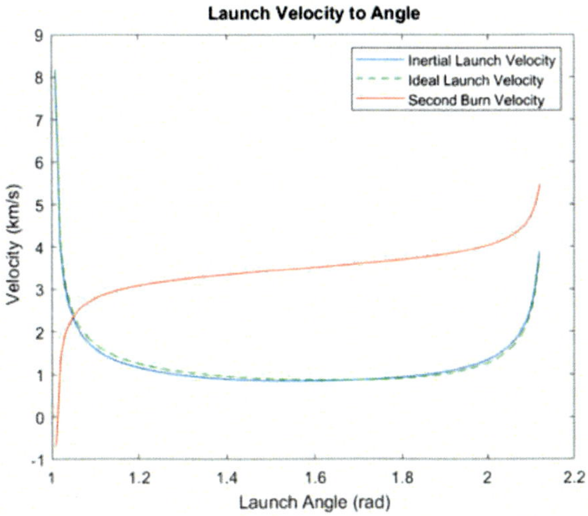

Fig. 4 Relation between Inertial, Idealistic values of Launch Velocity, second burn Velocity to Launch Angle

Table 1 Values for Nozzle Characteristics

Sr. No	Nozzle characteristic	Value
1	Specific Heat Ratio	1.1642
2	Chamber Length (m)	0.16
3	Throat Length (m)	0.68
4	Nozzle Length (m)	0.9
5	Chamber Diameter (m)	0.8
6	Throat Diameter (m)	0.235
7	Nozzle Diameter (m)	1.3
8	Chamber Temperature (°C)	3000
9	Chamber Pressure (millibar)	100

4 Nozzle Characteristics

Reusable Module Nozzle is used to general thrust such that it can lift-off mass and also achieve a successful descent after deployment of the Payload Module. With conditions being different in Martian Environment, nozzle design differs from that of Earth. The thrust generated by a rocket engine depends highly on Rocket Nozzle Characteristics, as indicated in Eq. (1). The selected nozzle for this mission, which is appropriate for the selected fuel, LCH4/LOX, must be SpaceX's Raptor [16, 17] (Table 1).

Mass flow rate for the mission $\dot{m}$. . is hence calculated as 734.8 kg based on rocket engine nozzle characteristics. Based on values mentioned in the table, plots for variation in pressure temperature, Mach number (on the Martian surface), and area of the nozzle can be represented as shown in Fig. 5.

5 Mass Calculations

Being a reusable ascent Launch Vehicle with payload module being able to achieve better heights, it is appropriate to take an educated guess regarding the reduced amount of propellant required for mission. This section concentrates on mass budgets for each subsystem of the launch vehicle, to have an appropriate estimate of payload mass and dry mass of the launch vehicle. The payload mass includes the three crew members, their life support system, and radiation protection jackets, and since the payload is merged with the second stage, it must also house a combustion chamber and propellant. At the point of payload separation, the launch vehicle should already be at an altitude free from drag and gravity losses. Since most of the propellant and dry mass parameters that include structure for propellant storage and thrust generation are variable and dependent on each other, the important step is the calculation of each mass budget of each subsystem.

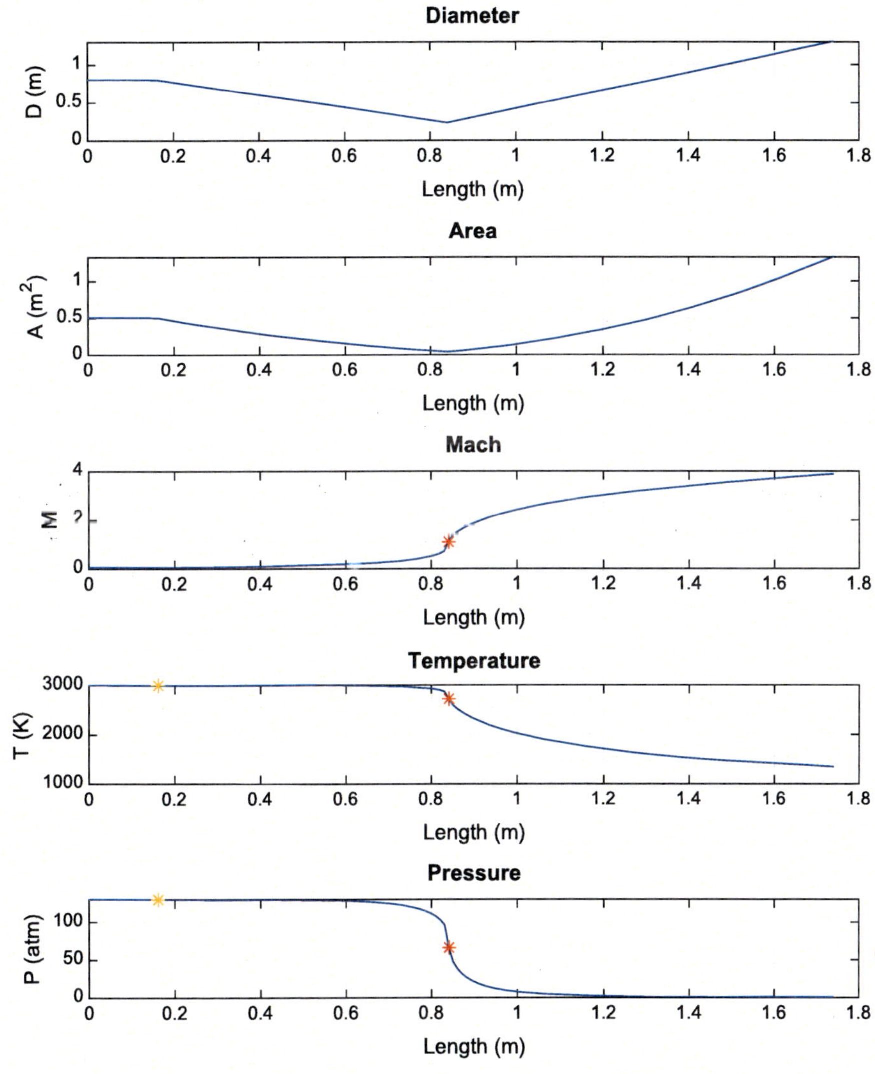

Fig. 5 Variation of diameter, mach number, temperature, and pressure inside the nozzle

5.1 Payload Module Mass Calculations

The Payload Module is the second stage of the proposed launch vehicle design. The net mass of each member of the human crew, and life support systems for each member crew have been reduced drastically due to very short flight duration. After the payload module docks into the Mars Space Station, the crew should be able to go inside and avail facilities for life support inside the space station (Table 2).

Table 2 Payload mass calculations

Sr. No	Element	Mass (in kg)
1	Human Crew	80 × 3 = 240
2	Radiation Protection Jackets [18]	27 × 3 = 81
3	Thermal Stability Mechanism	875
4	Oxygen (for the crew) [19]	3
5	Life Support System	500
6	Miscellaneous	301
Total		2000

Presently, the payload module is designed to be thermally stable, with appropriate air pressure systems, radiation protection systems, and most importantly, the appropriate amount of oxygen in the module for a period of 1 h despite the burn duration and flight time being less than 10 min under extreme circumstances. The remaining life support not only serves the purpose of redundancy in case some aspect of the mission goes wrong, and it takes less than an hour to rescue the crew members, but it also keeps the systems intact for a return flight of the next crew to land on the surface of Mars.

Human Crew consists of 3 passengers including tourists and a trained professional for any kind of problems faced during the mission. This number must be increased to at least 6 in further prospects of the mission to be able to carry at least 2 trained professionals and 4 other passengers along.

5.2 *Dry Mass and Structural Mass Calculations*

The reusable module of MAV is designed to be landed taking into account the combustion chamber, fuel tanks, and the relevant required vacant space. The tanks are filled with fuel and oxidizer, the engine is turned on, the thrusters are activated, and the MAV lands on Mars. The tanks are emptied and the MAV is ready for the next mission. This configuration includes four retractable legs to help facilitate the reuse of the reusable part for several missions on Mars.

Using three engines in the reusable module and one engine in the higher stage of the vehicle, the engines chosen for the first stage reusable and the top stage of the vehicle are able to generate an average propulsion force of 100 kN each on average. The method used to determine the required tank volumes is outlined as follows (Table 3):

1. The ratio of fuel to oxidizer, which is 3.2 in our case, is the available data.
2. Considering that the total propellant is 30 tons, we can calculate the masses of both the fuel and the oxidizer with the aid of the ratio.

Table 3 Fuel mass and volume characteristics [11]

Parameter	Attribute
Mass of fuel	4142 kg
Mass of oxygen	13,255 kg
Density of fuel	424 kg/m^3
Density of oxygen	1140 kg/m^3
Volume of LOX	20.05 m^3
Volume of LCH4	**16.844**

3. Knowing the density of each, we can figure out the minimum tank volume needed for both components.
4. It is critical to remember to set aside 5% of the tank's capacity to maintain the necessary pressure inside, or to allow for fluid motion and turbulence when storing liquid propellant.

6 Conclusion and Future Scope

It is necessary to find the appropriate launch windows for a reasonable, cost-effective delta v through a comprehensive, repetitive procedure. Alterations to the Delta v value directly affect structural measurements, capacity, and lift-off mass. This paper introduces a preliminary strategy for these figures, which must be reassessed. In the near future, a system engineering approach that looks at various potential strategies analytically will be presented as part of this project. Moreover, following the design of the MAV, a guidance, navigation, and control system in accordance with existing technologies will be developed for it.

NASA has ambitious plans for future missions on Mars. They plan to use a combination of launch vehicles, orbiters, satellites, and rovers to explore the Red Planet. The first mission will be a launch vehicle carrying an orbiter, a satellite, and a rover. The orbiter will map the planet's surface and relay data back to Earth for analysis. The satellite will monitor the environment and the rover will explore the terrain and collect samples. Future missions may include sample return missions, where a second launch vehicle will collect samples and return them to Earth for further study. Additionally, the mission may include a surface habitat where astronauts could stay, conduct research, and explore the planet's surface. These missions will be expensive, but they will greatly expand our knowledge of Mars and provide valuable data for future space exploration.

References

1. Higgins M, Benaroya H (2020) Utilizing the analytical hierarchy process to determine the optimal lunar habitat configuration. Acta Astronaut 173:145–154
2. Meyers C, Toutanji H (2007) Analysis of lunar-habitat structure using waterless concrete and tension glass fibers. J Aerosp Eng 20(4)
3. Crusan JC, Craig DA, Herrmann NB (2017) NASA's deep space habitation strategy. In: IEEE aerospace conference, Big Sky, Montana, USA
4. Howard R (2019) Opportunities and challenges of a common habitat for transit and surface operations. In: IEEE aerospace conference, Big Sky, Montana, USA
5. Hoerner SV (1962) The general limits of space travel. Science 137, 3523, 18–23
6. Bell D, Parker M (2009) Special issue: space travel and culture: from apollo to space tourism. Sociol Rev 57(s1):1–223
7. Fendall M, Denkins P, Fisher S, Oluseyi H, Yarborough P (2014) Human mars exploration research objectives. In: NBSE aerospace conference
8. Lele A (2014) Mars missions: past, present and future. In: Mission Mars, SpringerBriefs in Applied Sciences and Technologies, pp 85–92
9. Muirhead BK, Nicholas A, Umland J (2020) Mars sample return mission concept status. In: IEEE aerospace conference, Big Sky, MT, USA
10. SpaceX (2021) SpaceX-Mission Mars. https://www.spacex.com/human-spaceflight/mars/
11. Pawar A, Soota M, Swaroop BM, Aggarwal R (2022) Systems engineering and design of a martian launch vehicle with a human crew, Big Sky, Montana
12. Glenn Research Center (2021) Mars Atmosphere Model (Metric Units). https://www.grc.nasa.gov/www/k-12/airplane/atmosmrmhtml
13. PDS: Planetary Data System (2022) Gas constant. https://pds-atmospheres.nmsu.edu/education_and_outreach/encyclopedia/gas_constant.htm
14. Rogers R (2015) Offshore gas hydrates. Gulf Professional Publishing, Boston
15. Svitak A (2022) SpaceX's gwynne shotwell talks raptor, Falcon 9, CRS-2, Satellite Internet and More. http://aviationweek.com/blog/spacexs-gwynne-shotwell-talks-raptor-falcon-9-crs-2-satellite-internet-and-more. Accessed 5 Jun 2022
16. Belluscio AG (2022) SpaceX advances drive for mars rocket via Raptor power. http://www.nasaspaceflight.com/2014/03/spacex-advances-drive-mars-rocket-raptor-power/. Accessed 5 Jun 2022
17. AIAA (2019) AstroRad: vests for radiation protection. https://aerospaceamerica.aiaa.org/iac-2019-anti-radiation-vests-headed-for-iss/
18. Loer SA, Scheeren T, Tarnow J (1997) How much oxygen does the human lung consume. Anesthesiology 86(3):532–537
19. Wilcox B, Nguyen T (1998) Sojourner on mars and lessons learned for future planetary rovers. In: International conference on environmental systems